INTERACTIVE DECISION MAPS

KLUWER ACADEMIC PUBLISHERS
Boston / Dordrecht / New York / London

Applied Optimization
Volume 89

Series Editors:

Panos M. Pardalos
University of Florida, U.S.A.

Donald W. Hearn
University of Florida, U.S.A.

INTERACTIVE DECISION MAPS
Approximation and Visualization of Pareto Frontier

by

Alexander V. Lotov
State University – Higher School of Economics, Russia,
Dorodnicyn Computing Centre of Russian Academy of Sciences, Russia,
Lomonosov Moscow State University, Russia

Vladimir A. Bushenkov
University of Evora, Portugal

Georgy K. Kamenev
Dorodnicyn Computing Centre of Russian Academy of Sciences, Russia

KLUWER ACADEMIC PUBLISHERS
Boston / Dordrecht / New York / London

Distributors for North, Central and South America:
Kluwer Academic Publishers
101 Philip Drive
Assinippi Park
Norwell, Massachusetts 02061 USA
Telephone (781) 871-6600
Fax (781) 871-6528
E-Mail <kluwer@wkap.com>

Distributors for all other countries:
Kluwer Academic Publishers Group
Post Office Box 322
3300 AH Dordrecht, THE NETHERLANDS
Telephone 31 78 6576 000
Fax 31 78 6576 474
E-Mail <orderdept@wkap.nl>

 Electronic Services <http://www.wkap.nl>

Library of Congress Cataloging-in-Publication

Lotov, Alexander V./ Bushenkov, Vladimir A./ Kamenev, Georgy K.
Interactive Decision Maps: Approximation and Visualization of Pareto Frontier
ISBN 1-4020-7631-2

To our Teachers

Contents

Foreword

Multiple criteria optimization (MCO) is an important problem area which has challenged researchers and practitioners for several decades. Its importance derives from the prevalence of multiple criteria problems in virtually every sphere of policy and decision making. Its challenge results from computational complexity, as well as the inherent difficulty of making decisions and crafting policy when there is a large number of choices and many conflicting objectives.

Now, with the publication of this book, we have a welcome addition to the MCO literature and to the toolkit of the MCO practitioners. Alexander Lotov, whose work I have followed and admired for thirty years, is a leader in MCO and multiple criteria decision making (of which MCO is an important branch) — leadership which has been recognized with the prestigious Edgeworth–Pareto Prize awarded by the International Multiple Criteria Decision Making Society. Dr. Lotov and his colleagues, Dr. Bushenkov and Dr. Kamenev, have recorded here the major methods that they have developed and applied over several years.

It is especially noteworthy that the book and its original methods cover all of the major aspects of MCO problems. We have a method for approximating the Pareto frontier — the first and fundamentally important problem in MCO. Though many techniques and treatments would stop at this point, this book continues on to the crucial (but often ignored) problem of visualizing the Pareto frontier. This more comprehensive approach makes the methods covered in the book all the more valuable, but Lotov and his colleagues do not stop there. They also explore ways to use their methodology to support the decision maker's search for a preferred solution from among the many (likely infinite) Pareto solutions.

This book also presents many applications of the techniques. This is important, for MCO is a field that has developed and evolved in response to our understanding of real problems and the demands they place on real decision makers. Dr. Lotov has been especially attentive to this, taking on tough problems, particularly in environmental and energy planning. He and his colleagues have also been early proponents of web-based methods — another notable feature of this book.

Having watched the development of the early versions of the work presented here, it is particularly pleasing to see this book published. I am happy for Alexander Lotov and his colleagues and for you, who will now have access to this important work.

> Jared L. Cohon
> President
> Carnegie–Mellon University
> Pittsburgh

Preface

The book is devoted to application of computer visualization in the framework of multi-criteria optimization (MCO), which is a mathematical theory of methods related to decision making in the case of conflicting objectives. Decision problems with conflicting objectives, that is, conflicting criteria of decision selection, can be met extremely often. For example, they arise in management, finance, machinery design and everyday life. If the cost of a wrong solution to a problem is high, one needs computer support in searching for a reasonable solution. Such support is especially important for the process of the design of large projects that may have negative environmental consequences and influence by this the lives of many people.

MCO provides a mathematical basis for various decision support techniques applied in those decision problems where a mathematical model of the decision situation can be constructed and the values of decision criteria can be related to the variables of the model. Though methods of MCO develop further the single-criterion optimization methods, there is a substantial difference between these two groups of methods. In single-criterion optimization, a single solution to a problem is usually found: the optimal criterion value and the decision that provides it. In contrast, a solution of a MCO problem is given by two related varieties: the variety of non-dominated (Pareto-optimal) criterion vectors, known as the Pareto frontier, and the variety of decisions that result in vectors of the Pareto frontier. The latter variety is known as the Pareto-efficient or Pareto-optimal decision set. The absence of a single solution is the result of the conflict between the criteria.

It is important to note that these two varieties (or sets, speaking mathematically) may contain a very large or even infinite number of elements. A human being (so-called decision maker) has to settle a balance between the criterion values and select a single solution from the mathematically equivalent Pareto optimal solutions. However, to select a solution consciously, the decision maker needs to know the structure of these sets. Various multi-criteria decision support techniques inform the decision maker on Pareto-optimal sets (or at least on the Pareto frontier) in different ways, supporting by this the selection of a single Pareto-efficient decision. MCO theory studies properties of the Pareto frontier and Pareto-efficient decision sets.

Different decision support techniques use information on the Pareto frontier to a different extent and in different ways. One group of such techniques, known as Pareto frontier generation techniques (Cohon, 1978) provides decision makers with information on the Pareto frontier. This information supports direct identification of a preferred Pareto criterion point by decision maker; then, the associated Pareto-efficient decision point can be computed. In the case of more than two criteria, information on the Pareto frontier is usually provided in the form of a list of a large number of criterion points that approximate the Pareto frontier (see, for example, Zeleny, 1974; Krasnoshchekov, Morozov and Fedorov, 1979; Steuer, 1986). However, selecting an alternative from large lists is too complicated for human beings (see Larichev, 1984; Larichev, 1992). For this reason, methods based on informing decision makers on the Pareto frontier have not found broad recognition in the case of more than two criteria.

An effective approach to resolving this deadlock can be based on computer visualization. Visualization of information, i.e., transformation of symbolic data into geometric figures, can support human beings in forming a mental picture of the symbolic data. Computer visualization on the basis of graphic user interface proved to be a convenient and effective technique that helps people to assess information. Enormous effectiveness of visualization is based on the fact that about one half of the human brain's neurons is associated with vision. Due to this fact, visualization of information can be considered as a direct way to its understanding.

Important attempts have been made in the field of transforming the Pareto frontier generation techniques into visualization-based procedures. The most known one is the Pareto Race technique (Korhonen and Wallenius, 1988; Korhonen and Wallenius, 1990), which is based on a user-controlled movement along the Pareto frontier that applies successive display of single criterion points that belong to the Pareto frontier.

However, there is another concept of computer visualization for informing a decision maker on the Pareto frontier. This idea has been introduced as early as in the 1950s for the case of two criteria (Gass and Saaty, 1955). Saul Gass and Thomas Saaty noted that, in the case of linear bi-criterion problems, the Pareto frontier can be computed and depicted on the criterion plane. Bernard Roy, who is the leading French specialist in multi-criteria decision aid, has stressed that the display of the Pareto frontier is sufficient for decision support in bi-criterion decision problems (Roy, 1972). In this book, we describe methods that develop the same idea for the case of a larger number of conflicting decision criteria, that is, three, four, five and more.

To visualize the Pareto frontier for three and more decision criteria, we have developed a special technique, called the Interactive Decision Maps (IDM) technique. The main feature of the technique consists in approximation of the variety of feasible criterion vectors (Feasible Criterion Set, FCS) and further interactive visualization of its Pareto frontier. The Pareto frontier is provided in the IDM technique in the form of decision maps, that is, collections of frontiers of differently colored bi-criterion slices of the FCS (or broader sets that have the same Pareto frontier). Decision maps are displayed in the interactive mode (on-line). As a part of the interactive display of decision maps, animation of them is used.

Before visualization of a Pareto frontier is started, the task of approximating FCSs (or broader sets) must be solved. In the case of more than two criteria, methods for solving the extremely complicated task of approximation of multi-dimensional sets must be developed. So, the approach discussed in this book can be described in one sentence as visualization of the Pareto frontier on the basis of computational methods for approximation of multi-dimensional sets (both convex and nonconvex). Development and analysis of such methods constitute the main mathematical result described here. However, we do not discuss the mathematical aspects of the approach in the preface, since users of our visualization technique do not need to worry about them at all — the approximation methods have already been developed and coded. They compute the result without human intervention. Therefore, we concentrate first on discussing how visualization of the Pareto frontier can be applied in decision support tools. However, before we start, we want to stress an important difference of our approach from sporadic attempts to construct cross-sections of the Pareto frontier in advance, which are not effective in the case of more than three criteria: we approximate a FCS (or broader a set) in advance, but the visualization of a Pareto frontier (including animation) is provided in human-computer interaction, which is extremely important in the case of more than three criteria.

However, there is another important merit of the approximation of a FCS (or a broader set) and visualization of the Pareto frontier as its frontier: it helps to avoid complications that can arise in the case of instability of the Pareto frontier to disturbances of data, which are inevitable in real-life decision problems. Conditions of the stability of the Pareto frontier that are given in the book (Sawaragi, Nakayama and Tanino, 1985) show that a Pareto frontier fairly often can be not stable, and one cannot know about it before the Pareto frontier is constructed. In contrast, the sufficient conditions of the stability of an FCS can be easily tested and they are usually satisfied in models of real-life problems. Such conditions are considered in this book in Chapter 9.

The Pareto frontier describes the limits of what is possible in terms of decision criteria. Moreover, it informs how an improvement of one particular criterion results in worsening of the values of other criteria — one says that the Pareto frontier describes the efficient (criterion) tradeoffs. Visualization of criterion tradeoffs informs the decision maker on the properties of the decision problem. Therefore, it is consistent with the exploratory nature of soft operations research, in particular, of soft systems methods (Checkland, 1982).

In addition to informing the decision maker on criterion tradeoffs, visualization of the Pareto frontier can be used in the process of searching for the preferred decision. Such a search can be based on a combination of Pareto frontier visualization with practically any decision support technique and can result in various graphic decision support tools. In this book, however, we concentrate on integration of Pareto frontier visualization with the goal programming. Due to such an integration, the goal programming methods are transformed into graphic tools.

Goal programming is a well known and widely used approach to decision problems with conflicting interests. In goal programming procedures, the decision maker has to identify the desired criterion values (or, in other words, goal vector). Then a decision is computed so that its output is as close to the goal as possible. However, if the goal is distant from the feasible criterion vectors, the computed decision depends mainly upon the concept of distance, but not on the identified goal. Therefore, it is desirable to use the goals that are close to the feasible criterion values. Decision makers may need a support to identify such goals, since both underestimation and overestimation of the opportunities may result in the above negative consequences.

Visualization of the Pareto frontier solves this problem, since it provides the limits of what is possible and the conflict between the criteria in an understandable form. Due to it, the decision maker can identify the goal vector consciously. The information on the Pareto frontier helps the decision maker to identify a goal that is not only feasible, but maximizes his/her preferences. Therefore, the method described in this book provides further development of goal programming by visualization of a Pareto frontier for supporting decision makers, experts or other users in the process of goal identification. Visualization simplifies the procedure of goal identification − to select a goal, it is sufficient to click the computer mouse on a preferred point of the Pareto frontier. As usual in goal programming, the identified goal vector is used for computing a feasible decision alternative; however, due to the feasibility of the goal vector, the computer can find a decision alternative, which output coincides with the identified goal vector. Such a goal method based on identification of the preferred goal in a graphic display of the Pareto frontier is named Feasible Goals Method (FGM). It is important that the FGM is transparent − visualization of the Pareto frontier can help explain why a particular goal was selected.

A further development of the FGM is the Reasonable Goals Method (RGM) that is based on visualization of the Pareto frontier for the envelope of feasible criterion vectors. Visualization of the envelope instead of the variety of feasible criterion vectors simplifies the picture. Though the identified goal vector may be not feasible in the RGM, it is close to the variety of feasible criterion vectors − so, it is reasonable (Lotfi, Stewart and Zionts, 1992). The RGM can be applied in the case of alternatives provided in large tables (the simplest form of relational databases), which may contain millions of alternatives. Application of the FGM and RGM on the basis of the IDM technique (FGM/IDM and RGM/IDM techniques) turned out to be especially effective.

It is worth noting that the goals identified by users of the IDM technique can be considered as the aspiration levels expressed after studying the Pareto frontier. Therefore, the IDM technique can be considered as a tool that helps to form personal aspiration levels. Such an interpretation is very helpful sometimes.

The FGM/IDM and RGM/IDM techniques can be applied in negotiation support. Indeed, different criteria can be related to interests of different negotiators. By visualizing the criterion tradeoffs, the FGM/IDM and RGM/IDM techniques can support pre-negotiation strategic planning. In negotiation preparation , the IDM technique helps to accomplish the main concepts of Principled Negotiations , which were developed in the Negotiation Program of Harvard University (Fisher and Uri, 1983). The main idea of Principled Negotiations is that the negotiations must focus on interests rather than on particular positions such as being for or against a particular decision. Then, the search for mutual gains among the variety of possibilities with respect to all recognized interests should be provided; only then a coordinated balanced decision should be constructed, starting with the negotiated balance of interests (such interpretation of Principled Negotiations in environmental decision making is given in (United Nations, 1988)). In this aspect, Principled Negotiations are contrasted to position-oriented negotiations, inefficiency of which was proven by theory and practices (Raiffa, 1982).

Since the IDM technique displays tradeoffs between criteria that represent the interests of negotiators, the positions (decisions) are hidden and are not considered at all. A negotiator who applies the IDM-based methods can find a decision that is in line with his interests, but is still satisfactory for other negotiators. Application of the IDM technique on the negotiation preparation stage may turn out to be a decisive advantage in a negotiation process. Moreover, such a negotiator may use the IDM technique to prove to other negotiators the advantages of the decision proposed by him/her.

It is important that the IDM technique can be used for negotiation support via computer networks (say, on the Internet). Actually, the IDM technique can be used via computer networks for supporting any form of multi-criteria decision making. It is important that non-professionals can easily access the IDM technique. Due to application of visualization, a computer-literate user can master the IDM-based Web tools fairly quickly. Several forms of IDM-based Web resources can be proposed; two of them are considered in this book. The first one is based on the RGM/IDM technique and is aimed at supporting e-commerce: it supports clients in the process of searching for preferable goods or services given in large lists provided on the Web (such as lists of real estate, tourist tours, second hand cars, etc.). On the other hand, it can support business-to-business relations, say, selecting several potential partners in the process of supply chain design. Another possible application is graphic on-line support of technical analysis of electronic stock markets.

The second, perhaps more important Web application of the IDM technique is related to public decision problems. Web application of the IDM technique can be used for on-line visualization of Pareto frontiers for the variety of alternative solutions of a public decision problem. This information can help ordinary people to understand the tradeoffs in such decision problems. Due to it, a Web implementation of the IDM technique can be applied in the framework of new paradigms of environmental decision making, which require involvement of non-experts into decision processes by supporting them in their active computer-based preparation for legal and political actions.

Once again, we have to stress that though the IDM technique is based on complicated computational methods that provide approximation of an FCS (or a broader variety), a user does not need even to know about the approximation procedures, since they are stable and are carried out automatically (except for very large problems). For this reason, we arranged the content of our book in parts that require different levels of mathematical background.

Part I contains a non-formal introduction to the IDM technique and describes its applications in decision support procedures. It is written in a simple form and can be understood by any computer-literate person interested in application of visualization methods in decision making. Part I will be of interest to specialists and students in various fields related to decision making, including environment, management, business, engineering, etc.

Part II is devoted to computational methods used for approximation of an FCS or of a broader set, an Edgeworth–Pareto Hull (EPH) of an FCS, which is the maximal set that has the same Pareto frontier as an FCS. Though the methods are introduced in Part II in a relatively simple form, certain mathematical background is required here. Part II will be of interest to specialists and students in the field of applied optimization, operations research and computer science.

Two important topics of the mathematical basis of the approximation methods are considered in Part III. Reading of Part III requires certain knowledge of differential geometry and theory of Banach spaces. It is written for specialists and students in applied mathematics interested in the theoretical basis of modern optimization.

Due to this structure of the book, its parts can be read independently. Say, students interested in applications could restrict themselves to Part I and the Epilogue. In contrast, those who are interested in computational methods can skip Part I and read Part II only. Finally, specialists interested in the theory of approximation of multi-dimensional convex sets or in estimation of disturbances of polyhedral sets can read the corresponding chapters of Part III.

The book consists of nine chapters. Chapters 1–5 constitute Part I. The IDM technique is introduced in Chapter 1 on the basis of an example problem of regional environmentally sound economic planning. Then, the concepts of the FGM and RGM are considered and Web application of the IDM technique is outlined. The second and the third chapters describe applications of the FGM/IDM technique. Chapter 2 shows that the technique can be applied in a broad range of environmental problems, starting with local problems of water management and culminating with global climate change. Chapter 3 is devoted to real-life applications of the technique, which include supporting the national goal choice at the State Planning Agency of the USSR in the 1980s and decision screening in the process of water quality planning in Russia in the 1990s. Several applications of the RGM/IDM technique are described in Chapter 4. Chapter 5 is devoted to the IDM technique in the non-linear case and to several other new developments, which include decision support under risk and visualization in Data Envelopment Analysis.

Chapters 6 and 7 constitute Part II of the book. In Chapter 6, methods for approximating FCS and its Edgeworth–Pareto Hull are described. Three groups of methods are considered:

- methods based on convolution techniques for inequality systems;

- methods for polyhedral approximation of multi-dimensional compact convex bodies; such methods are based on combination of convex optimization with the convolution techniques; and

- methods for approximation of multi-dimensional compact sets in the non-convex case; such methods are based on hybridization of local optimization with stochastic simulation of random feasible decisions and filtering of associated criterion vectors.

Chapter 7 is devoted to methods for approximation of FCSs and EPHs for dynamic systems described by ordinary differential equations and for systems in partial derivatives. In this chapter much attention is given to the problems of approximation of reachable sets for dynamic systems.

Chapters 8 and 9 constitute Part III. They are devoted to two important topics that belong to the theoretical basis of the computational methods considered in Part II. Chapter 8 describes the asymptotic theory of methods for polyhedral approximation of multi-dimensional compact convex bodies. In Chapter 9, the problem of stability of an FCS is considered in the case of linear systems, for which an FCS is given as the image of the solution set of a linear system of equalities and inequalities. Their perturbations are estimated.

The Epilogue is devoted to application of the IDM technique via the Web by non-professionals in the process of preparation for political actions on environmental topics.

The book is partially based on a shortened and updated translation of books in Russian (Lotov, Bushenkov, Kamenev and Chernykh, 1997) and (Lotov, Bushenkov and Kamenev, 1999). A part of the material of the book has already been published in (Lotov, Bushenkov and Kamenev, 2001). The additional material related to the book (full color pictures that are given by black-and-white copies in the text of the book, the downloadable software and Web resources that implements the IDM technique) can be found at the Web site

http://www.ccas.ru/mmes/mmeda/book6.htm

The first six chapters of the book were written by the authors together. Chapters 7, 9 and the Epilogue were written by A. Lotov, and Chapter 8 was written by G. Kamenev.

Acknowledgments

The book describes the results of studies carried out at Dorodnicyn Computing Center of the Russian Academy of Sciences (CC RAS). Authors are grateful to researchers of CC RAS for their important advice, constructive critical remarks and moral support in the process of the studies. We express our sincere gratitude to Professor Alexander A. Petrov, Head of the division of CC RAS where the research was carried out, for his patience and help during the long years of the research. Director of CC RAS, Professor Yuri G. Evtushenko proposed the idea to publish the book in the Applied Optimization series of Kluwer, and Professor Panos M. Pardalos from University of Florida, Gainesville, USA, editor of the series, supported this idea. We are grateful to them for it.

The authors would like to stress the influence on their research of two distinguished Russian scientists Nikita N. Moiseev (1917–2000) and Germogen S. Pospelov (1914–1998). As early as in the 1960s, they stressed the need for visualization of the variety of feasible criterion vectors. Their ideas provided a starting point of our research. An enormous support was provided to us during the last 20 years by the head of the Russian decision science school, Professor Oleg I. Larichev (1934–2003). We never forget his advice and care.

An important role in the development of the methods described in the book was played Dr. Oleg L. Chernykh (1954–1996). Oleg Chernykh belonged to the team of researchers who developed the methods. He was the author of an effective robust algorithm for constructing the convex hulls of multi-dimensional points. He coded the software that implements his algorithm, providing by this a computational basis of our research in the case of convex problems. He was one of the co-authors of the algorithm for computing of large series of two-dimensional slices of convex polytopes; this algorithm helped to develop efficient visualization software. He took part in a large number of the case studies described in this book.

The authors would like to express their gratitude to the members of the International Society for Multiple Criteria Decision Making for their permanent support. We are especially grateful to Presidents of the Society Stan Zionts, Ralph Steuer, Pekka Korhonen and Valerie Belton. Bestowing A. Lotov with the Edgeworth–Pareto Award of the Society in 2000 is another example of the support provided by the MCDM Society to our research.

Jared L. Cohon, President of the Carnegie–Mellon University, was the first who recognized the potential of our visualization techniques as a multi-criteria decision support method. The authors are grateful to him for his concepts that played an important role in the progress of their research.

For 20 years, the authors have been involved in collaboration with Finnish researchers supported by joint projects of the Russian Academy of Sciences and the Academy of Finland. The authors like to stress the importance of this collaboration, which provided an effective bridge to the research of the Western decision science community. In the 1980s, our research was partially supported by the International Institute for Applied Systems Analysis (IIASA) located in Laxenburg, Austria. We are grateful for this opportunity to former members of the IIASA Prof. Andrjei Wierzbicki (Poland), Prof. Janush Kindler (Poland) and Dr. Sergei Orlovski (Russia).

The authors are grateful to researchers who spent much time collaborating with us in the process of writing joint papers, material of which was used in this book: Professors Olavi Hellman (University of Turku, Finland), Matti Pohjola and Jyrki Wallenius (Helsinki School of Economics, Finland), Hannele Wallenius (Helsinki University of Technology, Finland), Pete Loucks (Cornell University, USA), Piotr Jankowski (University of Idaho, USA), Antonio Camara (New University of Lisbon, Portugal), Dr. Kaisa Miettinen (University of Jyväskylä, Finland) and Dr. David Soloveichik (Ministry of National Infrastructures, Israel). Professor O. Hellman was the first to include our methods in his course of lectures outside of Russia.

A computer-based educational tool was developed on the basis of the IDM technique. The educational tool was developed in the beginning of the 1990s for students of the Institute (University) of Physics and Technology, Russia. Since 1995 it has been used at Lomonosov Moscow State University, Russia, and University of Idaho in Moscow, Idaho, USA. The authors would like to thank Professor Alexander Kurzhanski and Professor Piotr Jankowski for their desire to teach students our visualization technique. We are thankful to Professor PingSun Leung from University of Hawaii in Manoa, Hawaii, USA, for his job aimed at dissemination of our technique.

A large part of the job related to writing of this book was carried out during the stay of one of the authors at the University of Siegen, Germany, and University of Jyväskylä, Finland. We are grateful to Professor Manfred Grauer from University of Siegen and Dr. Kaisa Miettinen from University of Jyväskylä for these opportunities. Fraunhofer Institute for Autonomous Intelligent Systems, Germany, partially supported coding of the demo version of the Web application server described in this book. We are grateful to Drs. Hans Voss, Natalia and Gennady Andrienko.

We want to thank unknown referees of our papers for their valuable comments.

We are grateful to Dr. Lioubov Bourmistrova who carefully read the book and corrected multiple typos and vague statements. We would like to stress her role in the development of the algorithms and software. We would like to thank former Ph.D. students S. Ognivtsev, S. Utkin and N. Zezulinski who were involved in our research as early as in the 1980s. Former and current graduate and Ph.D. students V. Berezkin, A. Biryukov, N. Brusnikina, A. Chernov, D. Gusev, R. Efremov, A. Kistanov, D. Kondratiev, A. Pospelov and A. Zaitsev took part in our research and coding of the software. We would like to express our gratitude to them.

Our book includes results that were published in papers supported by Russian Foundation for Basic Research (projects No.95-01-00968, No.98-01-00323, No.01-01-00530), Russian State Program for supporting of leading scientific schools (project No.00-15-96118), Programme Perspective Information Technologies of Ministry for Science and Technology of Russian Federation (project No.1235), and Linkage Research Grant of NATO Scientific Affairs Division (No.ENVIR.LG-931565).

Russian Foundation for Basic Research (grant No.01-01-00530), Russian State Program for supporting of leading scientific schools (grant No.NS-1843.2003.1) and Program of the Presidium of Russian Academy of Sciences (grant No.2-31) have supported writing of this book. We express our deep gratitude for this support.

I

ENVIRONMENTAL, BUSINESS AND OTHER APPLICATIONS OF INTERACTIVE DECISION MAPS

Chapter 1

INTRODUCTION TO INTERACTIVE DECISION MAPS

In this chapter the Interactive Decision Maps (IDM) technique is introduced and its simplest application in the framework of the goal approach is described. First the screening phase of the decision making process is considered and the role of visualization in it is outlined. Then, the Pareto frontier, Edgeworth–Pareto hull, decision maps and related notions of MCDM theory are introduced using a simple regional environmental model.

The IDM technique for linear models is introduced in Section 3. Then, the Feasible Goals method (FGM) is described and discussed. Finally, computational aspects of the IDM technique are outlined. The mathematical formulation of the FGM/IDM technique is given in Chapter 6.

1. Visualization in decision screening

In his classic book "The New Science of Management Decision" (Simon, 1960), Professor Herbert Simon, a Nobel prize winner, split the decision making process into four main phases: *intelligence, design, choice* and *review*. The intelligence phase concentrates on identification of the decision problem and collection of related information. The second phase, design, is concentrated on developing a relatively small number of decision alternatives that must be studied in detail. The choice phase consists in detailed analysis of decision alternatives prepared at the design phase and choice of the most preferred one. At the final phase, review, the selected decision alternative is implemented. Additional experience is obtained in this process.

So, the decision is made in the framework of two phases:

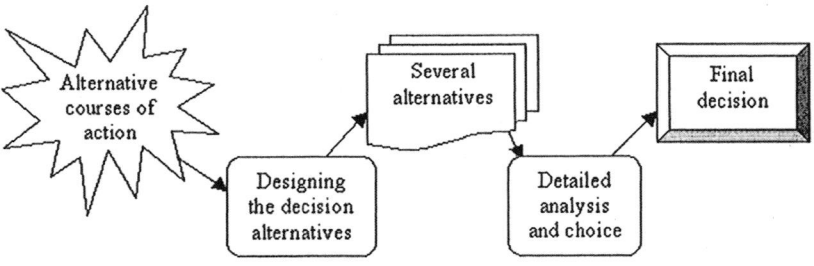

Figure 1.1.   Phases of the decision making process

- *designing* a list that contains a relatively small number of decision alternatives, and

- *final choice* of a single decision alternative from the small list of designed alternatives.

These two phases are totally different − their goals, information and methods differ drastically (see Figure 1.1). On the phase devoted to developing a relatively small number of decision alternatives (this phase is often denoted as *early screening* of decision alternatives), one has to take into account all possible courses of action. In contrast, on the phase of detailed analysis and final choice, decision makers restrict themselves to a short list of prepared alternatives, but apply a detailed exploration of the decision alternatives.

Computer visualization tools play now an important role in supporting of decision processes. However, their application is mainly concentrated on detailed analysis and final choice. Detailed analysis is often based on simulation of comprehensive mathematical models combined with visualization of simulation output. Multimedia tools and geographic information systems, which provide decision makers with exciting opportunities for rapid graphic assessment of one or a few decision alternatives, step-by-step find their proper place in the real-life decision processes.

In contrast, decision design is usually not supported by modern visualization tools. This does not seem to be wise since screening of decision alternatives plays an extremely important role in decision processes. Indeed, the decision alternatives excluded at the screening phase cannot be revived later. Traditionally, decision makers have to design decision alternatives by themselves guided by their experience and feelings.

Often, experts are asked to develop a small number of decision alternatives for further detailed exploration. Expert involvement saves time for decision makers, but introduces additional complications related to the fact that the alternatives developed by experts usually reflect their experience, perceptions, and goals, which may differ from those of decision makers. It can result in a deadlock during the final phase since decision makers are forced to choose among decision alternatives that do not reflect their opinions or interests, since the courses of action that could satisfy the decision maker's interests were excluded in the screening phase. Therefore, new tools are needed that can involve decision makers in the screening phase and amplify their experience and intuition. Visualization must play an important role in the use of these tools.

It is important to note that the final phase of decision making is often a negotiation process that involves decision makers with different experience, interests and goals. In this case, the design phase plays the role of the *negotiation preparation*. In this phase, negotiators need to find such decision alternatives that are preferable for them and acceptable for other negotiators. This requires a direct involvement of negotiators in the design phase.

Pre-negotiation activities are often separated from negotiations. This means that the time requirements may be not so restrictive as in the negotiation process. Say, the screening phase may take months and even years, especially in such public undertakings as environmental planning. Due to this time lag, multiple stakeholders, independent institutions and political groups may have the opportunity to take part in the designing and screening activities by developing, selecting, modifying or even rejecting possible decision alternatives. Computer methods applied by professional decision makers must be transparent and easy to use. Involvement of ordinary people, which has become more and more important in public decision making (especially in environmental problems), only amplifies this requirement.

Modern psychology asserts that human beings make their decisions on the basis of their *mental models of reality*. Mental models usually comprise several levels (Figure 1.2). The upper, rational level is based on logical inference. The second level includes images, relations of which (in contrast to the upper level) may be not precise. The third level contains vague subconscious relations. All levels interact, and the process of coordinating them is permanently under way, however, certain discordance between the levels is a natural feature of human thinking processes. The decision processes incorporate activities of all levels, and it is very complicated to estimate which level contributed more to the final choice. One has to take into account that the imaginary and subconscious elements of the decision process are usually camouflaged by the logical inference

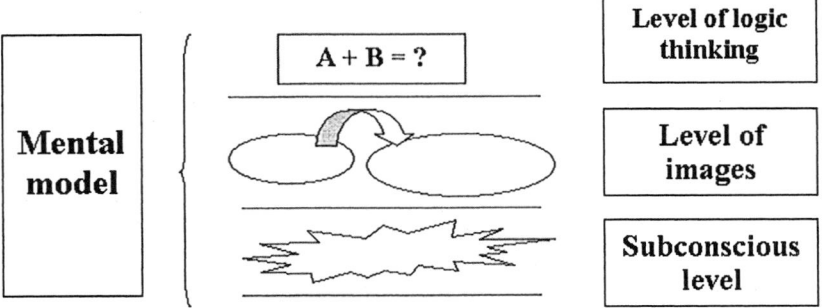

Figure 1.2. Levels of a mental model.

that is used to justify a decision after it has been made. Therefore, it is important to have a possibility to influence all levels of mental models.

The need to influence mental models by the output of mathematical models is related to the fact that mental models are often not only rough, but not true in many aspects. In particular, people often underestimate or overestimate the feasibility of the goals they identify. Therefore, one of the effective forms of decision and negotiation support may consist in correcting the mental models. Computer-based visualization has a chance to refine logical constructions, images and subconscious relations simultaneously. If the graphic information is given in a relatively simple form, it may help a human being to assimilate it not only consciously but on subconscious levels, too. In the framework of the methods described in this book, decision information is visualized in a form that is supposed to influence all levels of mental models. These methods do not force anyone to answer complicated questions; instead, they visualize information that is hidden in equations and parameters of mathematical models and databases.

The visualization technique described in this book is based on the following assumptions:

- All important *conflicting interests* that must be taken into account have already been identified and included into the list of decision criteria;

- Original knowledge and information on the problem have already been transformed into a *mathematical model* that describes the variety of feasible decision alternatives and their relation to decision criteria; by this, the problem of decision design is transformed into the problem of selecting a small number of alternatives from a given large (or even infinite) variety of all feasible alternatives (decision screening);

- Users would rather prefer to *receive information* on the decision problem than to answer questions concerning their preferences;

- Users are eager to know about the *limits of what is possible and how conflicting interests can be exchanged* in a reasonable way; and

- Users would rather prefer to receive some kind of *visual information* than usual data tables.

These assumptions are taken into account in the *Interactive Decision Maps* (IDM) technique described in this book. The IDM technique is applied in the case of several (two and more) conflicting criteria. First we describe visualization of the Pareto frontier for the case of two and three criteria. The latter case is used for introduction of decision maps. Only then we do discuss the IDM technique in general.

2. Visualization of the Pareto frontier in a regional environmental problem

We describe visualization of the Pareto frontier on the basis of a simple water-related model of regional production. The original version of the model, which was elaborated in the beginning of the 1980s at the International Institute for Applied Systems Analysis (IIASA), Austria, and described water-related problems of the South-Western Scone region, Sweden, was explored using visualization of the Pareto frontier (Lotov, 1981b; Bushenkov, Ereshko, Kindler, Lotov and de Mare, 1982). Later, the model was modified to make the conflict between production and environment more acute, and so now it differs a lot from the original problem. The modified model was used for introduction of Pareto frontier visualization in (Lotov, 1984a), and since then it has permanently been used for this purpose. It is applied now in computer educational tools in several universities around the world. An educational environmental computer game called LOTOV_LAKE (Lotov, Bushenkov and Chernykh, 1992) and the demo Web resource considered in this chapter are based on exploration of the same model.

Regional environmental problem. A region with intensive agricultural production is considered. The region is located in the basin of a river that runs through a lake and then flows into a sea (Figure 1.3). The lake serves as the municipal water supply, and it is an important environmental and recreational site.

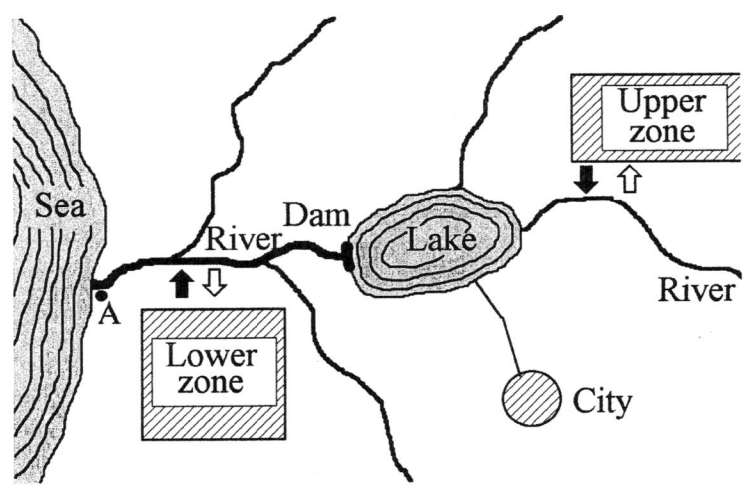

Figure 1.3. Map of the region

The problem of economic development of the region is studied. If the agricultural (to be precise, grain-crops) production increases, it would spoil the environmental situation in the region. This is related to the fact that the increment in the grain-crops output requires irrigation and application of chemical fertilizers. It may result in negative environmental consequences, namely, a part of the fertilizers may find its way into the river and the lake with the withdrawal of water. Moreover, shortages of water in the lake may occur during the dry season.

Two agricultural zones are located in the region. Irrigation and application of fertilizer in the upper zone (located higher than the lake) may result in a drop of level of the lake and an increment in water pollution. Irrigation and application of fertilizer in the second zone that is located lower than the lake may also influence the lake. This influence is, however, not direct: irrigation and application of fertilizer in the lower zone may require additional water release from the lake into the river (the release is regulated by a dam) to fulfill the requirements of pollution control at the monitoring station located in point **A** (Figure 1.3).

A finite number of grain-crop production technologies are considered in the model. Intensive technologies are related to high levels of water consumption and fertilizer application. The technologies that are related to low water consumption and fertilizers application levels are characterized by low production. Several technologies use moderate amounts of water and fertilizers; they result in a moderate production output.

Reasonable combinations of production technologies and water release from the lake should be found. Several economic and environmental performance indicators characterize the production and release strategies. The indicators represent different interests: farmers are interested in grain-crop production while recreational business is interested in keeping the level of the lake, and the inhabitants of the city are mainly interested in water quality. So, we consider three performance indicators that are used as screening criteria in the process of selecting a reasonable strategy:

- agricultural production (to be maximized),

- level of the lake (to be maximized),

- additional water pollution in the lake (to be minimized).

Mathematical description of the model is given in Appendix 1.A.

Pareto frontier: schematic introduction. Here, we provide a schematic introduction of the Pareto frontier (Figures 1.4–1.7). Then, we introduce the concept of decision maps (Figure 1.8). Only then we do display the real decision maps for the regional model (Figures 1.9–1.11).

Let us start with the case of two criteria: agricultural production and level of the lake. Any feasible (possible) decision concerning the production technologies and water release results in certain values of production and level of the lake (say, point **Q** in Figure 1.4). Such a combination of the criterion values is called *feasible*. Often, a feasible combination of criterion values is denoted as a "feasible criterion vector". In the case of two criteria, all feasible criterion vectors may be displayed on a plane. In Figure 1.4, the variety of feasible criterion vectors is given by its frontiers.

It is important that, for any point of the variety of feasible criterion vectors, one can find a feasible strategy of regional agricultural production and water release that results in this criterion point. Say, a strategy can be found that results in point **Q**. Therefore, if a point of the variety is considered as a goal, a strategy does exist that satisfies the goal requirements. In contrast, no feasible strategy exists that results in point **R** that is outside the variety. For this reason, the variety of feasible criterion vectors displayed in Figure 1.4 may be denoted as the *variety of feasible goals*.

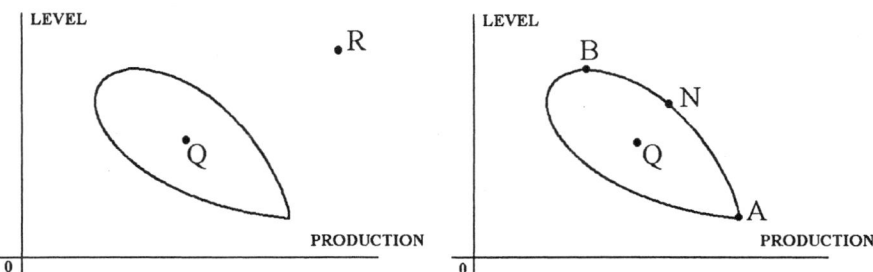

Figure 1.4. A feasible point **Q** and non-feasible point **R**.

Figure 1.5. Dominated **Q** and non-dominated **N** combinations of production and level of the lake.

Identification of goals is a well-known decision support procedure. It is a part of goal programming (Charnes and Cooper, 1961; Ignizio, 1985; Steuer, 1986; Romero, 1991, etc.). Usually, the variety of feasible goals is not displayed in the goal procedures. This results in a sophisticated question: What decision should be provided if the goal identified by the user (say, point **R** in Figure 1.4) is not feasible? To solve this problem, a feasible criterion vector closest to the identified goal is usually computed. If the closest feasible combination is fairly distant from the identified goal, the user may be disappointed with the result. Moreover, the notion of the "closest" feasible criterion vector may depend more upon what is understood by the distance between points than on the goal itself. Therefore, the computed strategy may disregard preferences of the user. Visualization of the variety of feasible criterion vectors helps to get rid of such problems, since visualization helps users to understand what is feasible and to identify only feasible goals.

Let us consider additional features of the variety of feasible criterion vectors. In Figure 1.5, in addition to the feasible point **Q**, feasible points **A**, **B** and **N** are marked. It is clear that the point **N** is better than the point **Q** since both criteria, production and level of the lake, are higher in point **N** than in point **Q**. In such a case, one says that **N** dominates **Q**. In contrast to **Q**, a feasible point that dominates **N** does not exist. Feasible criterion points of this kind are called *non-dominated points*. The non-dominated points are displayed in Figure 1.5 by curve **AB**, which is a part of the frontier of the variety. Since we are interested in maximizing both criteria (production and level of the lake), the non-dominated points are those located in the top right part of the frontier. A frontier of this kind is called a *Pareto (non-dominated, non-inferior)* *frontier*. Associated strategies are denoted as *Pareto* (or simply *efficient*) *strategies*. In the framework of the model, reasonable strategies are associated to the non-dominated points only. Therefore, the decision maker who wants to find a reasonable strategy has to choose one of the points of the Pareto frontier.

The Pareto frontier plays an important role during decision making and negotiations. Let us consider an example. Figure 1.6 displays a Pareto frontier **AB** (the same as in Figure 1.5) with two additional non-dominated points, **P** and **M**. If one moves along the Pareto frontier from point **B** to point **P**, just a small decrement in the level of the lake is needed for a substantial increment in production. Vice versa, if one moves along the

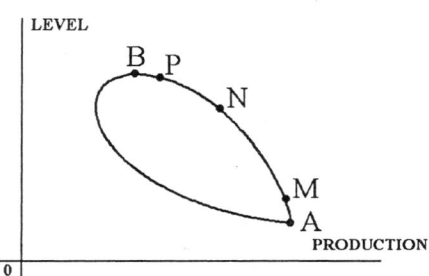

Figure 1.6. Points of Pareto frontier.

Pareto frontier from point **A** to point **M**, only a small decrement in production results in a substantial increment in level. So, the Pareto frontier shows how agricultural production is transformed into the level of the lake if efficient strategies are used.

To provide a qualitative description for the transformation of a criterion into another one (or other criteria) while moving along the Pareto frontier, the notion of efficient (criterion) tradeoff was introduced. To be precise, several forms of the notion of efficient tradeoff do exist. Let us denote by $L_\mathbf{A}$ and $L_\mathbf{M}$ the levels of the lake in points **A** and **M** of the Figure 1.6. Let $P_\mathbf{A}$ and $P_\mathbf{M}$ be the values of production in these points. Since $P_\mathbf{A}$ does not equal to $P_\mathbf{M}$, one can consider the value

$$\lambda(\mathbf{A}, \mathbf{M}) = (L_\mathbf{A} - L_\mathbf{M}) / (P_\mathbf{A} - P_\mathbf{M}).$$

This value is called the *criterion tradeoff* for level and production between points **A** and **M**. In turn, the inverse ratio

$$(P_\mathbf{A} - P_\mathbf{M}) / (L_\mathbf{A} - L_\mathbf{M})$$

is called the *criterion tradeoff* for production and level between points **A** and **M**.

Though the criterion tradeoff can be considered between any feasible criterion points, usually the criterion tradeoff between non-dominated points is considered. In this case the criterion tradeoff is known as the efficient tradeoff, too. Since for non-dominated points any increment of production must be compensated by a loss of level, the value of the efficient tradeoff $\lambda(\mathbf{A}, \mathbf{M})$ is negative.

It is important that the criterion tradeoff is feasible (available). It exists objectively and does not depend on preferences of the decision maker. Thus, criterion tradeoffs differ from value tradeoffs, which describe preferences (value tradeoffs are studied, say, in Keeney and Raiffa, 1976).

In the case of a smooth Pareto frontier, the notion of efficient criterion tradeoff in a single non-dominated point can be introduced. Let point $\mathbf{A}$ approach point $\mathbf{M}$. The value of $\lambda(\mathbf{A}, \mathbf{M})$ in this case tends to a negative value $\lambda(\mathbf{M})$ that is denoted as the *efficient (criterion) tradeoff* for level and production in point $\mathbf{M}$. Roughly speaking, the value of $\lambda(\mathbf{M})$ equals the slope of the tangent line to the frontier in point $\mathbf{M}$. Sometimes the value $\lambda(\mathbf{M})$ is called the *tradeoff rate in point* $\mathbf{M}$. The terms *transformation rate* and *substitution rate* are used sometimes, too.

In the case of a kink point of the Pareto frontier, the efficient tradeoff can not be defined in this way. In this case one usually speaks about efficient tradeoffs to the left and to the right from such point. These values are not equal at the kink point, i.e., the efficient tradeoff changes drastically. Such information may be very important for a decision maker. Because of the importance of the notion of efficient tradeoff, a bi-criterion Pareto frontier is often called the *tradeoff curve*.

Decision Maps and the Edgeworth–Pareto Hull. The Pareto frontier of the variety of feasible criterion vectors cannot be displayed so easily in the case of three, four or a greater number of criteria. The *Interactive Decision Maps* (IDM) technique was developed to display it. Here, we introduce the concept of decision maps, and the IDM technique is considered in the next section.

Let us first modify Figure 1.6 slightly. Since the user is interested in the Pareto frontier of the variety of feasible criterion vectors and other frontiers only spoil the picture, it is reasonable to get rid of useless frontiers and thus simplify the picture. It can be done by broadening the variety of feasible criterion vectors. However, it should be taken into account that the new variety must have the same Pareto frontier. The broadest variety of this kind can be obtained by incorporating all dominated criterion vectors into the variety.

In Figure 1.7, the original and the broadened varieties are displayed. The additional non-feasible points are shaded in the picture. In accordance with the proposal of Stadler (Stadler, 1986), we denote the broadened variety as the *Edgeworth–Pareto Hull* (EPH) of the variety of feasible criterion vectors. It is clear that the dominated frontiers of the variety of the feasible criterion vectors disappear in the EPH. The frontier contains two additional rays that do not belong to the Pareto frontier (the vertical ray that starts from the point **A** and goes down and the horizontal ray that starts from the point **B** and goes to the left). However, the user can recognize these rays easily, since, for the points of the vertical ray, the criterion tradeoff for production and level equals zero, and, for the points of the horizontal ray, the criterion tradeoff for level and production is zero. Therefore, these rays do not hinder analysis of the Pareto frontier. Display of the EPH instead of the original variety plays a minor role in the case of two criteria, but it is extremely important in the case of a larger number of criteria.

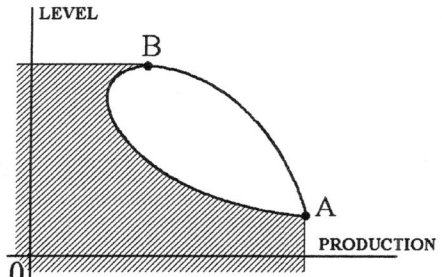 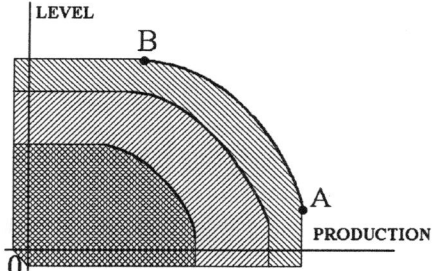

Figure 1.7. The Edgeworth–Pareto Hull of the variety of feasible goals.

Figure 1.8. Superimposed bi-criterion EPH (decision map).

Let us consider the third criterion of the above regional environmental problem, namely pollution of water in the lake. To display the Pareto frontier for all three criteria, one can consider several bi-criterion EPHs related to different constraints imposed on water quality (pollution is not higher, than...) and superimpose these bi-criterion EPHs (see Figure 1.8). Figure 1.8 provides an example of a decision map. Generally speaking, a *decision map* is a picture that displays several Pareto frontiers for any two criteria (tradeoff curves) while several constraints are imposed on the value of the third criterion. Though the tradeoff curves are given in Figure 1.8 in bold, usually it is not needed since the user can, as we have said already, easily recognize the non-dominated parts of the frontier − they are neither vertical, nor horizontal.

The decision map in Figure 1.8 informs the user on the Pareto frontier for all the three criteria. Indeed, a tradeoff curve informs the user on the tradeoffs between production and level for a particular constraint imposed on pollution. A decision map helps the user to compare two neighboring tradeoff curves. This comparison helps the user to realize how efficient is a possible increment of pollution that helps to improve two other criteria. Though the notion of efficient (criterion) tradeoff has not been defined for three criteria in a precise way, the concept is clear qualitatively and one can use it. Efficient tradeoff between three criteria can be understood as the transformation rate for the value of one of the criteria in the process of its transformation into the values of the rest of the criteria, while moving along the Pareto frontier in three-criterion space. In the same way one can speak of an efficient (criterion) tradeoff between four, five and more criteria.

Decision maps for the regional problem. Now let us consider decision maps related to the above regional problem. In the following pictures the values of production are measured in percents of its maximal feasible value, the level of the lake is given in percents of the gap between its maximal and minimal values, and pollution is measured in milligrams of pollutant per one cubic decimeter of water.

In Figure 1.9 the tradeoff curves (production versus level of the lake) are depicted for several constraints imposed on pollution. Production is given in the horizontal axis, and level of the lake is given in the vertical axis. The constraints imposed on pollution are specified directly in the figure. Any tradeoff curve displays non-dominated values of two criteria for a given constraint imposed on the value of the third one. The curve defines the limits of what can be achieved, say, it is impossible to increase the values of agricultural production and level of the lake beyond the frontier.

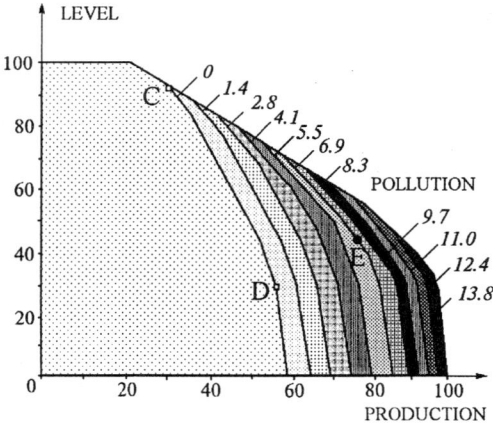

Figure 1.9. A decision map for the regional problem (production versus level of the lake)

The internal tradeoff curve (marked by points **C** and **D**) is related to minimal, i.e., zero pollution. It shows how the level of the lake must be decreased to increase production while keeping zero level of pollution. For small values of production (about 20%), the maximal level (100%) of the lake is feasible. Then, with the increment in production, the maximal feasible level of the lake starts to decrease more and more abruptly (especially, after point **C**). The maximal (for zero pollution) value of production (a little bit less, than 60%) is related to the minimal level of the lake. Note that it is necessary to exchange a substantial drop of the level (about 30% starting at point **D**) for a small increment in the production needed to achieve its maximal value.

Other tradeoff curves have a similar shape. Note that as the allowable level of additional pollution increases, the possible production level increases as well. The outer curve in Figure 1.9 is related to the situation when constraints on the pollution permit maximal non-dominated concentration of 13.8 mg/l. Note that if the level is reasonably high, the tradeoff curves are close to each other. This means that for these levels of the lake even a substantial increment in pollution does not result in economic advantages.

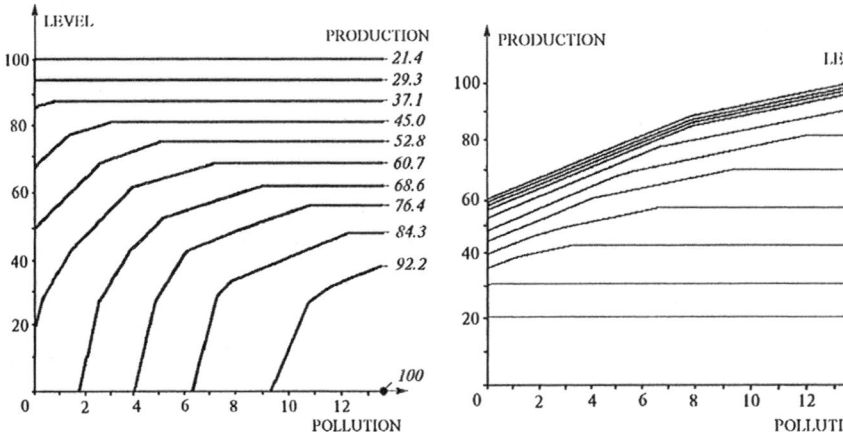

Figure 1.10. Pollution versus level.

Figure 1.11. Pollution versus production.

In Figures 1.10 and 1.11, two different decision maps (pollution versus the level of the lake and pollution versus production) are depicted. They cannot provide any new information on the problem since all information is already displayed in the decision map given in Figure 1.9, but some features of the problem are displayed on the additional decision maps in a more convenient form.

Figure 1.10 shows the tradeoff curves for pollution and level of the lake for different constraints imposed on production. Note that it is desirable to minimize pollution and maximize the level of the lake. Therefore, the Pareto frontier is provided now by the top left frontiers of slices. If production is maximal (100%), the only non-dominated point that is possible is the point with the minimal level of the lake (zero) and the maximal pollution (13.8 mg/l). If production is decreased, the conflict among pollution and levels arises. If, say, production is relatively high (92.2%), one can choose, say, among 40% level and maximal pollution, 30% level and 11 mg/l, or minimal level and 9.3 mg/l. If production is less than 30%, then the conflict among pollution and levels disappears − one has simply to select zero pollution and the level of the lake that is defined by the selected production.

Figure 1.11 displays the tradeoff curves between pollution and production for several constraints imposed on the level of the lake. Here, it is desirable to minimize pollution and maximize production. Therefore, the top left frontiers of slices are of interest. If the level of the lake is high (90 − 100%), there is no conflict between pollution and production − one has to choose minimal (zero) pollution and the production value that is defined by the level of the lake. If the level of the lake is decreased, then the conflict between pollution and production arises. If the level of the lake is less than 30%, its value influences the tradeoff curves for pollution and production to a minimal extent.

3.   Interactive Decision Maps

As we have mentioned already, the idea to display a bi-criterion Pareto frontier in decision problems was introduced by S. Gass and T. Saaty (Gass and Saaty, 1955). They showed that the Pareto frontier of a linear model can be computed and displayed using standard parametric linear programming. One simply has to compute the non-dominated vertices of the variety of feasible criterion vectors and depict them along with the line segments connecting the neighboring vertices.

Application of parametric linear programming, however, is not so simple if the number of criteria is more than two. The linear multi-criteria methods, which develop the idea of Gass and Saaty in a straightforward way, usually construct the list of all non-dominated vertices and provide it to the user (see Zeleny, 1974; Steuer, 1986). Since vertices may not be located on the Pareto frontier regularly, the set of non-dominated vertices may fail to describe the frontier accurately. For this reason, sometimes, the non-dominated faces are provided to the user. The non-dominated faces are actually the multi-criteria analogues of the line segments of the bi-criterion frontier. However, this information provided in the form of large lists of multi-dimensional vectors, is extremely complicated and usually people fail to assess it. Visualization of such information is very complicated, even in the case of three criteria. That is why decision maps and other possible pictures are used extremely seldom. An application-oriented paper (Louie, Yeh and Hsu, 1984, p.53, Figure 7) where tradeoff curves for the case of three criteria are displayed provides one of the rare examples.

The IDM technique develops the idea of Gass and Saaty in an alternative way — it provides on-line visualization of the Pareto frontier in the form of decision maps on the basis of approximation of the feasible criterion set.

Concept of Interactive Decision Maps. It is clear that, in the case of three criteria, one can construct a bi-criterion tradeoff curve by imposing a constraint on the value of the third criterion and using a parametric linear programming code for computing the non-dominated vertices. Therefore, the decision maps displayed in Figures 1.9–1.11 could be constructed with the help of parametric linear programming. Such an approach, however, is effective for constructing a single decision map, since it may take a lot of time to construct a decision map using parametric methods, especially in the case of models with hundreds or thousands of decision variables. Application of the IDM technique for more than three criteria requires exploration of a large number of decision maps requested on-line, and so one cannot hope that a user would wait until the parametric method computes all non-dominated vertices for all tradeoff curves of all requested decision maps. Animation applied in the framework of the IDM technique also requires fast computing of a large number of decision maps. Therefore, to make decision maps practical, some kind of preprocessing is needed that can speed up on-line computing of the decision maps. It is shown later in this chapter that such preprocessing is also important in Internet applications of the IDM technique.

An effective preprocessing procedure developed by us is based on approximating of the EPH for the entire list of decision criteria (from three to seven). It is important that frontiers of bi-criterion slices of an EPH provide the tradeoff curves that constitute decision maps. To be precise, they display Pareto frontiers related to constraints imposed on the values of all criteria except two criteria given on axes. Thousands of slices of a given approximation of an EPH can be constructed in seconds even by a personal computer. Therefore, hundreds of decision maps can be computed and depicted extremely fast. This is why it is possible to display decision maps on request and even animate them on-line. In the

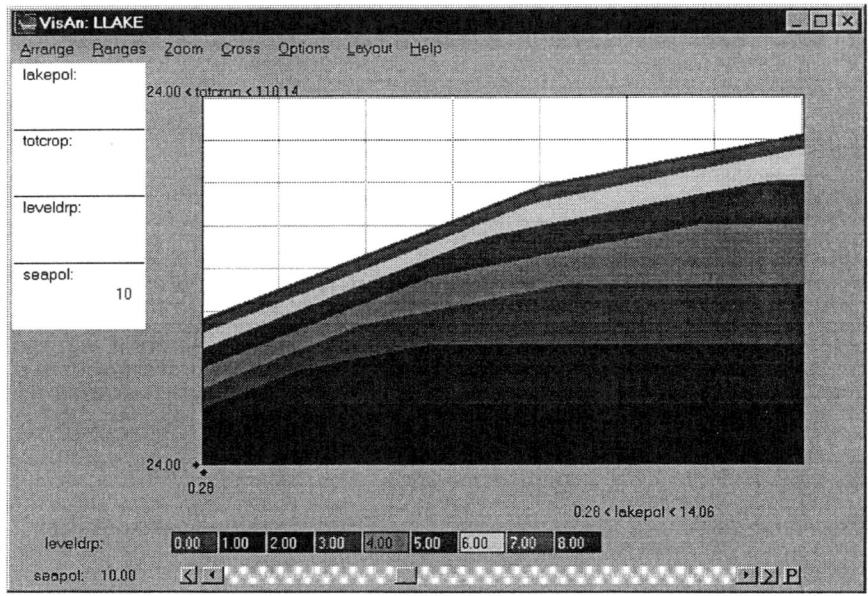

Figure 1.12. Black and white copy of color display for four criteria

case of three criteria, for example, one may want to request any of the three decision maps, or change the number of the tradeoff curves in a decision map, or zoom a part of the map. In the case of more than three criteria, the values of the fourth, fifth and other criteria can be given by sliders of scroll-bars. The position of a scroll-bar specifies the value of, say, the fourth criterion and its movements result in a fast change of the decision map. Let us consider an example.

In Figure 1.12, the black and white copy of a color computer display for the case of four criteria is provided. The criteria are related to the above problem of regional water management. Production (denoted as "totcrop") and pollution of the lake (denoted as "lakepol") are given on axes. Drops in the level of the lake (denoted as "leveldrp") are given by shading. The constraint imposed on the fourth criterion, water pollution at point **A** in Figure 1.3 (denoted as "seapol"), is given by the slider of the scroll-bar located under the decision map

This decision map is very close to the decision map drawn in Figure 1.11. However, two major and several minor differences do exist between this decision map and the map given in Figure 1.11. Among the minor differences one can mention the drop of the level, which is used as the criterion instead of the level of the lake. Moreover, natural units are used to measure the criteria — the drop of the level is measured in feet and production is measured in thousand tons.

The first major difference is related to the fact that bi-criterion slices of the EPH are given on the decision map instead of the frontiers of the slices in Figure 1.11. The slices are depicted in different colors on display that are transformed into different shadings in the black-and-white copy given in Figure 1.12. The second difference consists in the presence of a scroll-bar in the picture. The slider of the scroll-bar can be moved manually. In Figure 1.12 the value of pollution in point **A** is restricted by 10 mg/l. To explore the influence of this constraint, the user can change this value by moving the slider — when it moves, the decision map changes immediately.

The scroll-bar can be used for animation of the decision map, too. Animation of a decision map is based on an automatic movement of the slider, i.e., on a gradual increment (or decrement) in the constraint imposed on the value of the fourth criterion. Fast replacement of decision maps provides the effect of animation.

It is clear that any reasonable number of scroll-bars can be located on the display. Due to this, one can explore the influence that the fifth, sixth and seventh criteria have on a decision map by using manual movement of sliders or animation. Since the preprocessing (approximating of the EPH for the whole list of criteria) is usually completed in advance, various forms of animation can be used, say, a simultaneous movement of several sliders. However, we avoid such effects that may be too complicated for the user. Animation of only one slider at once is used now. Positions of the other sliders during animation can be arbitrary, but they must be fixed.

A collection of animation snap-shots (actually, collection of decision maps) can be displayed in a row. Such a row can be displayed without animation at all. In the case of five criteria, a matrix of snap-shots for several animation runs can be displayed. The snap-shots of the matrix can be selected manually by the user or automatically. The maximal number of rows and columns in such a matrix depends exclusively on the desire of the user and on the quality of computer display. Several examples of matrices of decision maps are given in Chapters 2 and 3. It is important to add that animation of an entire matrix of a decision map is possible — in this case, the value of the sixth (or seventh) criterion that is related to a scroll-bar is changed automatically.

Note that criteria in a decision map can be arranged in an arbitrary order, i.e., any criterion can be associated with an axis, a scroll-bar or the color palette. A related decision map (or even a matrix of decision maps) is displayed very quickly after the request. Ranges of criteria can be squeezed, so a more detailed decision map can be displayed immediately. These opportunities also use preprocessing, i.e., approximating of the EPH carried out in advance. One can consider the approximation of an EPH as a source of an infinite number of possible animation films. Therefore, application of matrices of decision maps and scroll-bars can be used for display of the EPH for any reasonable number of criteria. However, we usually recommend restricting the number of the criteria to seven, otherwise the display becomes too complicated for a human being.

Interactive decision maps as visualization technique. We would like to recall that visualization is a transformation of symbolic data into geometric information that must aid in the formation of mental pictures of the symbolic data. Three main qualities are required of visualization to make it effective (see McQuaid, Ong, Chen and Nunamaker, 1999):

- *simplicity* of visualization that measures the degree to which the picture is immediately understandable;

- *persistence* of visualization that measures the picture's propensity to linger in the mind of the beholder;

- *completeness* of visualization that measures the extent to which all relevant information in the data is given in the picture.

Do decision maps meet these requirements? To answer the question, we explore an interesting parallel among decision maps and topographic maps.

First of all, let us note that the tradeoff curves do not intersect in a decision map (though they may coincide sometimes). Due to this, they look like contour lines of topographic maps . Indeed, a value of the third criterion (related to a particular tradeoff curve) plays the role of elevation value related to a contour line of a topographic map. Say, a tradeoff curve describes such combinations of values of the first and second criteria that are feasible for a given constraint imposed on the value of the third criterion (like "places higher, than..." or "places lower, than..."). Moreover, one can easily estimate, which values of the third criterion are feasible for a given combination of the first and of the second criteria (like "elevation of this particular place is between..."). If the distance between tradeoff curves is small, this could mean that there is a steep grade, i.e., a small move of the tradeoff curve is related to a substantial change of the value of the third criterion. Such information concerning the conflict among three criteria is very important; it means that one has to pay with a substantial change of the third criterion value for a small improvement of the values of the first two criteria.

So, decision maps are fairly similar to topographic maps. For this reason, one can use topographic maps for the evaluation of the effectiveness of visualization in the form of decision maps. Topographic maps have been used for about two hundred years, and educated people usually understand information displayed in them without any problem. Experience of application of topographic maps shows that they are

- *simple enough* to be immediately understood;

- *persistent enough* not to be forgotten by people after the exploration is over; and

- *complete enough* to provide elevation information.

Analogy between decision maps and topographic maps makes us assert that decision maps satisfy the above requirements. In particular, decision maps are complete since they can display information on Pareto frontiers with any desired precision.

Comment concerning the term "decision maps". It must be noted that pictures that we display in the book differ a bit from standard decision maps that are collections of bi-criterion cross-sections of a multi-criteria Pareto frontier (see, for example, Haimes, Tarvainen, Shima and Thadathil, 1990). So, they may be denoted as modified decision maps. Though the modified decision maps look quite similar to standard decision maps, they have several advantages, which are related to computational aspects of the IDM technique. First, since the modified decision maps are given as frontiers of slices of an EPH, they can be provided on-line and animated. The second advantage is related to the fact that the modified decision maps are robust to small disturbances of parameters of mathematical models (see Chapter 9). It is well known that the Pareto frontier often does not possess this property. So, its two-dimensional cross-sections may also be not robust to disturbances. We do not discuss this sophisticated mathematical topic in detail, but it is clear that robustness is required for computation of any mathematical object. In this book we use the words "decision maps" in the sense of modified decision maps.

4. The FGM/IDM technique

The FGM/IDM technique is an application of Feasible Goals Method (FGM) with the help of the IDM technique. First we introduce FGM and then consider its application jointly with the IDM technique.

Introduction to FGM. Let us return to the regional environmental problem and consider screening the variety of feasible strategies on the basis of decision maps. As it was said earlier, the goal approach is applied in the book for decision screening. In this section we describe how the knowledge provided by decision maps can be used in the process of goal identification. The most important feature of the process consists in identification of a non-dominated feasible goal directly in a decision map.

Let us suppose that a user, by studying a sufficient number of various decision maps on-line, has got a proper understanding of Pareto frontier for all the three criteria, especially efficient tradeoff between them, and is ready to identify a preferred feasible goal. To do it, the user must first select one (actually, the most convenient) decision map. Then the user has to specify the value of the criterion given by color. In this way one of the tradeoff curves is selected. Finally, a preferable combination of the other two criteria must be identified at the chosen tradeoff curve (say, point **E** in Figure 1.9). It can be done with the help of the computer mouse. A strategy associated with the identified goal is computed then fairly fast. The user can identify several goal points if needed. In this case the same number of associated strategies will be computed. Several strategies of regional development are given in Table 1.1. Let us consider them.

There are three columns that are related to three different goal points. The first column of the table contains Strategy 1 that is related to a balanced goal point given by point **E** in Figure 1.9. The goal point is characterized by a fairly high production (77%), medium level of the lake (44%) and medium pollution (6.9 mg/l). Release of water through the dam is less than maximal − it is only 5.00 cubic meter per second (compare with possible 6.00 cubic meter per second). In the framework of the strategy, the fourth and sixth technologies are applied in the upper zone, and the fourth, sixth and seventh technologies are applied in the lower zone.

Let us compare the balanced strategy with two other strategies given in Table 1.1. These two strategies have been obtained as the best outcomes for particular interests. The second column contains Strategy 2, which is related to the interests of farmers − it means that production is maximal. The selected goal (production equal to 100%) is given by a single non-dominated point in Figure 1.10. The strategy is characterized by the maximal release of water through the dam and by the use of technologies that are related to an intensive application of water and fertilizers. In particular, the application of fertilizers in the upper zone is two times higher than under the balanced strategy. However,

Table 1.1.   Goal-related strategies.

Strategy	1	2	3
	Goal vectors		
Production	76.8	100	21.4
Level of the lake	44.3	0.00	100
Additional pollution	6.9	13.8	0.00
	Strategies		
Water release through the dam	5.00	6.00	4.50
	Upper zone		
Production	61.88	2.6	18
Production per hectare	2.06	2.75	0.60
Water application	3.60	4.66	0.00
Fertilizer application	1381	2754	0.00
Area distribution			
Technology 1	0.00	0.00	30.00
Technology 2	0.00	0.00	0.00
Technology 3	0.00	0.00	0.00
Technology 4	25.50	0.00	0.00
Technology 5	0.00	0.00	0.00
Technology 6	4.50	21.15	0.00
Technology 7	0.00	0.00	0.00
Technology 8	0.00	8.85	0.00
Technology 9	0.00	0.00	0.00
	Lower zone		
Production	24.52	9.8	6.1
Production per hectare	2.45	2.98	0.61
Water application	1.60	2.40	0.00
Fertilizer application	578	700	0
Area distribution			
Technology 1	0.00	0.00	10.00
Technology 2	0.00	0.00	0.00
Technology 3	0.00	1.25	0.00
Technology 4	5.54	0.00	0.00
Technology 5	0.00	0.00	0.00
Technology 6	1.10	0.00	0.00
Technology 7	3.36	8.75	0.00
Technology 8	0.00	0.00	0.00
Technology 9	0.00	0.00	0.00

productivity is only about 25% higher in the upper zone and about 20% higher in the lower zone.

The third column of Table 1.1 contains Strategy 3 related to the requirements of environmentalists who propose to plan the maximal level of the lake and the minimal pollution of water. This strategy is associated with the goal given in Figure 1.11 by the point with zero pollution on the frontier with 100% level of the lake. One can see that in this case only the first technology is used and production equals about 21% percent of the maximal.

So, one can see that different non-dominated goal points identified at the decision maps result in different efficient strategies of agricultural production. Additional details of decision maps for the regional problem are given in the books (Lotov, Bushenkov, Kamenev and Chernykh, 1997) and (Lotov, Bushenkov and Kamenev, 1999).

Main steps of the FGM/IDM technique. Let us summarize the main steps of the FGM/IDM technique in the process of searching for preferred decision alternatives. It is assumed that a mathematical model that describes the decision problem was prepared and the variety of feasible strategies is given. The main steps include (Figure 1.13):

- approximating of EPH;

- interactive display of decision maps as collections of two-dimensional slices of EPH;

- identification of a feasible non-dominated goal at a decision map;

- computing and display of a feasible decision alternative that results in the identified goal.

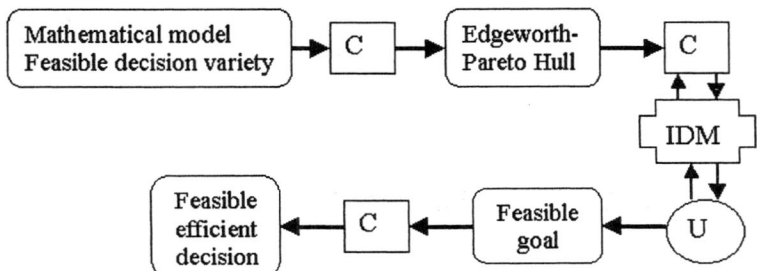

Figure 1.13. Main steps of the FGM/IDM technique. Computer processing is denoted by C, user activity is denoted by U

The computed feasible decision alternative, which output coincides with the identified goal, can be displayed in any convenient form. Many forms of multimedia and GIS may be used for it. In real-life problems, the user may want to apply the experience obtained during the process of searching for the preferred decision alternative − he/she can formulate a new problem with different screening criteria and constraints imposed on variables and performance indicators. Then, a new process of decision screening supported by the FGM/IDM technique can be started. An example of the closed-loop application of the FGM/IDM technique in a real-life DSS is described in Chapter 3.

The FGM/IDM technique in the framework of the MCDM. The FGM/IDM technique is one of multi-criteria decision making (MCDM) techniques. To describe its place among thousands of MCDM methods, we need first to classify these methods.

In accordance with the role of the user, which is the most important feature of any MCDM method, the methods can be classified into four main groups (Cohon, 1978; Steuer, 1986; Miettinen, 1999):

- Methods that do not take into consideration the user's preferences (no-preference methods);

- Methods that are based on development of the user's preference model before a particular variety of feasible decision alternatives is considered (*a priori* preference methods);

- Methods that combine step-by-step exploration of a variety of feasible decision alternatives with the step-by-step development of a user's preference model (interactive methods); and

- Methods that are based on some kind of approximation of the Pareto frontier and on subsequent informing the user of it (*a posteriori* methods).

The decisive role of user is recognized now by the MCDM society, and so methods of the first group are actually out of date. Methods of this kind can be applied in the case of a very large number of decision makers, who have got equal rights, and so their preferences must be aggregated in a fair way (in elections, for example). However, such a topic is out of the MCDM field.

The *a priori* preference methods are based on multi-attribute utility theory (MAUT), which is a mature discipline now. It has developed a lot of theory and algorithms aimed at solution of the MCDM problem (see Keeney and Raiffa, 1976). The theory proved that even if some extremely restrictive assertions are valid, the preference identification should be related to boring and complicated interactions, during which the user has to compare multiple pairs of criterion points. For this reason, the scope of real-life application of MAUT-based methods is rather limited.

Goal programming based on the single-shot identification of a goal (see Charnes and Cooper, 1961; Ignizio, 1985; Steuer, 1986) can be considered as an example of *a priori* preference methods, which are not based on MAUT. Though these methods have found a broad real-life application, the requirement to identify a goal without knowing any information on feasibility frontiers hinders further propagation of goal programming.

The number of the interactive methods is extremely large (see, for example, Miettinen, 1999). However, their applications usually meet the same difficulties as those of MAUT methods, i.e., they are too complicated and require too many boring comparisons of criterion points. For this reason, the scope of real-life application of the interactive methods is not so broad as one could hope.

Finally, the *a posteriori* methods that were started by the above paper (Gass and Saaty, 1955), continue to develop. In the book by Cohon (Cohon, 1978), the idea of Gass and Saaty was reformulated as a generic approach to MCDM problems that is based on Pareto frontier generating. The main principle of MCDM methods of this group consists in preprocessing the problem by constructing some kind of approximation of the Pareto frontier and in further informing on the frontier. It is important that the user is free to select any point of the Pareto frontier − a free search among non-dominated criterion values is assumed. Most of the MCDM methods of this group provide information in the form of large lists of non-dominated points or even of non-dominated multidimensional faces. As we have said already, it is extremely complicated to assess such information for more than two criteria. For this reason, the *a posteriori* methods in this form have not found a broad real-life application yet. To avoid this deadlock, P. Korhonen and J. Wallenius proposed the Pareto Race technique that applies consequent dynamic graphic display of points, which belong to the Pareto frontier, by movement along the Pareto frontier in accordance to the wishes of the decision maker (Korhonen and Wallenius, 1988; Korhonen and Wallenius, 1990). Successful application of the Pareto Race technique proves the need of visualization in the framework of the *a posteriori* methods. However, the IDM technique differs from the Pareto Race technique, since it is aimed at visualization of the Pareto frontier as a whole, but not at listing the non-dominated points or using a consequent display of them. Due to it, the IDM technique informs not only on non-dominated criterion points, but visualizes criterion tradeoffs, too.

It is important to recall that the first studies in the field of constructing of Pareto frontiers (Gass and Saaty, 1955), were aimed at visualization of the Pareto frontier as a whole. Therefore, the IDM technique seems to be a natural approach in the framework of the Pareto frontier methods – we continue the old tradition by visualizing the multi-criteria Pareto frontier in the form of interactive and animated decision maps. It is important to note that, by displaying the Pareto frontier and providing the user with the opportunity to identify a goal point in a decision map, the single-shot goal programming is shifted from the *a priori* group to the *a posteriori* group of MCDM methods.

By the way, free movement along the Pareto frontier used in the Pareto Race technique can be supported by visualization of tradeoff information provided the IDM technique, as driving a car can be supported by a road map. Such an idea was proposed in (Lotov, Bushenkov, Chernykh, Wallenius and Wallenius, 1998).

Internet applications of the IDM technique. In the IDM technique, user interface is based on visualization, and so computer-literate people can master its network applications. Due to it, Internet applications of the IDM technique can provide new opportunities to a large number of users. Several features of the IDM technique simplify its implementation on computer networks. First of all, approximating of a EPH, which requires about 99% of the computing efforts needed, is separated from human exploration of decision maps. Secondly, methods of approximating an EPH are robust and do not require human involvement. Therefore, the approximating process can be performed automatically on a server. Finally, since exploration of decision maps is related to minor computing efforts, it can be fulfilled by means of Java applets on the user's computer.

One Internet application of the IDM technique is devoted to Web-based support of Internet users in their independent search for solutions to public problems. It is clear now that the Internet helps millions of its users to receive information directly from the sources, independently from mass media which inevitably screen (i.e., distort) the information. In particular, Internet users have access to data concerning various public problems. These data are now collected and located on the Internet by national, regional or local authorities as well as by private people. For example, special Web servers that contain various facts concerning particular environmental problems are gradually being established. However, free access to data on recent situations is not sufficient for un-

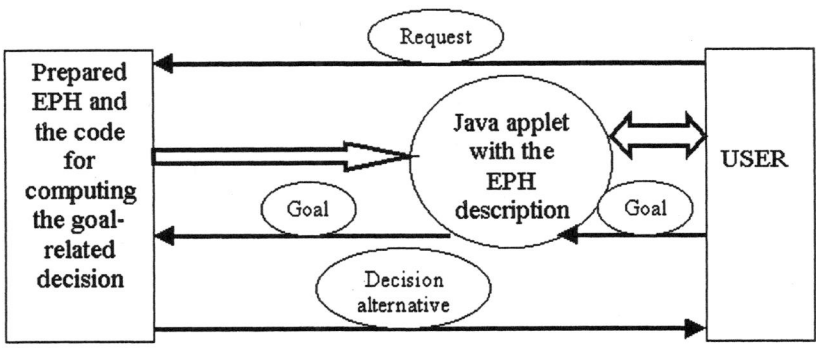

Figure 1.14. Scheme of the demo Internet resource

derstanding of all possible environmental strategies and sifting a preferable one. Special Internet-based methods must be applied for it. The IDM technique can help in this field.

The first version of a Web resource of this kind was started as early as in 1996. It was developed on the basis of Common Gateway Interface scripts that are out of date now. In 2000, a new Web resource was established that is based on application of Java applet technology. The regional problem described in Section 2 is studied in the Web resource. The scheme of the new resource is given in Figure 1.14.

Approximating of the EPH has been carried out in advance; its description was located on a server. After the request, the Java applet and EPH approximation are transmitted to the user's computer. The Java applet provides an opportunity to obtain various decision maps using one of the Web browsers. The user can select a view, i.e., arrange the criterion location, and change ranges of the criteria. The ranges can be changed in small steps, and so the user receives an animated picture of zooming.

Finally, the user can identify a goal. Goal identification, however, is a bit different than in the IDM software described above. Now the preferred goal can be located at any point of the decision map using a special moving tradeoff curve associated with the slider of the scroll-bar related to the color. A cross located at the curve can be moved along the curve by the computer mouse. After fixation of a position of the cross, i.e., identification of the goal, the related information is transmitted from the user's computer back to the server. It computes a related decision alternative and transmits it to the user's computer. The user may receive the alternative after several seconds or minutes, depending on the connection quality. Reader can have a look at the decision maps provided by the applet directly at the Web site (see Lotov, Bushenkov, Kistanov and Chernov, 2000).

A new advanced Java applet that implements scroll-bars has been developed recently. It helps to apply the IDM technique for exploration of decision problems described by more than three decision criteria. The advanced Java applet was used in a Web application server that implemented the Reasonable Goals method (RGM) for databases described in Chapter 4. RGM is a development of FGM, and it applies the IDM technique, too. Therefore, both FGM and RGM can be applied now in the framework of Web resources. An example of a Web application server based on the RGM/IDM technique can be used for selecting of preferable items from large lists through the Internet (see Section 5 of Chapter 4). In the Epilogue it is shown that both FGM/IDM and RGM/IDM techniques can be used in the Web resources used for supporting the new democratic paradigm of environmental decision making.

5. Computational aspects of the IDM technique

The main computational problem in the framework of the IDM technique consists in approximating the variety of feasible criterion vectors or its EPH. Here we outline the problem just to give an idea of the mathematical methods that were elaborated in the process of development of the IDM technique.

Three groups of methods for approximating the variety of feasible criterion vectors or its EPH were developed. Methods of the first group deal with linear models with a relatively small number of decision variables. These methods are based on direct application of the classic method of convolution of linear inequality systems proposed by J.B. Fourier in the first part of the 19th century (Fourier, 1826).

The methods of the second group are based on approximating the varieties of feasible criterion vectors by polytopes. They can be used if the variety of feasible criterion vectors is a compact convex multidimensional body. In particular, the methods can be applied in the case of linear systems with a large number of decision variables (maybe, thousands of them). Such methods can be used for approximating of an EPH even in the case of non-linear systems if the EPH is convex. However, the number of criteria should be not too great (as a rule, not larger, than seven). It is interesting that the psychological theory states that a normal human being can operate not more, than seven objects (see, for example, Solso, 1988). So, the number of criteria studied by the second group of methods seems to be sufficient.

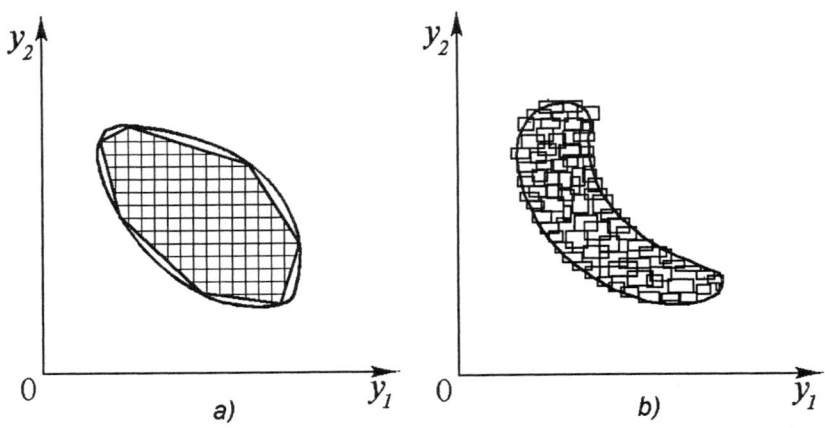

Figure 1.15. Approximation of a convex variety by a polytope (a) and of a non-convex variety by a collection of boxes (b)

The methods of the third group are used for approximating the non-convex varieties of feasible criterion vectors or their EPH. Such methods are extremely important since the approximated varieties are usually non-convex for non-linear models. The difference between approximating of convex and non-convex varieties is illustrated for the case of two criteria in Figure 1.15. A convex variety can be approximated by a single polytope (Figure 1.15a), and so its EPH can be approximated by the sum of a polytope and of the cone of dominated criterion points. In contrast, approximating of non-convex varieties requires more complicated methods that apply approximating by collections of boxes (Figure 1.15b). For approximating of a non-convex EPH, collections of cones may be used.

If methods of the first two groups are applied, visualization of the Pareto frontier is based on display of collections of bi-criterion slices of the approximation. These slices can be computed very fast since the approximating body is a polytope. Slices of collections of boxes can be constructed relatively fast, too. However, approximating by collections of boxes is a more time-consuming procedure. For this reason, it is desirable to transform a problem into a convex formulation, if possible. Two ways of such transformation are illustrated in Figures 1.16 and 1.17.

First, the EPH can happen to be convex even for a non-convex variety of feasible criterion vectors. An example is given in Figure 1.16a. Another opportunity to get a convex variety is based on the application of the concept of a partial Edgeworth–Pareto Hull, which is close to the concept of the EPH, but applies information concerning the improve-

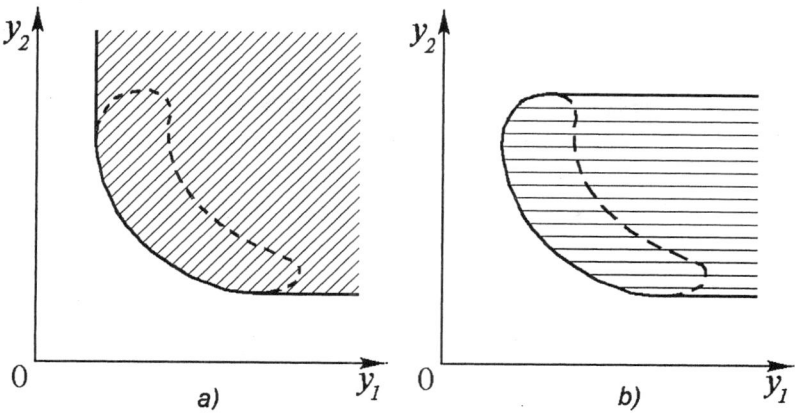

Figure 1.16. The EPH (a) and the partial EPH (b); it is desirable to decrease the value of the first criterion only

ment direction only for a part of the criteria (see Figure 1.16b). The partial EPHs are used in Chapter 2.

Another approach is based on exploration of the convex hull (envelope) of a non-convex variety of feasible criterion vectors. Moreover, the convex hull of an EPH can be approximated (Figure 1.17). Due to the enveloping, the IDM technique can be applied for exploration of non-linear systems. Application of the IDM technique for exploration of the Pareto frontier of the envelope (convex hull) of a feasible criterion set and identifying the goals at the envelope is called the Reasonable Goals method (RGM).

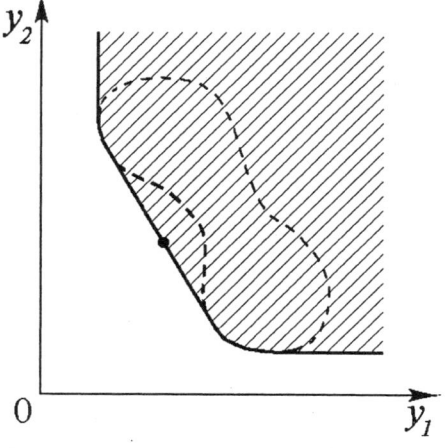

Figure 1.17. Envelope of the EPH of a non-convex variety.

As we have said already, the RGM/IDM technique is a further development of the FGM/IDM technique. An important particular case of the RGM/IDM technique is provided by its simplest modification aimed at the visualization of large relational databases (tables) of decision alternatives. Such tables may contain millions of alternatives. In the case

of the convex hull of an EPH, the user obtains "averaged" dependence
among the criteria. Due to this, the picture is much simpler and it can
be assessed much easier. However, the user has to pay for it — a selected
goal usually is not feasible. Fortunately, it belongs to the envelope,
points of which are fairly close to the variety of feasible goals. In the
RGM, several decisions that are close to the identified goal (in various
senses) are provided to the user. A detailed description of the RGM
is given in Chapter 4, including Internet application of the RGM/IDM
technique.

The IDM technique is a particular form of the Generalized Reachable
Sets (GRS) method (Lotov, 1973b; Lotov, 1975a; Lotov, 1984a; Lotov,
1989). The GRS method was developed for exploration of non-closed
mathematical models, i.e., models with input variables. It consists in
approximation and display of the variety of attainable output vectors
for a given variety of feasible input vectors. In the MCDM context,
the GRS method results in mapping the feasible decision set into the
criterion space. This job is done by methods for approximation of multi-
dimensional sets. Such methods that constitute the basis of the IDM
technique are described in Chapter 6.

Reachable sets for dynamic systems (varieties of possible states of a
dynamic system at given time-moments) provide another example of the
varieties of attainable output vectors. Therefore, computational methods of the GRS method can be used for approximation of reachable sets.
This topic in outlined in Chapter 7 of this book.

Other applications of the GRS method and approximation techniques
include aggregation of economic models (Lotov, 1982), coordination of
decisions using such economic models (Lotov, 1983; Lotov and Ognivtsev, 1984), constructing of aggregated production functions (Dzholdybaeva and Lotov, 1989), identification of parameters of models (Kamenev,
1998; Lotov, Bushenkov and Kamenev, 1999), etc.

The first detailed description of the GRS method in English was given
in Chapter 18 of the book (Lieberman, 1991).

APPENDIX 1.A: Mathematical model of regional development

In this appendix, the mathematical model of the regional development used in
Section 2 is described.

The production in an agricultural zone is described by a technological model, which
includes N agricultural production technologies. Let x_{ij} be the area of the j-th zone
where the i-th technology is applied. The areas x_{ij} are non-negative

$$x_{ij} \geq 0, \quad i = 1, 2 \ldots, N, \quad j = 1, 2, \tag{1.A.1}$$

and restricted by the total areas of zones

$$\sum_{i=1}^{N} x_{ij} = b_j, \quad j = 1, 2. \tag{1.A.2}$$

The i-th agricultural production technology in the j-th zone is described by the parameters a_{ij}^k, $k = 1, 2, 3, 4, 5$, given per unit area, where
a_{ij}^1 is production,
a_{ij}^2 is water application during the dry period,
a_{ij}^3 is fertilizers application during the dry period,
a_{ij}^4 is volume of the withdrawal (return) flow during the dry period,
a_{ij}^5 is amount of fertilizers brought to the river with the return flow during the dry period.

Then, one can relate the values of performance indicators for a zone to the distribution of the area among technologies

$$z_j^k = \sum_{i=1}^{N} a_{ij}^k x_{ij}, \quad k = 1, 2, 3, 4, 5, \quad j = 1, 2, \tag{1.A.3}$$

where
z_j^1 is production in the j-th zone,
z_j^2 is water application during the dry period,
z_j^3 is fertilizers application during the dry period,
z_j^4 is volume of the withdrawal (return) flow during the dry period,
z_j^5 is amount of fertilizers brought to the river with the return flow during the dry period.

The water balances are fairly simple. They include changes in water volumes during the irrigation period. The deficit of the inflow into the lake due to the irrigation equals $z_2^1 - z_4^1$. The additional water release through the dam during the dry period is denoted by d.

Let T be the length of the dry period. The level of the lake at the end of the dry period L_T is assumed to be approximately given by

$$L_T = L - (z_2^1 - z_4^1 + d)/\alpha \tag{1.A.4}$$

where L is the level without irrigation and additional release, and α is a given parameter. It is assumed that the release d and water applications are constant during the dry season. Then, the flow in the mouth of the river near monitoring point **A** denoted by v_A equals

$$v_A = v_A^0 + \left(d - z_2^2 - z_4^2\right)/T$$

where v_A^0 is the normal flow at point **A**. The restriction is imposed on the value of the flow

$$v_A \geq v_A^*$$

where the value v_A^* is given. So, the following restriction is included in the model

$$v_A^0 + \left(d - z_2^2 - z_4^2\right)/T \geq v_A^* \tag{1.A.5}$$

The increment in pollution concentration in the lake denoted by w_L is assumed to be equal to

$$w_L = z_5^1/\beta, \tag{1.A.6}$$

where β is a given parameter. This means that in the formula for pollution concentration we neglect the change of the volume of the lake in comparison with the normal volume.

The pollution flow at monitoring point **A** is calculated as

$$z_5^2/T + q_A^0$$

where q_A^0 is the normal pollution flow. This means that we neglect the influence of fertilizers application in the upper zone on pollution concentration in the mouth. Then, the concentration of pollution at point **A** denoted by w_A equals to

$$w_A = (z_5^2/T + q_A^0)/v_A.$$

Taking into account the above expression for v_A , we obtain

$$w_A = (z_5^2/T + q_A^0)/\left(v_A^0 + \left(d - z_2^2 - z_4^2\right)/T\right).$$

The restriction

$$w_A \leq w_A^*$$

on pollution concentration at point **A** is imposed where the value w_A^* is given. So, the following restriction is included in the model:

$$(z_5^2/T + q_A^0)/\left(v_A^0 + \left(d - z_2^2 - z_4^2\right)/T\right) \leq w_A^*$$

or

$$(z_5^2/T + q_A^0) \leq w_A^* \left(v_A^0 + \left(d - z_2^2 - z_4^2\right)/T\right). \tag{1.A.7}$$

It is important that the restriction (1.A.7) is linear, too. Due to this, all expressions of the model (1.A.1)–(1.A.7) are linear.

Criteria:

The first criterion, production is the sum of productions in both zones

$$y_1 = z_1^1 + z_2^1.$$

The second criterion is the final level of the lake L_T given by (1.A.4)

$$y_2 = L - (z_2^1 - z_4^1 + d)/\alpha.$$

And the third one, the additional pollution in the lake w_L, is given by (1.A.6)

$$y_3 = z_5^1/\beta.$$

Chapter 2

ACADEMIC APPLICATIONS

Several examples of applications of the FGM/IDM technique are considered in this chapter. They show how the technique can be applied in various environmental decision problems and provide by this a new visualization-based methodology for the development of efficient environmental strategies. First we describe the search for efficient ocean waste disposal strategies based on the example of the New York Bight. In Section 2, the technique is applied in the process of searching for efficient strategies of regional agricultural development while environmental characteristics are taken into account. Section 3 is devoted to exploration of long-term development of a national economy. In this study, pollution and unemployment are considered in addition to economic growth. Section 4 describes an FGM-based software tool that supports searching for efficient regional strategies for trans-boundary air pollution abatement. Section 5 outlines the process of searching for efficient strategies aimed at the abatement of the global climate change on the basis of the FGM/IDM technique.

Simplified mathematical models of environmental systems are used in these studies. The simplified models are usually based on integration of simplified descriptions of sub-systems of the environmental systems. The methods for constructing such models are described in the next chapter where they are exemplified by the process of developing a simplified integrated model used in environmental water management. In this chapter, we use simplified models without detailed description of the process of their construction.

1.   Ocean waste management decisions

In this section we illustrate the FGM/IDM technique by an ocean waste disposal example, requiring difficult decisions concerning cost and pollution. We reconsider the old problem of choosing sewage sludge disposal sites in the New York Bight. This section is based on the paper (Lotov, Bushenkov, Chernykh, Wallenius and Wallenius, 1998).

Sewage Sludge Disposal Problem.   Contamination of the New York Bight has been a concern of the USA Environmental Protection Agency (EPA), and the neighboring municipalities for many years. Concern for water quality in the Bight region is long standing, particularly for waters in the inner portion of the Bight. Highly publicized pollution-related episodes that have occurred over the past decades have had a lasting impact on public opinion. Being concerned about the contamination of the inner Bight region, in 1985 the EPA ordered New York City and the remaining users of the inner Bight region to begin shifting their dumping operations to the 106-mile site. (Figure 2.1).

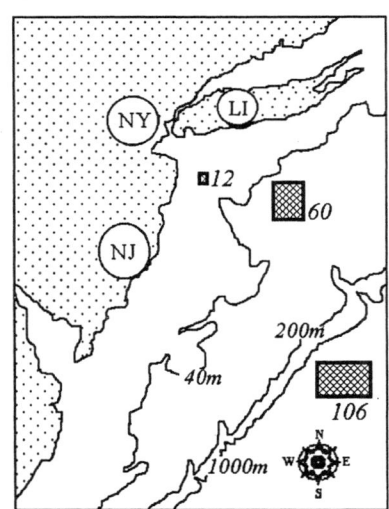

Figure 2.1. Map of the New York Bight region.

In this study we reexamine, following (Wallenius, Leschine and Verdini, 1987) and (Leschine, Wallenius and Verdini, 1992), the EPA decision in a way, which permits simultaneous multiple-site dumping.

Three alternative disposal sites were considered in the model: the 12-mile site, the 60-mile site, and the 106-mile site. We assumed that a combination of the above sites was a possibility, such that all three sites could be used at the same time in different portions. In the model all sludge was assumed to be produced in New York City, NY, (where 52% is produced), New Jersey, NJ, (41%), and Long Island, LI, (7%). Production of sludge was assumed to be constant from year to year. Two types of vessels were used for the transportation of the sludge: towed barges, and self-propelled barges. The constraint set of the model contained four parts:

1 constraints to ensure dumping of all generated sludge;

2 constraints on annual dumping capacity of barges;

3 constraints on amount dumped at each site;

4 constraints related to the ocean's assimilative capacity.

The following three criteria were used to evaluate different sludge disposal strategies:

- total cost of sludge disposal operation (in millions of US$);

- pollution level at inshore monitoring station (pollution concentration, in percent of a given value);

- pollution level at offshore monitoring station (pollution concentration, in percent of a given value).

The decision variables included the number of self-propelled/towed barge trips from source (NY, NJ, LI) to site (12-, 60-, 106-mile sites). A formal description of the model is given in (Leschine, Wallenius and Verdini, 1992).

Application of the IDM technique. On the basis of the model outlined above, the EPH for the three criteria was approximated. By using the approximation, various decision maps were displayed and explored. To begin with, we consider only two criteria. Let us restrict total cost by some value, say $15 million. Then all feasible values of inshore pollution and of offshore pollution are given by the variety of feasible criterion vectors on the criterion plane (see Figure 2.2 where the variety is shaded).

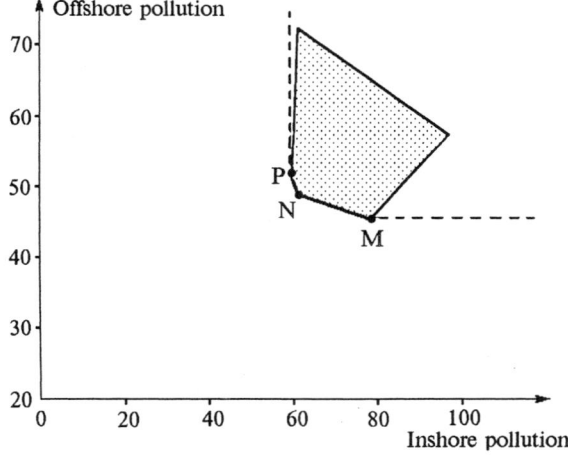

Figure 2.2. The variety of feasible criterion vectors and its EPH

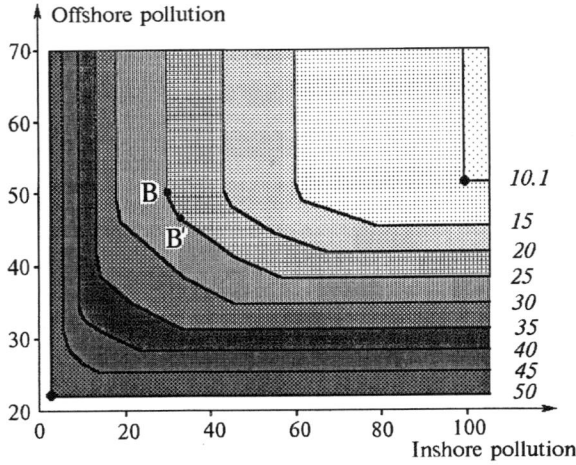

Figure 2.3. A decision map for several fixed costs

The frontier of its EPH is depicted by a dashed line. Once again, the variety of feasible criterion vectors and its EPH have the same Pareto frontier.

Since it is preferable for users to decrease both inshore and offshore pollution, we are interested in its left lower frontier, which is the Pareto frontier. Note that in the vicinity of point **M**, a small decrement in the offshore pollution requires a substantial increment in the inshore pollution. Vice versa, in the vicinity of point **P**, just a small rise in the inshore pollution results in a sharp decrement in offshore pollution. One can easily understand how the offshore pollution is transformed into the inshore pollution if efficient strategies are used. In other words, the criterion tradeoff for inshore and offshore pollution is displayed in a clear way in Figure 2.2.

The decision map in Figure 2.3 is constructed as collections of bi-criterion slices of the three-criterion EPH. Pareto frontiers of the slices (tradeoff curves for inshore and offshore pollution) are related to different constraints imposed on values of total cost. These constraints are given directly in the figure. The decision map helps to understand the influence of an increment in the total cost on the improvement of the environment (i.e., a reduction in the inshore and/or the offshore pollution). The constraints on the value of cost are between the minimal \$10.1 million and the maximal \$50 million. Pareto frontiers in Figure 2.3 are drawn with heavy lines. They have the following important feature: there is

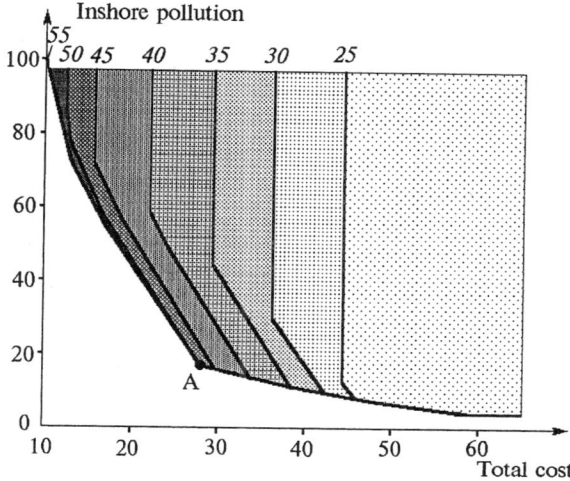

Figure 2.4. Cost versus inshore pollution

a conflict between inshore and offshore pollution, except for the $10.1 million and $50 million frontiers, which consist of just one point. It is important that the criterion tradeoff changes in process of moving along Pareto frontiers, and it depends on cost, too.

Let us compare the distances between pairs of Pareto frontiers related to different costs. The $10.1 million and $15 million frontiers are quite distant while the distance between the $15 million and $20 million frontiers is obviously smaller. This means that the extra $5 million investment has much more impact if the cost equals $10.1 million rather than $15 million.

This phenomenon can be easily explored from another angle of view with an alternative decision map (Figure 2.4). Pareto frontiers on the "Cost versus inshore pollution" decision map are given for several constraints imposed on values of offshore pollution. They range from 25% to 55% of the maximum value of offshore pollution. One can see that every frontier has a kink, where the tradeoff between inshore pollution and cost changes drastically − it is much smaller to the right of the kink than to the left of it. If offshore pollution equals 55% of its maximum value, the kink occurs at point **A**. It is interesting to note that point **A** is very close to another frontier that corresponds to 50% of the maximum value of offshore pollution.

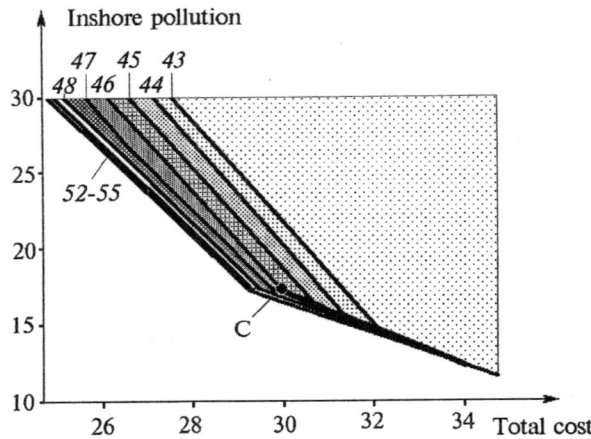

Figure 2.5. Cost versus inshore pollution with the goal point **C**

Figure 2.5 provides a zoomed part of the decision map "Inshore pollution versus total cost" for which the constraint imposed on offshore pollution varies between 43% and 55% of the maximum value.

It is interesting to note that, for relatively small values of inshore pollution (between 15–20%), a growth of the offshore pollution beyond 46% is practically useless – it results only in a small move of the tradeoff curve. For this reason, the following feasible goal (point **C**) may be of interest:

- cost equals $30 million,

- inshore pollution equals 17.5%,

- offshore pollution equals 46%.

Suppose that the user decided to identify the point **C** as the goal. The associated decision provided by the computer is the following: transport all the waste to the 60-mile site. Strategies associated with other points of Pareto frontiers could be found as well.

2.   Search for efficient water-related strategies of regional agricultural development

The results of the study carried out at the International Institute for Applied Systems Analysis (IIASA) in the middle of the 1980s is described in this section. The study was performed in the framework of the project "Regional water management strategies" headed by Dr. S. Orlovski. The research was based on experience obtained in the process of application of the FGM/IDM technique in the regional study described in Chapter 1. However, this time a real-life region was considered. This section is based on the paper (Kamenev, Lotov and Walsum, 1986).

Introduction. The Southern Peel Region in the Netherlands was studied. The region has intensive agricultural production, which includes both crop and livestock. Future agricultural development of the region may result in negative environmental consequences. The problem is related mainly to livestock production, which results in animal slurries as by-products. Animal slurries can be used for fertilization. To do this, slurries produced during the summer and the winter must be temporarily stored in tanks till the next spring and only then applied to the land. However, the storage is restricted by storage capacities, and for this reason a part of slurries may be stored outside of the tanks. In turn, this may result in groundwater pollution. Intensive application of animal slurries for fertilization may have the same result. Groundwater pollution may have not only negative environmental consequences; negative consequences in the field of population health may arise, too. It is related to the fact that local population uses the water from deep aquifers. Water quantity problems may arise, too: groundwater level is very important in several natural zones located in the region.

Model. We used an aggregated version of the model of the water resources and agricultural production in the region developed in (Orlovski and Walsum, 1984). The Orlovski–Walsum model (OW-model) linked submodels of agricultural production, water quantity and quality processes and soil nitrogen processes. The Southern Peel region was divided into 31 subregions. The division was based on classes of groundwater conditions and soil physical units. In the framework of the OW-model, a year was split into two parts:

- "summer" which starts on April 1, and

- "winter" which starts on October 1.

For any region the groundwater levels in the beginning and at the end of the summer were considered. A model that relates deviations of the groundwater levels from their natural levels to water supply extractions was developed by means of linear parameterization of a complicated model of ground water flow developed at the Institute for Land and Water Management, Wageningen, the Netherlands. Influence of water extractions in particular sub-regions on groundwater levels in all sub-regions was described on the basis of two influence matrices: winter influence matrix and summer influence matrix. The influence matrices constructed in simulation experiments that were carried out in the Netherlands (see Orlovski and Walsum, 1984, for details). Several constraints on the groundwater levels were introduced during the parameterization process to describe hydro-geologic circumstances. Environmental constraints on groundwater levels were imposed in some subregions as well. Leaching of nitrate to groundwater was also approximated in the aggregated model by a linear form.

Two types of water extractions were considered: agricultural extractions and public water supply. Demands of public water supply in winter and in summer were given. Agricultural extractions depended on agricultural production, which was described in the model. The agricultural production was described by means of technologies. A technology is actually a combination of agricultural activities involved in growing and processing of a certain crop or livestock. Technologies differ from each other by their inputs and outputs. Two types of agricultural technologies were considered: technologies that use land and technologies that do not use it. It turned out to be convenient to further divide technologies into the technologies involving livestock and technologies not involving livestock. The following types of inputs (resources) characterize a technology: labor, capital, and water. Land-using technologies are additionally characterized by the input of nitrogen supplied by fertilization. Each technology is also characterized by the output (crop yields, livestock products). Technologies that involve livestock are additionally characterized by outputs of animal slurries produced as by-products.

Application of agricultural technologies was described in terms of their intensities. For land-using technologies, the intensities were measured in areas of land allocated to these technologies. For technologies that do not use land and that involve livestock, intensities were measured in number of livestock-heads. For other technologies, intensities were measured in, say, amount of slurry transported to outside the region.

Inputs of such resources as labor and capital were given by corresponding quantities per unit of intensity of a technology, for example, by amount of labor per unit area of land for a technology. It was also assumed that the water inputs for technologies, which do not use land, are quantified in the same way (amount per unit intensity). However, in the case of water inputs and yields of land-use technologies, the description is different. One reason for this difference is that both the water availability and the output of land-use technologies depend on weather conditions. Another reason is that the availability of water is also influenced by activities in the region, especially pumping of groundwater. In order to take into account the respective possible variations in the performance of land-use technologies, a finite number of options for each such technology were considered that covered a suitable variety of typical water availability situations in each subregion. The term "sub-technologies" was used to describe such options.

Each sub-technology in the model was characterized by crop productivity, by the corresponding seasonal averages of soil moisture and of actual evaporation, as well as by the total nitrogen requirement per unit area of land. Actually, the "demand" for soil moisture was considered in the model. Satisfaction of the demand together with satisfaction of the requirement for nitrogen guaranteed obtaining of certain crop productivity. Each of ten land-using technologies was divided into three sub-technologies.

The crop production technologies included greenhouse horticulture, intensive field horticulture, extensive field horticulture, production of potatoes, cereals, maize with low nitrogen application, maize with medium nitrogen application and maize with high nitrogen application. The animal production technologies included grassland with high cow density, grassland with low cow density, beef calves breeding, pigs for feeding, pigs for breeding, egg-laying chickens and broilers production.

Soil moisture that played an important role in the model of crop production was related to precipitation and to sprinkling. The sprinkling volume was restricted by sprinkling capacities. Sprinkling from surface water was connected with irrigation. The surface water supply capacity was limited. The moisture content of the root zone in summer was related to the moisture content at the beginning of summer and to the level of groundwater at the end of winter.

As it was said earlier, the technologies that involve livestock produce animal slurries as by-products that can be used as fertilizers for land-using technologies. Five kinds of slurries were described in the model: cattle slurry, beef calf slurry, pig slurry, chicken slurry and broiler manure. Two directions of slurry application were considered: on the arable land and on the grassland.

Labor requirements were satisfied by the labor of the region and the labor hired from outside of the region.

In our research, the original OW-model was aggregated. The region was divided into five economic clusters, each of them consisting of a number of subregions (see Figure 2.6). Though the explored model was aggregated, it was still sufficiently large: it had 460 variables and 672 constraints.

Exploration of the model. Two environmental and two economic criteria were studied in the research. The environmental criteria included

- nitrates concentration in deep aquifers, C, measured in milligram per liter of water; and

- groundwater level in natural zones, h, characterized by the maximal value of the fall for all zones and measured in centimeters.

The economic criteria were:

- investment in agricultural production, I; and

- additional yearly pure income, W.

Values of both economic criteria were measured in millions of Dutch florins. It is clear that one is interested in minimizing both environmental criteria. As to economic criteria, increment in income is preferable. Investment in agricultural production, however, has no certain improvement direction.

In this study, we applied approximation and visualization of the frontier of a Partial Edgeworth–Pareto Hull (PEPH). It means that we specified improvement direction (maximization) only for one criterion (additional income W). However, it turned to be sufficient to make the PEPH convex for this model (in contrast, the variety of feasible criterion vectors is not convex). Due to this, it was possible to approximate it using polyhedral sets.

Let us consider several slices of the PEPH. One of the slices is given in Figure 2.7. Additional yearly pure income W is located in the horizontal axes, and concentration of nitrates in deep aquifers C is located in the vertical axes. The values of the other two criteria satisfy the constraints $I < I_{max} = 250$ million florin, $h < h_{max} = 20$ cm.

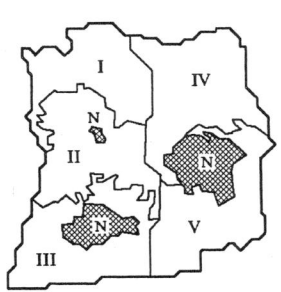

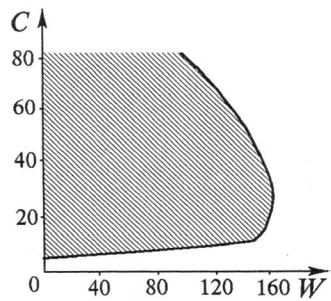

Figure 2.6. Five economic clusters of the region. Natural zones are depicted by **N**.

Figure 2.7. The frontier for two criteria

The lower right frontier of the PEPH (tradeoff curve) is of interest. The value $W = 0$ is related to nitrates concentration of $C = 8$ mg per liter. The increment in income W requires first a moderate growth of pollution level, but after the kink concentration of nitrates starts to grow very fast (is clearly visible in Figure 2.7). The kink is characterized by income of about 150 million and pollution about 15 mg per liter. Maximal income of 164 million is related to the concentration of 25–30 mg per liter. So, the relatively small (about 10%) increment in income to its maximal value requires doubling the pollution level! Once again, it proves how it is dangerous to maximize economic criteria without properly considering the tradeoff between them and the environmental criteria. Further movement along the frontier is not interesting − it proves only that the increment of pollution requires resources as well.

Now let us consider the influence of two other criteria, i.e., of investment and groundwater level in natural zones. This information is given in Figures 2.8–2.9. In Figures 2.8a and 2.8b, three frontiers for additional income W and concentration of nitrates C are given for three constraints imposed on the fall of groundwater level. Investment is zero in Figure 2.8a, and it is maximal in Figure 2.8b. The largest of these slices was displayed already in Figure 2.7.

One can see that increment in investment makes the frontiers move, but it practically does not change their shape. However, increment of h makes the shape change. The change is visible, say, in Figure 2.8a. It is clear that the increment of h from 5 cm to 10 cm results in a substantial growth of income, especially if the pollution is 20 mg per liter and higher, but its increment from 10 cm to 20 cm gives practically nothing.

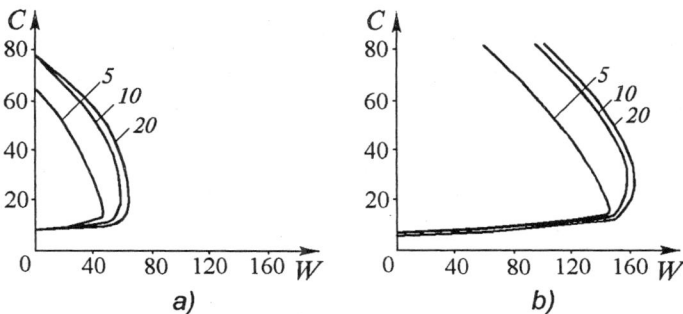

Figure 2.8. Income versus pollution for three constraints imposed on the fall of groundwater level: a) investment is zero; b) investment is maximal

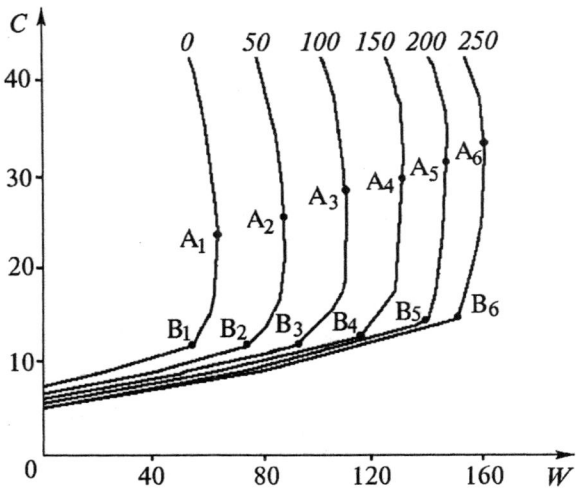

Figure 2.9.   Income versus pollution for several values of investment ($h \leq 10$ cm)

Now let us consider the influence of the investment in detail. Figure 2.9 provides the most interesting parts of the slices depicted for the same criteria as in Figure 2.8, but several constraints imposed on investment are specified in the figure. The value of h is not greater than 10 cm. Once again, it is visible that increment in investment makes the frontiers move, but does not change their shape. One can see how the kink points denoted by **B1, B2,...,B6** depend on investment. Let us compare the kink points with the points of the maximal income **A1, A2,...,A6**. First of all, it is clear that the locations of points **A1, A2,...,A6** are not stable, a minor change in the data can cause points to move in the direction parallel to the vertical axis.

This example shows that location of points related to maximal criterion values may not be stable. It seems that such a situation is a normal feature of real-life problems. It means that the end-points of Pareto frontiers, which coincide with **max**imal criterion points, are also not stable. For this reason, one has to display the whole slice of the EPH (or PEPH) as we did in the framework of the FGM/IDM technique.

Figure 2.9 proves that it is not reasonable to increase the income beyond the kink level. However, since the above instability of points **A1, A2,...,A6** is not of great importance: struggling for the maximal income and disregarding environmental aspects is simply stupid in this case. Figure 2.8 shows that the same inference is true for all values of groundwater level. Note that we do not recommend applying the points

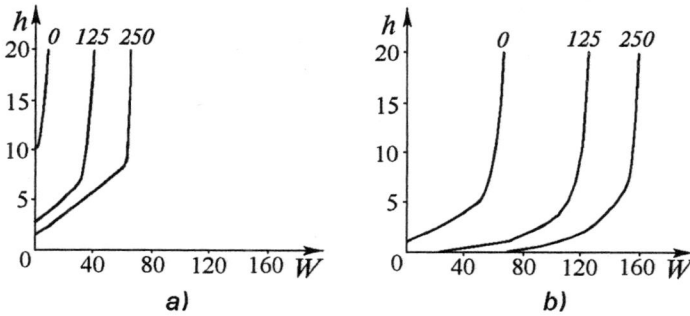

Figure 2.10. Income versus groundwater level for several investments: a) $C \leq 8$; b) $C \leq 25$

B1, B2,...,B6; the user has to decide about the feasible goal. It may happen that further water quality improvement is preferable.

Figure 2.10 displays slices of the PEPH from a different angle of view: frontiers for income versus groundwater level are given depending on constraints imposed on investment (they are specified in the figure). In Figure 2.10a, nitrates concentration is not higher than 8 mg per liter, and it is not higher than 25 mg per liter in Figure 2.10b. One can see that the frontiers do not touch the horizontal axis in Figure 2.10a. It means that it is impossible to provide the initial level of groundwater for this value of nitrates concentration. In contrast, Figure 2.10b shows that for higher nitrates concentration a sufficiently large investment can help to avoid a fall in the groundwater level. Non-zero investment means in turn that a change of intensities of agricultural technologies is not sufficient for supporting the groundwater at its original level. So, a non-zero investment in new technologies is needed also from the environmental point of view.

It would be interesting to consider different collections of slices and discuss the regional strategies related, say, to points **B1,B2,...,B6**. However, we cannot do it because of the complexity of the description of a strategy in the model (recall that it is described by 460 variables). Several strategies and other details of the research are discussed in (Kamenev, Lotov and Walsum, 1986).

Note that a moderate value of investment can solve environmental problems of this region − it can be sufficient for conservation of the groundwater level in natural zones. In this aspect, this region differs drastically from the example region discussed in Chapter 1 where environmental and economic criteria were in a conflict that could not be settled so easily.

3. Analysis of long-term development of a national economy

In this section, the FGM/IDM technique is applied in the framework of analysis of long-term strategies for development of a national economy. A simple model, which describes a transition process from an "old" polluting technology to a "modern" clean one, is studied taking into account levels of consumption, pollution and unemployment. The section is based on the paper (Lotov, Chernykh and Hellman, 1992).

Models of economic growth are usually studied in order to gain a better understanding of the general problems of long-term development of a national economy. The problem of distribution of the national income between consumption and investment is usually of special interest. Here, we demonstrate that multi-criteria approach can bring new important results, especially if environmental and social aspects are taken into account.

Description of the model. The potential opportunities for development of a national economy are studied here on the basis of a simple, illustrative model. We consider only one product produced by several technologies that have different economic and environmental features. The product is used both for investment and consumption. Output of a technology is restricted by the capacity of funds that implement the technology. Note that the funds can be used only partially. There exists another constraint that is common for all technologies − the constraint related to the given labor force. If the labor force in the country is used only partially, it is considered as unemployment. Pollution is assumed to be a result of production activities, and pollution destruction is assumed to be proportional to the pollution level. The initial capacities of technologies and the initial pollution are given. A sufficiently long time-period is considered (about 50 years).

Three criteria for strategies screening are studied in this section:

- consumption criterion C^*;

- unemployment criterion U^*;

- pollution criterion Z^*.

For the pollution and unemployment criteria their maximal values during the time-period under consideration are used. The consumption criterion is based on the deviation of computed consumption from a given reference trajectory. For the reference trajectory we apply the trajectory of exponential growth. Consumption and pollution are measured in their

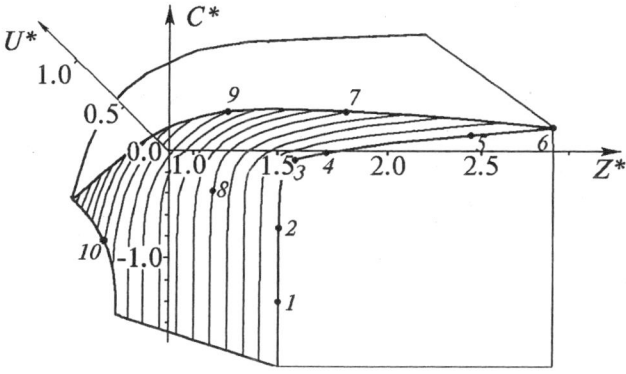

Figure 2.11. The EPH for three criteria. The Pareto frontier is hatched.

initial values, and unemployment is measured in parts of the labor force. The values of the coefficients of the model were selected to be typical for a national economy.

Exploration of the Pareto frontier. Here a model with two production technologies is analyzed. The first one represents the "old" technology that is cheaper than the second, "modern" one. The modern technology is less polluting than the old one. Moreover, the new technologies require less labor force. The requirements for labor force decrease in time for both technologies. It is clear that the increment in the consumption criterion and the decrement in the pollution and unemployment criteria are needed.

Figure 2.11 displays the three-criterion EPH for the model under study. The Pareto frontier is hatched. This picture provides a good understanding of the general structure of the EPH and its Pareto frontier, but at the same time it seems that the picture cannot help to understand the criterion tradeoffs or recognize coordinates of non-dominated points. For example, it is fairly complicated to evaluate the criterion tradeoffs related to the points identified in the picture, except points 1–6 that belong to the plane (Z^*, C^*). The decision maps can help in this case.

A decision map is given in Figure 2.12. The tradeoff curves are given for pollution and consumption while the value of unemployment criterion is restricted for any frontier. The constraints imposed on unemployment are given in Figure 2.12; they vary between 10% and 70% of the labor force. One can see that the decision map provides tradeoff information in a clearer form than the three-dimensional picture given in Figure 2.11.

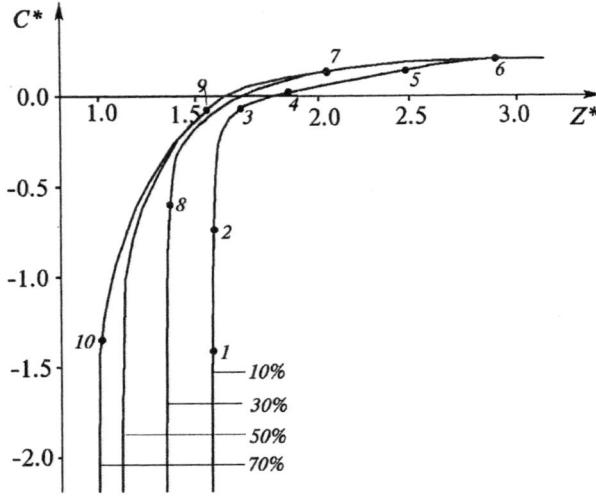

Figure 2.12. Tradeoff curves

One can easily understand how pollution is transformed into consumption on Pareto frontiers. The points, which were identified in Figure 2.11, are marked in the decision map, too. In contrast to the three-criterion picture, the user can easily understand the criterion values at these points and the criterion tradeoff in the decision map.

All four tradeoff curves of the decision map given in Figure 2.12 have the same shape. Three well-defined zones can be seen:

1 the first zone, in which a small increment in pollution results in substantial growth of consumption;

2 the second zone, where consumption growth is related to a substantial increment in pollution;

3 the intermediate (or elbow) zone.

The intermediate zone is fairly small for the curve related to the 10% constraint imposed on unemployment, but it gets broader when admissible unemployment increases to 70%.

In the second zone, values of consumption are pretty close to the reference trajectory (C^* is close to zero). The tradeoff curves between pollution and consumption are close one to another. This means that the constraint imposed on the unemployment has a small influence on the frontier in this zone. In contrast, an increment in unemployment can help to decrease the pollution value in the first zone (consumption is low, and C^* is less than -0.5). For example, if $C^* = -1$, the increment in maximal unemployment from 10 to 30% helps to decrease pollution from 1.5 to 1.3. Surely, this is a very high price to pay for a relatively small decrement in the pollution value. Pollution decrement is even less in other points. Therefore, the tradeoff curve seems to be of interest, which is related to an unemployment level that is not greater than 10%. Since the intermediate zone is fairly small for this tradeoff curve, it has a "kinked" form. Point *3* is located near the kink. It is related to a small deviation from the reference trajectory ($C^* = -0.1$), while pollution is about 1.7. To make pollution a bit less, one needs to decrease consumption drastically (to $C^* = -0.7$ in point *2*). An increment in consumption at this point requires substantial increment in pollution (to $C^* = 0.05$ and $Z^* = 2.5$ in point *5*). For these reasons, point *3* seems to be fairly reasonable.

Exploration of strategies. Strategies associated to points in Figures 2.11–2.12 were computed. Let us consider the strategy that results in point *3*. The strategy is given in Figure 2.13, where five graphs are depicted that describe the time dependence of the variables. Ten time periods are considered, each of them corresponds to five years. The following variables are depicted in Figure 2.13: pollution Z, unemployment U, capacities of first (old) technology M_1 and of second (new) technology M_2, output X_1 of the first technology (output of the second technology coincided with its capacity).

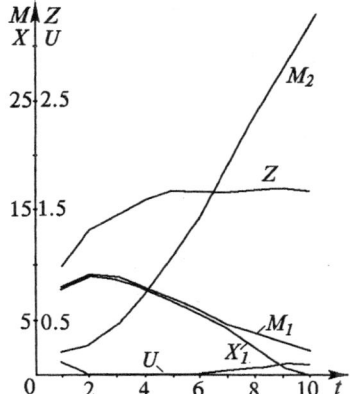

Figure 2.13. Strategy with a balanced consumption/pollution.

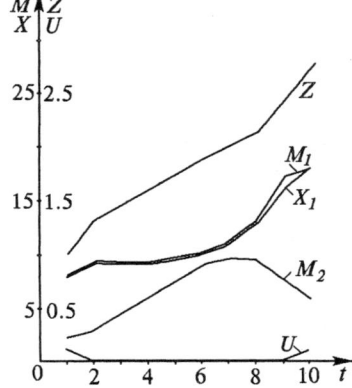

Figure 2.14. Strategy with maximal consumption.

One can see in Figure 2.13 that at the very end of the time interval, the second (new) technology substitutes the first (old) one. However, the substitution is fulfilled not at once, but gradually. During the first part of the interval (about 10 years), the capacities of the first technology are growing. Only after it, they start to decline. Full utilization of the capacities of the first technology stops only at the very end of the interval. So, the transformation of the economy takes about 50 years. The unemployment is small in the middle of the time interval. Its reappearance at the end of the time interval seems to be related to high efficiency of the second technology. Pollution gradually approaches the level of 1.7 that seems to be natural for the second technology.

Now let us compare the above strategy with the strategy that is associated with point *6*, which is characterized by maximal consumption. The same five graphs are depicted in Figure 2.14. The main feature of Figure 2.14 is evident − substitution of the first technology by the second technology does not happen! The capacities of the second technology start to rise first, but then they decline − the product is used for consumption, and so it is impossible to construct capacities based on the new technology that is more expensive than the old one. A permanent substantial growth of pollution is clearly related to the extensive application of the first technology.

Strategies related to other points from Figure 2.12 are given in the paper (Lotov, Chernykh and Hellman, 1992), where several additional features of the study are discussed.

It is extremely interesting to compare the development strategies that are associated with points *6* and *3*. Though point *6* is related to maximal consumption, the difference between point *6* and point *3* is not so large in this aspect − in point *6* it holds $C^* = 0.1$ instead of $C^* = -0.1$ in point *3*. However, this small difference results in extremely negative effects − the strategy associated with point *6* brings the nation into a catastrophic environmental situation and economic deadlock. In contrast, the strategy associated with point *3* results in the complete modernization of the economy and in a stable, relatively small, pollution level.

This example is very educational. It proves that a single-criterion optimization (say, optimization of consumption) cannot be sufficient in the process of searching for reasonable development strategies − the single-criterion optimization may result in inappropriate values of other performance indicators. We will meet the same effect in other problems studied in this book.

4. Searching for trans-boundary air pollution control strategies

In this section the FGM/IDM technique is applied in the framework of software that supports searching for strategies aimed at the abatement of long-distance air pollution. Users of the software can apply screening of possible strategies using an integrated model that consists of two parts: a model of regional sulfur abatement cost and a model of sulfur transportation between regions (countries). Screening criteria may include sulfur deposition rates and abatement costs in some or all of the regions. This section is based on the paper (Bushenkov, Kaitala, Lotov and Pohjola, 1994).

Introduction. Acid rain is one of the major environmental concerns in Europe. It is known to render lakes incapable of supporting aquatic life, to threaten forest and agricultural productivity and to damage statuary and other exposed materials. Airborne concentrations of sulfate particles can also increase morbidity and even premature mortality. Sulfur and nitrogen oxides stay aloft for one to three days and are transported by the wind over distances ranging from 50 to 2000 kilometers. This fact makes the environmental problem transnational. Cooperation among the countries concerned is a natural solution to the problem. In fact, a natural approach to studying the economics of acid rain consists in its formulation as an international decision problem.

A key problem in international cooperation on pollution control is the allocation of abatement resources among the countries involved. Where and how should these scarce resources be directed so as to maximize the benefit from abatement reduction? In (Maeler, 1990; Kaitala, Pohjola and Tahvonen, 1992) it was demonstrated that cooperation may entail side payments, which are a stark reality in international cooperation.

The main goal of the software described here is to provide new opportunities for studying international acid rain problems and for elaborating reasonable abatement strategies that may be related to software described here. As usual in our technique, collections of bi-criterion slices of the variety of feasible criterion vectors are displayed.

In this section, acid rain in Finland, Estonia and a part of Russia is studied. The problem has a long history. As early as in 1989 the governments of Finland and the former USSR signed an action plan for the purpose of limiting and reducing the deposition and harmful effects of air pollutants emanating from areas near their common border. Since the political events of 1991 have made this agreement obsolete, Finland has responded to the changes by seeking new forms of cooperation with the new nations. In particular, the government of Finland has decided to develop a program of environmental investment in areas that are the sources of trans-boundary air pollutants deposited in Finland. Supporting the decision making is required in the framework of the program.

FGM-based software can help to solve this kind of problem. Users, e.g., experts of the governments of the countries involved, can study the feasibility frontiers of sulfur deposition rates and abatement costs in the individual countries as well as the tradeoffs among these criteria. This information is obtained from a sulfur transportation model and from estimated abatement cost functions. The software supports a search for appropriate feasible combinations of criterion values and associated abatement strategies.

Data and the model. In 1988 the Finnish-Soviet Commission for Environmental Protection established a joint program for estimating the flux of air pollutants emitted close to the border between the countries.

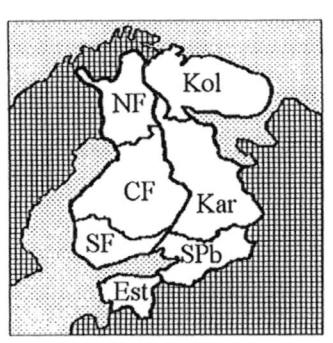

It consists of the estimation of emissions, model computing of trans-boundary transport of pollutants, analysis of observational results from measurement stations and conclusions for emissions reductions. The emissions inventory includes sulfur, nitrogen and heavy metals. Depositions were calculated in (Tuovinen, Kangas and Nordlund, 1990) by applying the latest version of the long-range transport model for sulfur developed at the Western Meteorological Center of the European Monitoring and Evaluation Program (EMEP). Emission data approved by the

Figure 2.15. Map of the region

Finnish-Soviet Commission were used as inputs of our model. Finland is divided into three sub-regions (Figure 2.15): Northern (NF), Central (CF) and Southern Finland (SF). The areas close to the eastern border of Finland are divided into four areas: Estonia (Est) and three areas in Russia: Kola peninsula (Kol), Karelia (Kar) and St. Petersburg (SPb). The annual sulfur depositions per square meter range from 0.5–0.6 gram in Northern and Central Finland as well as in Karelia to 1.2–1.3 gram in Southern Finland and Estonia.

In (Tuovinen, Kangas and Nordlund, 1990), an annual sulfur budget between these seven regions was estimated for the year 1987. It can be used to formulate a sulfur transportation matrix indicating how the emission in one area is transported in the atmosphere for deposition in another. The columns of Table 2.1 specify the distribution of deposition of one unit of sulfur emitted in each area between the regions. The large numbers on the diagonal show that their own sources of pollution play an important role in each region. The sums in a column (or a row) may be not equal to 1 since the regions both emit sulfur to and receive it from the rest of the world.

Table 2.1. Deposition distribution.

Emitting region	NF	CF	SF	Kol	Kar	SPb	Est
Receiving region							
Northern Finland *(NF)*	0.200	0.017	0.010	0.046	0.012	0.000	0.000
Central Finland *(CF)*	0.000	0.300	0.062	0.011	0.047	0.036	0.029
Southern Finland *(SF)*	0.000	0.017	0.227	0.003	0.000	0.027	0.038
Kola *(Kol)*	0.000	0.017	0.000	0.286	0.023	0.009	0.000
Karelia *(Kar)*	0.000	0.033	0.031	0.017	0.318	0.045	0.019
St. Petersburg *(SPb)*	0.000	0.017	0.031	0.003	0.012	0.268	0.058
Estonia *(Est)*	0.000	0.000	0.031	0.000	0.000	0.018	0.221

A sulfur transportation model was constructed on the basis of emission data and data provided in the table. Let E and Q denote the vectors of annual emission and deposition of sulfur, respectively, and let A stand for the matrix given in the table and B for the vector of exogenous deposition in 1987. The model can then be expressed in vector notation as

$$Q = AE + B.$$

To apply this model to the analysis of cooperation on trans-boundary air pollution between the three countries, we need information about both future emissions and sulfur abatement costs. The Finnish Integrated Acidification Assessment model (HAKOMA) provided estimates for emissions in the year 2000 (Johansson, Tahtinen and Amann, 1991). The Finnish estimates of future emissions were obtained by using the basic energy use scenario of the Ministry of Trade and Industry. The Russian and Estonian emissions were assumed to stay at their 1987 levels, as no other information was available.

The estimates for sulfur depositions were obtained from the above model by using the estimated emissions and by assuming that the man-made sulfur deposition originating from the rest of the world will be 50 per cent lower than in 1980. We justify this assumption by referring to the Helsinki protocol of the Convention on Long-Range Trans-boundary Air Pollution according to which the 21 signatories will reduce their sulfur emissions by 30 per cent from the 1980 levels. Moreover, about half of these countries have declared more ambitious cuts ranging from 40 to 80 per cent.

The sulfur abatement cost function for the regions were calculated for various sulfur reduction requirements ranging up to the maximal technologically feasible removal. The HAKOMA project at the Technical Research Center of Finland (VTT) has derived such cost functions for Finland and the nearby regions (see Johansson, Tahtinen and Amann, 1991). These piecewise linear functions (they are given in Bushenkov, Kaitala, Lotov and Pohjola, 1994) are used in the software, but other cost estimates can also be applied easily. The annual costs, measured in million Finnish marks, have been estimated on the basis of expected emissions for the year 2000, and they include both capital and operating costs.

The sulfur transportation model and the cost functions provide an opportunity to formulate performance indicators that can be used as possible screening criteria. Users can use the following six groups of performance indicators:

- sulfur abatement cost in each sub-region;

- abatement cost in Finland, in the nearby region of Russia, and in Estonia;

- total abatement cost for the whole territory;

- average sulfur deposition in each sub-region;

- maximal average depositions in the sub-regions of Finland, Russia, and Estonia;

- maximal average depositions in the whole territory.

Search for a trans-boundary acid rain abatement strategy. The search for preferable strategies includes the following main stages:

1 problem specification (defining the criteria and constraints);

2 compatibility test;

3 approximation of the variety of feasible criterion vectors;

4 visual exploration of the frontiers of bi-criterion slices of the variety of feasible criterion vectors;

5 identification of a feasible goal;

6 computing the goal-associated efficient strategy;

7 decision display.

Users must first specify criteria in the list of performance indicators (i.e., costs and deposition values), that includes:

1 regional cost related to the emission decrement;

2 national (for Finland or Russia) cost related to the emission decrement;

3 total abatement cost;

4 maximal rate (in gram per square meter) of sulfur deposition in any sub-region;

5 maximal rate of sulfur deposition in a country (Finland or Russia);

6 maximal rate of sulfur deposition in the whole territory.

Users must specify the improvement direction for the criteria. Originally, it was assumed that the user is interested in minimizing all the criterion values. However, then it turned out that the opposite type of interests are also possible (say, maximization of investment in a particular region). It may sometimes be the case that a user does not know about the improvement direction of a particular criterion. This situation can be specified as well. For this reason, a partial EPH may be constructed.

The user has an option to define constraints imposed on the values of performance indicators. Say, maximal rate of sulfur deposition can be specified in any sub-region, country or the whole territory. The initial values of pollution deposition are restricted from above by the values corresponding to the forecast for the year 2000. Since these values are treated in the model as the pollution depositions associated with zero cost, these constraints cannot be violated by any feasible decision. They are displayed to inform the user.

After criteria and constraints have been specified, the compatibility of the constraints is checked. If the constraints are not compatible, the user has to return to their definition. If they are compatible, the variety of feasible criterion vectors or a related set is constructed. As usual, visual exploration of the variety is based on displaying its two-dimensional slices. Then, a preferred feasible point should be identified on one of the slices. Afterwards, the efficient strategy associated with it is computed automatically. Any strategy obtained in this way can be studied by displaying it as a table, diagram or histogram.

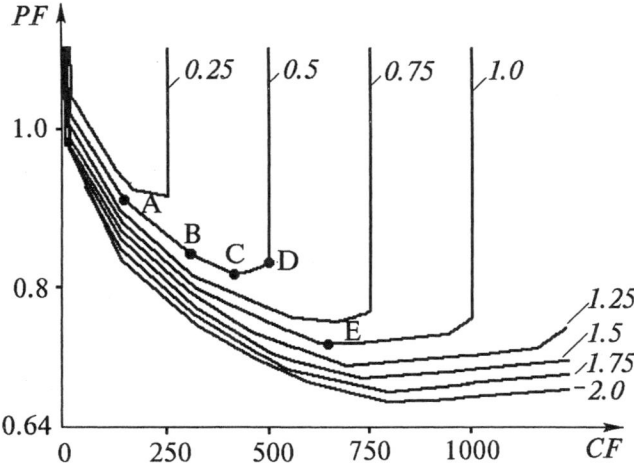

Figure 2.16. Abatement cost in Finland versus deposition in Finland (for several variants of total cost)

Let us consider an example. Suppose the user has specified three criteria:

1 abatement cost in Finland (CF, in million Finnish mark, FIM);

2 total abatement cost (TC, in billion FIM);

3 maximal deposition rate in Finland (PF, in gram per square meter).

It is assumed that the decrement in the second and the third criteria values are preferable. The direction of improvement of the first criterion is not specified. Let us assume that the following constraints were imposed:

1 the deposition rate in Northern Finland must be not greater than 0.4 gram per square meter;

2 the deposition rate in Central Finland must not be greater than 0.5 gram per square meter.

The constraints are compatible, and so the variety of feasible criterion vectors can be constructed and visualized. In Figure 2.16 several frontiers of the variety of feasible criterion vectors are depicted in the plane of the criterion CF (investment in Finland) and the criterion PF (maximum deposition rate in Finland) for several values of total abatement cost. The range of total costs in the whole region is between zero and 4.1 billion FIM. The following total cost values were selected:

TC = 0.25, 0.5, 0.75, 1.00, 1.25, 1.50, 1.75, and 2.00 billion FIM.

Let us consider the properties of the frontier TC = 0.5 that is related to the total cost of 500 million FIM. First of all, it is clear that the increment in the investment in Finland from zero to about 400 million FIM (points **A**, **B** and finally **C**) results in a gradual decrement in the deposition rate in Finland. Say, point **A** means that the 140 million of the total costs of 500 million will be invested in Finland. As a consequence, the maximal deposition rate in Finland is about 0.9 gram per square meter. Increasing the Finnish share of the total costs (points **B** and **C**) yields an even better result for Finland (the deposition rate decreases). However, increasing this share from 400 million FIM (point **C**) to 500 million (point **D**) does not yield a better result: the deposition rate in Finland increases. So, investment in point **C** is related to the minimal deposition rate in Finland (a little bit more than 0.8 gram per square meter). Note that the total cost remains the same. This means that, given the total cost of 500 million, all investments exceeding about 400 million FIM give better results when invested somewhere else.

So, investment of 400 million FIM in Finland from the total cost of 500 million FIM corresponds to Finnish interests. It is clear that 400 million FIM must be invested by Finland. Though the software informs where the investment of 100 million FIM must be applied (it is sufficient to consider the associated strategy), what country will pay the balance of 100 million FIM? The software is not supposed to answer this question − the answer must be a result of further negotiation. However, it is clear that the above allocation of total cost is related to the interests of Finland, and other countries may propose different allocations that may be found using the same software. So, negotiation results may depend upon the source of investment.

Comparison of the curve for 500 million FIM with the other curves given in Figure 2.16 informs the user concerning the influence of total costs. The curve for the cost of 250 FIM proves that Finland is interested in application of this investment on Finnish territory. This means that, in the case of small investment, Finland may be not interested in providing international aid.

In case the total investment is 1 billion FIM, it would be reasonable for Finland to select point **E** on the related curve: increasing the investment in Finland beyond this point (if the total costs remains the same) starts to increase the deposition rate in Finland. At point **E**, abatement cost in Finland is 650 million FIM. Thus, 350 million FIM is reasonable to invest in the neighboring regions.

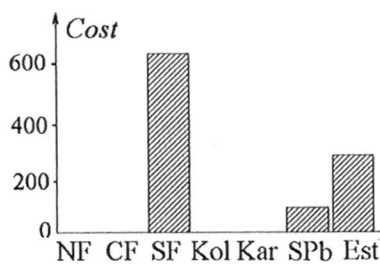

Figure 2.17. Cost allocation

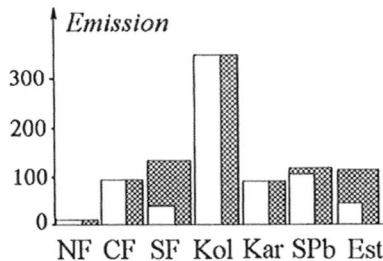

Figure 2.18. New versus old emissions

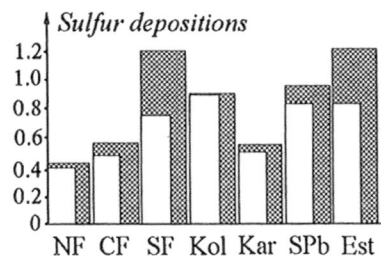

Figure 2.19. New versus old depositions

Let us consider the abatement strategy associated with point **E** in additional details. The performance indicators of the strategy are given in Figures 2.17–2.19. Figure 2.17 displays allocation of the total abatement cost among regions, Figure 2.18 displays sulfur emissions before and after the investment would be applied, and Figure 2.19 displays the sulfur depositions per square meter before and after the investment.

According to Figure 2.17 the optimal strategy consists in investment application mainly in Southern Finland. However, sulfur emissions must be reduced in St. Petersburg and Estonia, too. Estonia pollutes the atmosphere of Southern Finland heavily, so most of the investment in foreign regions (260 million FIM) must be applied there.

The rest (90 million FIM) must be applied in St. Petersburg. Perhaps the most surprising result in this case is that the emission in the Kola region must remain unchanged. As to the decrement in sulfur deposition, it is especially substantial in Southern Finland and Estonia, but in St. Petersburg and other regions it is visible, too.

It is important to note, that in the above study, the criteria were selected, frontiers were explored and the goal point was identified in the interests of a Finnish user. Absence of the investment in Kola is clearly related to the level of admissible pollution in Northern Finland. Exploration of additional criteria could make the study more fruitful. Surely a study of an Estonian or Russian user would result in different strategies. Let us add that the software turned out to be a prototype of several decision support systems.

5. Searching for smart strategies for abatement of global climate change

This section is devoted to an approach to supporting the analysis of global problems. The approach is based on application of the FGM/IDM technique, and it helps to develop efficient global strategies. Here we outline a demo version of the software tool that is devoted to the problem of global climate change. The research was started in the beginning of the 1980s (Moiseev, Alexandrov, Krapivin, Lotov, Svirezhev and Tarko, 1983; Bushenkov and Lotov, 1983) and completed in the beginning of the 1990s; a bit more detailed description is given in (Lotov, Bushenkov, Kamenev and Chernykh, 1997; Lotov, Bushenkov and Kamenev, 1999).

Introduction. Any system of models that is aimed at supporting the search for efficient strategies related to global climate change must contain at least the four following models:

- a model that describes the emission of carbon dioxide and other gases that seem to be responsible for global climate change; the model must describe the influence of economic and technological decisions on the emission level;

- a model of the global cycle of the above gases; the model must relate the concentration of the greenhouse gases in the atmosphere to their emission;

- a climate model that describes consequences of changes in the concentration of the gases; it is important that the consequences in different parts of the world must be estimated;

- a model that describes the influence of climate changes on economic development and standards of life in particular groups of nations.

In the demo version of the model, we decided to restrict the study to the problem of investment in the energy sector aimed at the decrement in carbon dioxide emission. The objective of the research was to develop a demo software tool that could show how the FGM/IDM technique can support the process of searching for smart cost-efficient strategies. The tool is based on the same ideas as the software for the abatement of trans-boundary acid rains described in the previous section.

Model. In the demo model, which was developed in the beginning of the 1990s, five groups of nations were considered (Figure 2.20):

1 post-industrial countries;

2 post-socialist countries;

3 new industrial countries;

4 Asian communist countries;

5 developing countries.

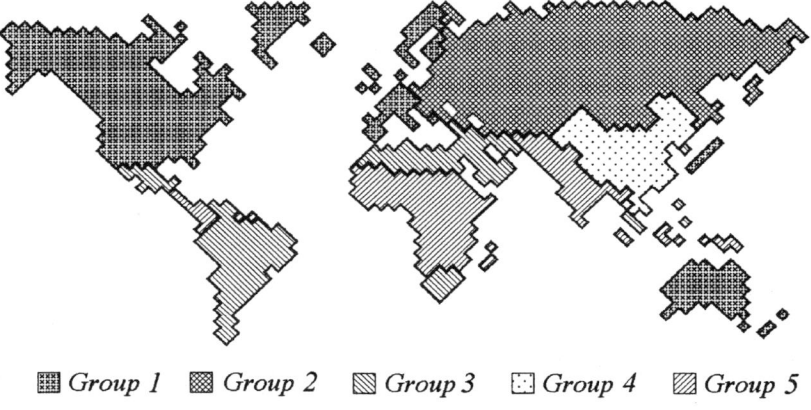

▦ *Group 1*   ▦ *Group 2*   ▨ *Group 3*   ⬚ *Group 4*   ▧ *Group 5*

Figure 2.20.   Five groups of nations considered in the demo model

This grouping is out of date now, but we use it here for demo purposes.

The climate change during the time period 1990–2050 was studied (actually, six decades were considered in the integrated model). In the simplest version, only one greenhouse gas was considered, namely carbon dioxide. The carbon dioxide emission was a function of energy consumption, which depended on the gross domestic product (GDP) and energy consumption per unit of the GDP. The energy consumption per unit of the GDP, in turn, was assumed to be able to decrease due to special kind of investment aimed at its decrement (energy-saving investment). The GDP for a group of nations was estimated by forecasting the population growth and the dynamics of the GDP per capita.

A linear dynamic model was used to describe the global carbon dioxide concentration in the atmosphere. Its coefficients were identified on the basis of existing data. The linear climate model used in the study was a parameterization of a global circulation model elaborated at the Computing Center of Russian Academy of Sciences in the 1980s (Moiseev, Alexandrov and Tarko, 1985). Simulation experiments with the global circulation model helped the authors of the model to develop its linear parameterization, in the framework of which average changes in solar radiation, precipitation and temperature in world regions were related to the global carbon dioxide concentration. Professor Alexander Tarko estimated the influence of the climate changes (changes of solar radiation, precipitation and temperature in world regions) on economic development and standards of life in particular groups of nations (to be precise, Professor Tarko estimated losses that result from climate change). We use this opportunity to express our gratitude to him.

The demo tool was coded for a PC. The user had to specify several decision screening criteria. Then, the EPH for these criteria was approximated, and the user was able to explore various decision maps.

Application of the software tool. Let us consider an example, in the framework of which five criteria were explored:

- $I1$, $I2$, $I3$ — discounted energy-saving investments during the whole time period (1990–2050) in the first, second and third group of nations, respectively;

- $I4$ — the same for both the fourth and fifth groups of nations;

- Ls — discounted investment-related decrement in global losses during the whole time period.

All values are given in billions of US$.

As usually, analysis was based on the exploration of various decision maps in interactive mode. A black and white copy of a decision map is provided in Figure 2.21. Here, discounted investment for the whole time period in the first group (post-industrial countries) is given on the horizontal axis. Discounted investment for the whole time period in the second group (post-socialist countries) is given on the vertical axis. The shading in Figure 2.21 (colors in display) is related to values of Ls.

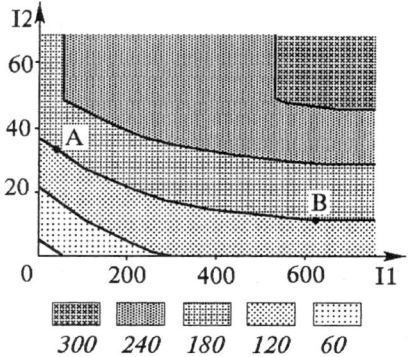

Figure 2.21. A black and white copy of a decision map.

The tradeoff curves are related to five

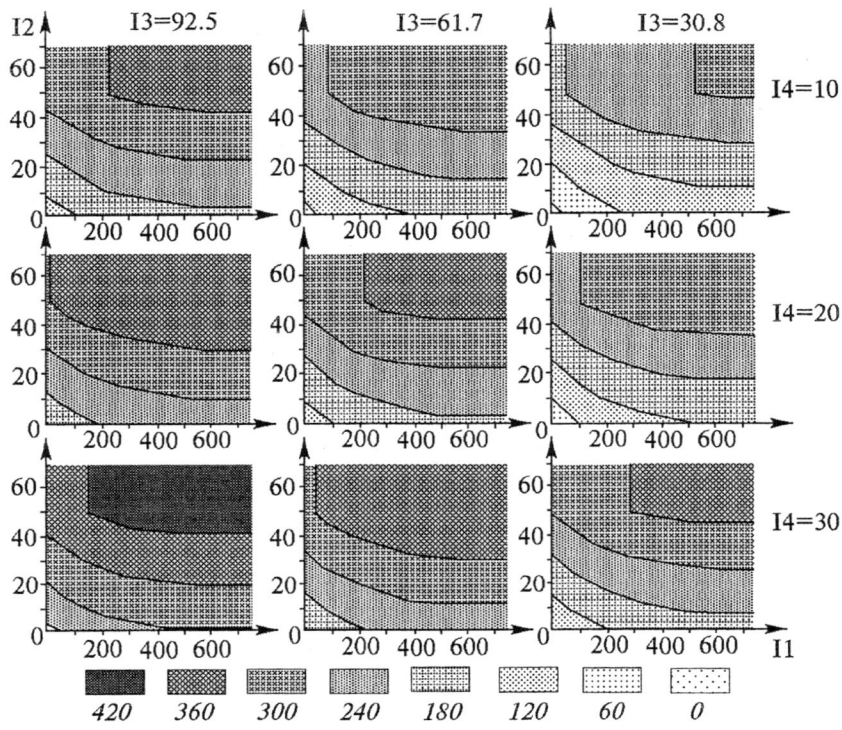

Figure 2.22. Black and white copy of a decision map matrix

values of Ls (300, 240, 180, 120 and 60). The values of $I3$ and $I4$ are not higher than 30 and 10 billion, respectively.

The tradeoff curves in Figure 2.21 provide extremely important information. Let us consider points **A** and **B** on the tradeoff curve related to $Ls = 180$ billion. By moving the goal from point **B** to point **A**, the user can substantively decrease the required energy-saving investment in the first group (less than 40 billion at point **A** instead of the shocking value of 630 billion at point **B**). This tremendous saving is provided by a relatively minor increment in investment in the second group (about 35 billion at point **A** instead of 15 billion at point **B**). So, additional investment of 20 billion in the second group can save about 600 billion of investment in the first group!

This impressive knowledge has a simple interpretation: statistical data prove that energy consumption per dollar of the gross domestic product is about three (!) times higher in the second group than in the first

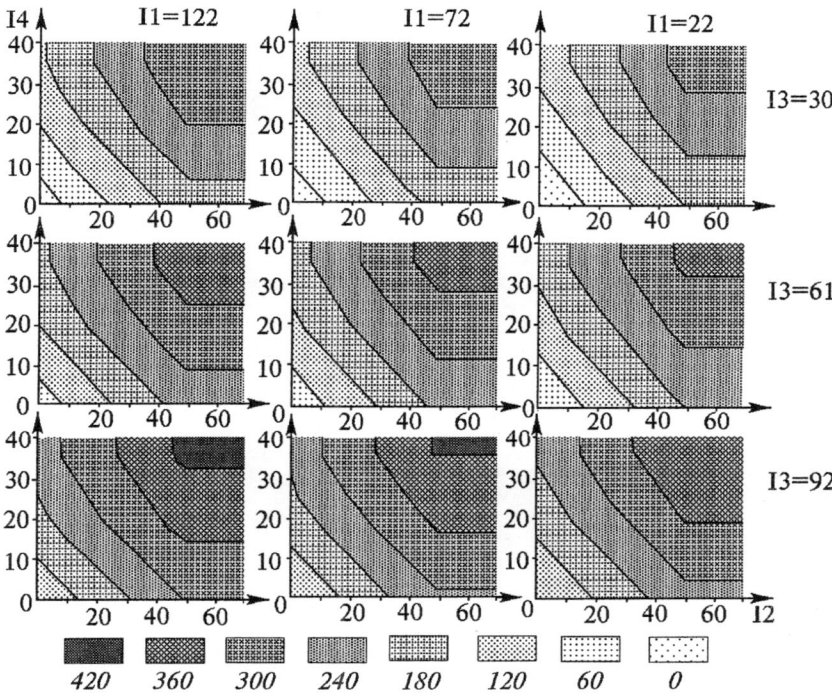

Figure 2.23. Black and white copy of a matrix of decision maps for an alternative arrangement of criteria

one. Therefore, a decrement in energy consumption (and the related decrement in carbon dioxide emission) is much more expensive in the post-industrial nations than it is in the post-socialist nations. This example shows how the tradeoff curves may help experts and politicians to find reasonable strategies for abatement of global climate change.

To study the influence of $I3$ and $I4$ on the relations among $I1, I2$ and global losses, one can use a matrix of decision maps, say, the matrix given in Figure 2.22. We recall that a matrix of decision maps can be considered as a collection of snap-shots obtained in the process of animating a decision map. So, the matrix given in Figure 2.22 can be related to animation of Figure 2.21. However, such interpretation of the decision map matrix is not obligatory.

Once again, discounted investments in the first ($I1$) and second ($I2$) groups are given in axes, and shading is related to values of Ls. Any column of the matrix in Figure 2.22 is associated to a certain value of $I3$

given above the columns, and any row is related to a value of $I4$ given to the right from the rows. Only three values of $I3$ and $I4$ are selected in Figure 2.22, and so only nine decision maps are given in the matrix. The decision map given in Figure 2.21 is located in the right upper corner of the matrix. Once again, the FGM/IDM-based software provides an opportunity to select larger numbers of decision maps in a column or a row, depending on the user's desire and quality of display.

Shading in Figure 2.22 is related to a broader range of decrement in global losses Ls than in Figure 2.21. The matrix of decision maps shows that the tradeoff curves for $I1$ and $I2$ have the same shape on all decision maps. So, the modification of the values of $I3$ and $I4$ does not influence our conclusions concerning energy-saving investment. Influence of the values of $I3$ and $I4$ on global losses is evident from the picture.

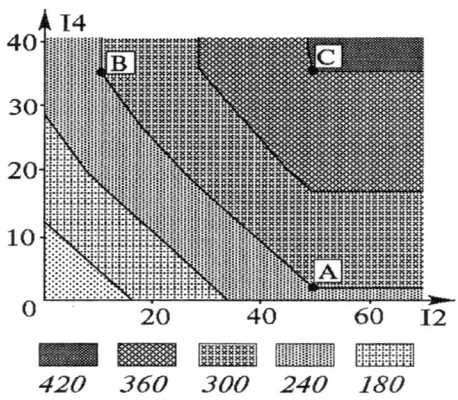

Figure 2.24. One of the decision maps for $I2$ and $I4$; $I1$ is not greater than 72 billion, and $I3$ is not greater than 92 billion.

Now let us have a look at the tradeoff curves for other pairs of criteria. It is interesting that the tradeoff curves for $I1$ and $I4$ look just the same (surely the particular numbers are different). For this reason we do not display the related decision maps here. In contrast, the matrix of decision maps, which is given in Figure 2.23 and contains the tradeoff curves for $I2$ (post-socialist countries) and $I4$ (Asian communist countries and developing countries), is fairly different. The shadings are related to the same values of Ls as in Figure 2.22. The values of $I1$ are related to columns, and the values of $I3$ are related to rows. One can see here that the tradeoff curves between $I2$ and $I4$ have a different shape than the frontiers between $I1$ and $I4$. Now it is impossible to save investment in the way depicted by points **A** and **B** in Figure 2.21. This fact has a simple interpretation: both groups of nations have the analogous pattern of energy consumption.

Let us consider this fact in details. For example, consider one of the decision maps, where $I1$ equals 72 billion and $I3$ equals 92 billion (Figure 2.24). Let us locate points **A** and **B** on the tradeoff curve related to $Ls = 300$ billion. At point **A,** it holds $I2 = 50$ billion and $I4 = 3$ billion. At point **B,** it holds $I2 = 11$ billion and $I4 = 36$ billion. So, the

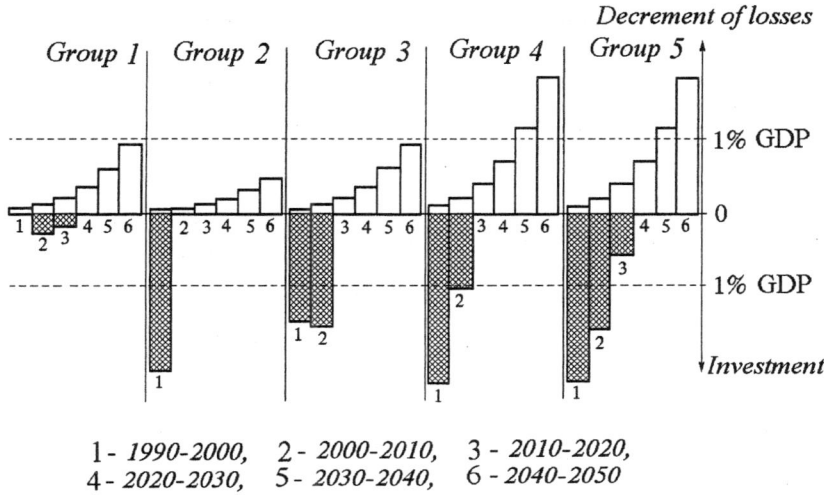

Figure 2.25. Time dependence of investments and losses for the groups of countries

sum of the energy-saving investments is approximately the same, and, in contrast to the case of the tradeoff curve for $I1$ and $I2$ in Figure 2.21, it is impossible to receive a substantial decrement in total investment by transferring investment from one group of nations to another. According to the matrix of decision maps given in Figure 2.23, the above property holds for all decision maps.

As usual in a FGM/IDM-based applications, search for a preferred strategy is based on the identification of a feasible goal. Point **C** in Figure 2.24, which is related to the maximal value of Ls in the decision map, was selected to be the goal. Time dependence of investments and decrements in losses (not discounted now) for the related strategy is displayed in Figure 2.25. In accordance to the number of the groups of nations in the model, there are five vertical sections in Figure 2.25. In any section, two column diagrams are given. The columns above zero display decrement in losses for the group (measured in percents of the GDP) in different time-periods (once again, six decades are considered). The shaded columns beneath zero display investment into the energy sector (measured in percents of the GDP, too). One can see that maximal investments (in percents of the GDP!) must be sufficiently large everywhere outside the post-industrial countries, where the investment is relatively low. However, taking into account that the GDP is much higher in post-industrial countries, real investments are not so different.

We do not discuss additional details because of the methodological nature of our research and low quality of the data, which are plausible, but not precise. Moreover, only part of the models (actually, the climate model only) can be considered as the result of a careful research, many features of the research are out of date. However, the main feature of the time-dependence of investments and losses seems to be clear − the main part of investment must be done at the beginning of the time period, while the result, that is decrement in losses, is expected mainly at the end of the period.

To apply the described approach in a real-life development of smart strategies for the global climate change abatement, one has to base the study on more detailed well-elaborated models. However, even the rough models used in this section prove that it is possible to discover smart abatement strategies that do not have destructive influence on the economic growth of particular countries and the world economy as a whole. The recent developments of international negotiations in the field of global climate change prove that this requirement is very important. They can be supported by the visualization technique described in this section. It may turn out to be even more important that the visualization techniques can help to involve large numbers of ordinary people in the discussion of possible strategies and the development of independent strategies. This topic is discussed in detail in the Epilogue.

Chapter 3

REAL-LIFE APPLICATIONS

In this chapter several real-life applications of the FGM/IDM technique are considered. First, the concept of a real-life application of a decision support technique is discussed. The discussion is illustrated with the application of the FGM at the State Planning Agency of the former Soviet Union in the first part of the 1980s. Sections 2 and 3 are devoted to two decision support systems that are used by water engineers for water quality planning in river basins of Russia. Screening of water quality plans is based on application of the FGM/IDM technique. The decision support systems (DSS) helps engineers to develop water quality improvement strategies that can be used as proposals in the process of final selection of water quality improvement plans.

1. On the real-life application of decision support techniques: identification of national goals

We start with considering a question that is extremely important for all developers of decision support techniques: what could a real-life application of such a technique mean? Often one comes across papers where authors argue that multiple-criterion decision support tools find real-life applications fairly seldom. It sounds a bit strange since hundreds of real-life applications of such decision support methods as Goal Programming (Charnes and Cooper, 1961; Ignizio, 1985; Romero, 1991) or the AHP (ANP) (Saaty, 1996) have been reported. On the other hand, a developer of a method may sometimes report a real-life application of the method even in a case when its application is restricted to a demonstration that the method can be adapted to some particular real-life problem. Surely, applications of the last type are only illustrative.

Moreover, one can find many examples where a decision maker is informed of a decision support tool that can be used in his/her problems, but does not use it. In such case, a developer may declare a real-life application of a method only because the decision maker has found time to listen to a description of the method and its possible application. Surely, one cannot consider such a situation to be a real-life application. It seems that a precise definition of a real-life application is needed.

Definition of a real-life application. Books and papers provide different definitions of the real-life application of a decision support method. Let us consider the definition used in the recent paper (Kasanen, Wallenius, Wallenius and Zionts, 2000) devoted to the discussion of real-life applications of MCDM techniques. An application of a decision support technique is said to be *a real-life application* if

- an actual problem of an actual organization is studied,

- using real data,

- in which decision makers participate, and

- the results of which have been implemented.

Though the above definition accurately reflects the modern state of common understanding of the topic, multiple real-life applications of goal programming and the AHP have nothing to do with the above definition. Decision makers, experts and other kinds of users apply these techniques without informing the developers about how they do it. Who knows what kind of problems they study? Are those problems "actual"? What kind of data do they use? Do decision makers participate in it or not? Or, perhaps, experts prepare decisions on the request of a decision maker without his/her participation? Or, perhaps, a university professor applies a technique for developing of educational examples? And what about the implementation? Do the techniques influence decision processes at all?

We cannot answer these questions: too many different institutions and private people use goal programming or the AHP in many different ways. It is known that in several cases, attempts to apply goal programming were not successful (one such case is described in this section). However, multiple successful applications of goal programming are known, too. Many users buy the software based on goal programming and the AHP or develop such software by themselves. Therefore, we can assert that goal programming and the AHP have found real-life application. It is important that in this case, the quality of an application depends on the user, but not on the developer.

So the above definition seems to be more related to the situation where a developer tries to test the applicability of a method, but not to its broad real-life application. For this reason, an alternative definition of real-life application of a decision support method is needed. In the framework of this book, we use the following definition.

A decision support technique is said to have real-life application if some experts or decision makers have mastered it and use it independently from the developers.

Both goal programming and the AHP comply with this definition. It seems that other techniques that pretend to have real-life applications must comply with it, too. This definition rules out many of alleged applications, in which the analysts cannot identify the organization that employs their results. On the other hand, it does not require involvement of the developers of a method in the decision process. It is important to stress that developers of methods cannot require such an involvement. Decision makers may not be willing to answer questions posed by developers (it is clear that a decision maker's openness may harm him/her). Therefore, decision makers may consider involvement of the developers to be undesirable.

Let us consider an example of real-life application of the FGM that illustrates the above ideas.

Example of real-life application of the FGM. In the beginning of the 1980s, a large research project was started at the State Planning Agency of the former Soviet Union. The aim of the project was to develop a computer-based decision support system for a medium and long-term national economy planning. The DSS was based on application of the hierarchical system of dynamic input-output models that described the development of the USSR economy with different levels of aggregation.

The most aggregated (upper level) model described possible development of the USSR production system on a time period of 15 years. It was a dynamic input-output model, elaborated by experts of the State Planning Agency, in the framework of which 17 production industries were considered. Yearly outputs of production industries were defined to be equal to the sum of investments, imports, exports, final consumption, and raw materials consumption of all other industries. Feasible labor statistics were projected, and the capacities of production industries depended upon investment. The delay between investment and the resulting capacity growth was given, its value depending on the industry. Decision alternatives in the model were related to distribution of the labor force among industries, production of industries, production investments, etc. (Matlin, 1978).

The upper level model was used for identification of long-term national social-economic goals. For the particular goals, values of several performance indicators of the national economic system were used. The performance indicators included consumption of several population groups, development of health care and educational systems, etc. In the early variant of the DSS, officials of the State Planning Agency had to identify the particular goals on the basis of their experience, without any computer support. As a rule, the goals identified by them were not feasible. Then, some optimization software was used to compute a feasible criterion vector as close as possible to the identified goal. Usually, such feasible criterion vectors were distant from the goals identified by the officials, and so it turned out that the identified goals had nothing to do with the reality. The officials were disappointed with such results. After several attempts, they refused to use the DSS. It seems that the officials regarded such results as undermining their prestige since it might be attributed to their incompetence. It was clear for the DSS developers that an additional decision support tool was needed to help officials to identify goals that are close to feasible performance vectors.

The developer of the upper level model, Dr. Ilya S. Matlin did know about the FGM and its application in the framework of methods for national goal feasibility analysis (Pospelov and Irikov, 1976; Lotov and Ognivtsev, 1980; Lotov and Ognivtsev, 1984). For this reason, he hired two of the authors of the book (namely A. Lotov and V. Bushenkov). Their obligations consisted in the development of an FGM-based software that could be used in the framework of the DSS. It was decided to use the simplified version of the concept proposed in (Lotov and Ognivtsev, 1980) and approximate the variety of feasible social-economic goal vectors for the upper level model.

At that time (at the very beginning of the 1980s) the State Planning Agency was unable to get personal computers, and so the authors had to approximate the feasible goals set by the mainframe computer and to print out a large number of graphs that contained collections of bi-criterion tradeoff curves (see Lotov, 1984a, for details). An album of such graphs was provided to Dr. I.S. Matlin. One of the graphs is given in Figure 3.1. Two of the criteria (payments per one employee, $C1$, and living room per capita, $C4$, both at the end of the planning period) are given on the graph's axes. The values of other criteria (fixed for a tradeoff curve) are given under the graph. Artificial values of the criteria were used − all the criterion values in the starting year of the planning period were set equal to 1. The tradeoff curves appear as segments of straight lines. This feature of the graphs is related to properties of the model that we do not discuss here.

Dr. Matlin informed the authors that their direct contact with the officials of the State Planning Agency would not be needed and that the officials would study the album of feasible social-economic goals by themselves with his help. Indeed, it turned out that the album of graphs worked sufficiently well without any support from the authors. Two years later, we met Dr. Matlin once again to discuss problems of further development of the DSS. He assured the authors that the album was used very intensively and helped the officials to apply the goal method (he even provided a proof of his words – the album that looked dirty, greasy and crumpled). So, it seems that the method indeed was used by the officials. Subsequent developments

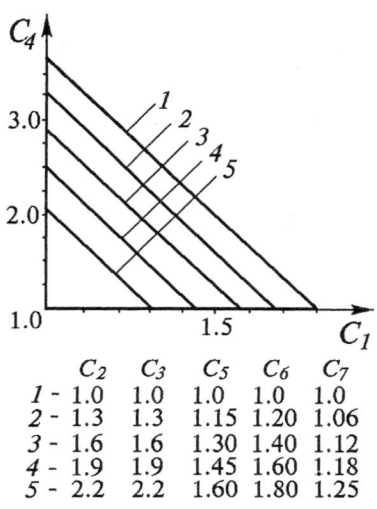

	C_2	C_3	C_5	C_6	C_7
1 -	1.0	1.0	1.0	1.0	1.0
2 -	1.3	1.3	1.15	1.20	1.06
3 -	1.6	1.6	1.30	1.40	1.12
4 -	1.9	1.9	1.45	1.60	1.18
5 -	2.2	2.2	1.60	1.80	1.25

Figure 3.1. A collection of tradeoff curves.

in the USSR ("perestroika") destroyed the planning system, and application of the DSS was halted. In total, the officials used the album for more than three years.

Strategy of real-life application of the FGM/IDM technique.
The above experience of a real-life application of the FGM helped us to develop our strategy for application of our decision support tools. First of all, we try not to be involved in decision procedures and even not to interact with decision makers and other stakeholders. Instead, we interact with experts who support the decision processes. We teach them to use our techniques, help experts to master them, and adapt our software to their particular requirements, if needed. Sometimes we help to develop the models, but usually we try to avoid it. This strategy for real-life application is illustrated in Figure 3.2 where we denote ourselves as "analyst".

In the following two sections, we describe utilization of this scheme in real-life application of the FGM/IDM technique in the framework of two DSSs. The first DSS described in Section 2 was used for water quality planning in several river basins in Russia in the first part of the 1990s. The DSS was developed at the request of Russia's State Institute

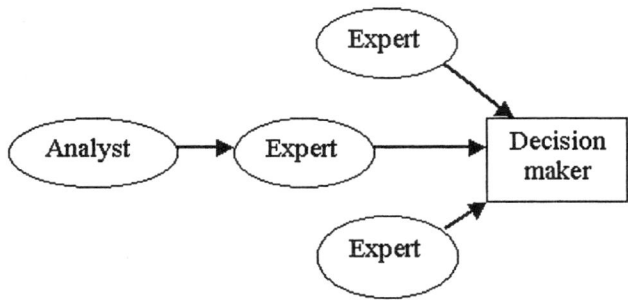

Figure 3.2.   Scheme of interaction between analyst and decision maker

for Water Management Projects (now the private Institute for Water Information Research and Planning, Inc.). The DSS described in Section 3 was developed at the request of the Russian Federal program "Revival of the Volga River". It is used now by engineers from the Engineering and Research Center on Water Management, Land Reclamation and Environment "Soyuzvodprojekt" who develop water-related projects on request from organizations in and outside of Russia.

2.   DSS for water quality planning in river basins

The DSSs described in this chapter were based on a special methodology of mathematical modeling for decision support in environmental problems. The methodology introduced in (Lotov, 1994) is discussed in detail in (Lotov, 1998; Lotov, Bourmistrova and Bushenkov, 1999). It is based on application of the FGM/IDM technique to integrated mathematical models of environmental systems. The DSS described in this section was first published in (Lotov, Bushenkov and Chernykh, 1997).

Methodology: short introduction.   Usually, DSSs for water management apply simulation of given decision alternatives as a main decision support technique. The FGM/IDM technique provides an additional tool that helps to select a small number of strategies for simulation. Multiple stakeholders, independent institutions and political groups that are involved in decision processes in water-related problems as well as experts associated with them can apply the FGM/IDM technique to assess all possible outcomes and screen possible strategies. Here, we restrict our investigation to water quality planning in river basins.

In water quality planning, decision screening requires integration of knowledge from a number of disciplines that provide information about different subsystems such as wastewater discharge, wastewater treatment, pollutants transport, the effect of pollutants on ecology, as well as economic impacts, environmental measures, and so forth. For this reason, simplified models must be used in an integrated mathematical model applied for screening. If an original mathematical description of a subsystem is provided, then a simplified model can be obtained from that description. In this case, a simplified model may have the form of an influence matrix. In different cases, expert judgments and empirical data may be used to help develop simplified models. As early as in the 1960s, Robert Dorfman stressed the importance of supporting decision screening in water management problems on the basis of simplified models (Dorfman, 1965). In this section, we describe a DSS for water quality planning in river basins that is based on the application of simplified models based on expert information. In contrast, Section 3 describes a DSS that applies a more sophisticated approach to parameterization of comprehensive simulation models of river basins.

Problem. To obtain even moderate investment, Russian environmental engineers have to prove to various stakeholders (federal and regional authorities as well as to owners and managers of industrial enterprises) that the investment would result in substantial improvement of the environment. Engineers have to prepare recommendations regarding wastewater treatment in industries and municipalities located in a river basin.

In the 1980s, engineers tried to apply single-criterion optimization procedures to obtain reasonable plans of wastewater treatment. They searched for plans related to minimal cost that would meet environmental requirements. Often it was impossible to find a feasible plan that met the requirements, and so the engineers had to change these requirements somehow. Even if such plans had been found, they were often too expensive to be implemented. Therefore, the engineers had to "improve" optimal plans by deleting several investment proposals from the plan in accordance with their experience. This resulted in inefficient strategies, which were sharply criticized.

For this reason, we had to elaborate a new decision support technology of water quality planning. In the framework of the technology, the measures devoted to water quality improvement were split into two phases:

- measures that had to be implemented immediately; they should be given by a water quality strategy that implements a balance between cost and pollution; and

- measures for final resolution of water quality problems.

A DSS was developed that supported the search for a strategy for the first phase. A search for such a strategy that implements a balance between cost and pollution was based on screening a myriad of feasible strategies. Along with the cost criterion, several water quality criteria were incorporated into the screening procedure. The FGM/IDM technique was used for computing the EPH and displaying the decision maps. Information on the criterion tradeoff helped the engineers to identify one or several reasonable feasible goals. Then, they received the related investment strategies, which were displayed both in graphical and table forms as well as in a specially developed GIS.

Model. A river under study was split into a finite number of reaches. It was assumed that monitoring stations observed pollutants concentrations at the downstream ends of reaches. The production enterprises in the river basin were grouped into industries, which included enterprises with analogous production technology and pollutants output pattern. Municipal services were grouped in the same way. Usually about 20 different types of industries and services were considered in the DSS. The production enterprises and municipal services were grouped in accordance to the reach they belonged to.

The problem was reduced to a search for an investment strategy for constructing wastewater treatment facilities. The investment (its volume was not given in advance) had to be allocated between production industries and municipal services in the reaches of the river. The current discharge was assumed to be known.

The integrated mathematical model that was used in the DSS consisted of two parts:

- a pollution transport model that helped to compute concentrations of pollutants at monitoring stations for any given discharge;

- two models of wastewater treatment that described the decrement in pollutant emission to the cost of wastewater treatment in an industry or a service.

The pollution transport model was based on empirical data and expert judgement. It was planned to substitute this model by another one based on parameterization of comprehensive simulation models of pollution transport. However, it turned out to be impossible to implement this desire at that time. It was implemented in another study that is described in the following section. Models of wastewater treatment were based on the description of wastewater treatment technologies. Decision variables described the fractions of wastewater treated by technologies in industries and services located in different reaches. A detailed description of the model is provided in (Lotov, Bushenkov and Chernykh, 1997).

Water quality criteria were based on pollution concentrations measured at monitoring stations. Since more than twenty pollutants were considered, the engineers applied aggregated environmental criteria. To be precise, they considered several pollutant groups. The value of a criterion that described a pollutant group was calculated in the following way. Relative pollutant concentrations (RPC) were used. An RPC is defined as the ratio of the actual concentration to the so-called maximum admissible concentration, which represents *a priori* environmental requirements. The sum of RPC values for the pollutants, which were considered as a group, was used as a water quality indicator. For the water quality criterion for a pollutant group, the maximal (i.e., the worst) value of water quality at monitoring stations was used. Desirable value was equal to 1.

It is important to stress that such criteria provide only one example from a large number of possible criteria that can be used (different criteria are described in the next section). The engineers provided the following grouping of pollutants:

- pollutants that lead to general degradation of water quality;

- pollutants that lead to degradation of fishing in the river;

- pollutants that lead to degradation of water quality related to the sanitation issues;

- pollutants that affect toxicological issues.

Therefore, four pollution criteria were used in the study.

Exploration of decision maps. As usual in the framework of the FGM/IDM technique, decision maps were used to display information on potential criterion vectors and on the efficient criterion tradeoffs. Figure 3.3 provides an example of a decision map used by the engineers.

The tradeoff curves in the figure display the frontiers for two pollution criteria, general pollution indicator, *GPI*, and fishing degradation indicator, *FDI*, for several values of cost. A black and white copy of the color display is given. The relation between the values of cost (millions of rubles for the year 1988 were used) and the shading is given in the palette. Note that, in accordance with Figure 3.3, the minimal cost here is 18 million rubles. This value corresponds to the darkest shading. For this value of cost, the combination of *FDI* and *GPI* values should belong

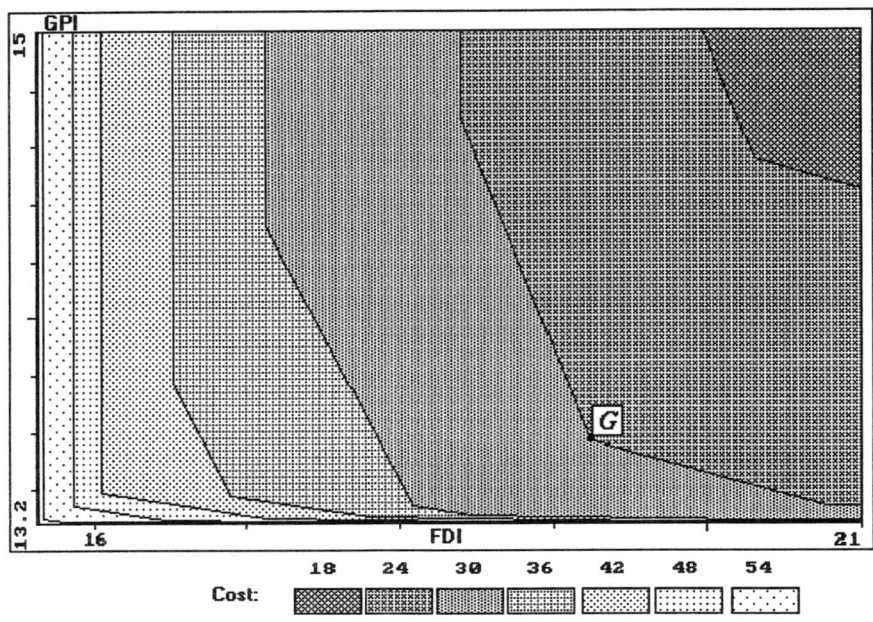

Figure 3.3. Tradeoff curves for *FDI* and *GPI* depending on cost

to the dark shaded variety placed in the upper right-side corner. In particular, the value of the *GPI* is not less than about 14.6, and the value of the *FDI* is not less than about 20. By adding 6 million rubles (24 millions in total), we provide a much broader variety of feasible values of *FDI* and *GPI*. In particular, 13.4 is now the feasible value of the *GPI*. The combination of *FDI* and *GPI* values identified by point **G** (it will be discussed later) is feasible as well. If cost is not less than 30 million rubles, the *FDI* can be decreased to about 17.25. Once again, the shape of the tradeoff curves helped engineers to assess how much the drop of the *FDI* is related to the increment in the *GPI*, and changing one tradeoff curve for another helped them to understand how the increment in cost results in the reduction of both pollution indicators.

Five criteria were explored in the DSS (in addition to *FDI*, *GPI* and cost, sanitation indicator, *SI*, and toxicological indicator, *TI*, were explored). The engineers drew decision maps for *FDI*, *GPI* and cost for different constraints imposed on the values of *SI* and *TI*. Matrices of decision maps were used, too (Figure 3.4). Once again, any decision map of the matrix is related to certain constraints imposed on the values of *SI* and *TI*. These values are given in Figure 3.4 above the columns and

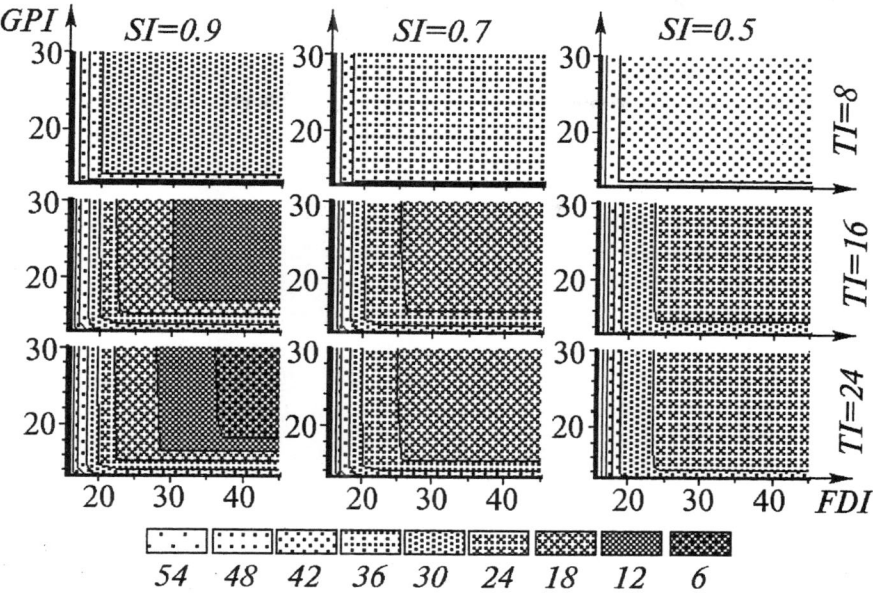

Figure 3.4. Matrix of decision maps

to the right of the rows of the matrix. These values may be chosen by the researcher or automatically.

By comparing the decision maps for properly chosen values of constraints, the engineers understood the influence of the fourth and the fifth criteria on the variety of feasible vectors for the first three criteria. As we have already said, any reasonable number of decision maps in a row or in a column may be presented (depending on quality of computer display).

Decision support system. The DSS consisted of five main subsystems:

- Data preparation subsystem;

- Subsystem for EPH approximation;

- Subsystem for exploration of EPH and identification of non-dominated goals;

- Subsystem for computing associated strategies; and

- Subsystem for display of strategies.

First, using the data preparation subsystem, the engineers prepared information concerning water balance and initial pollutant concentrations, parameters of pollution transport matrices, parameters of possible wastewater treatment facilities, etc. The subsystem provided a simple data compatibility test and converted initial files into an internal form of the software.

Constructing of the EPH was performed by the second subsystem. Then, engineers explored particular decision maps and their matrices. In the process of exploration, the engineers identified one or several feasible goals, say point **G** in Figure 3.3. Afterwards, the associated plan of wastewater treatment was computed automatically in the fourth subsystem. The fifth subsystem displayed the strategy in the form of column diagrams. Moreover, icons were placed on the map of a river basin and the diagrams were drawn on it when the corresponding icon was clicked. Reference information related to the problem was requested and provided in the same manner.

Unfortunately, it proved to be impossible to arrange the FGM/IDM-based negotiations among real decision makers at that time. For this reason, the engineers had to construct several variants of the project, which they provided to the decision makers who took this information into account. Hopefully, this improved their understanding of the situation. Therefore, the engineers played the role of experts who screened the whole variety of feasible decisions. The FGM/IDM technique gave them an opportunity to do it on the basis of information on potentialities of choice and efficient criterion tradeoff.

The system was implemented for water quality planning in several river basins. In particular, the project for a small river in the Moscow region, named the Nara River, was developed. The river was heavily polluted. In the framework of the old optimization procedure, the engineers failed to develop even a feasible plan, since the environmental requirements could not be met in this case. The DSS helped to solve this problem. The DSS was applied in the case of large rivers, too, say in the case of the Belaya River, which is one of the major tributaries of the Volga River.

As we have said already, the study described in this section was carried out in the first part of the 1990s. At that time, implementation of the projects developed by the engineers was facing institutional difficulties in Russia. The responsibility of the federal government for environmental quality was gradually shifting (jointly with financial resources) to regional and local authorities. Therefore, it was not clear what institution had to support regional environmental projects. Moreover, privatizing processes in Russia that were under way at that time complicated the problem drastically.

Now, the studies of this kind are carried out by the private institution Engineering and Research Center on Water Management, Land Reclamation and Environment "Soyuzvodprojekt". In the framework of collaboration with this institution, a new DSS for water quality planning in river basins was developed on the basis of the experience received in the process of application of the DSS described in this section.

3. DSS for screening of water quality improvement plans

The DSS water quality management which is described in this section was calibrated for the Oka River. The Oka River is one of the largest tributaries of the Volga River. The length of the Oka River is about 1,500 kilometers. The river has multiple tributaries. Its flow changes from 10 cubic meters per second in the upper part of the river (during the dry summer period) up to 1000 cubic meters per second in the lower Oka River. Geometric characteristics, roughness and other parameters of the riverbed vary substantially along the river. In Figure 3.5, the map of the river basin is provided. Along with the Oka River and multiple tributaries, frontiers of seven regions located at the main flow of the river can be seen. One can see that Moscow is located in the Oka River basin.

In the study, the riverbed was split into fourteen segments that approximately describe membership of banks to the regions. Pollution concentration was studied at the downstream ends of the segments, which are given by numbers in Figure 3.5.

The simplified integrated model used in the DSS for screening the strategies consists of three submodels:

- a wastewater discharge submodel that describes the current discharge attributed to particular regions, river segments, industries or services; structure of the wastewater discharge is provided, too;

- a wastewater treatment submodel that relates the decrement in wastewater discharge to the construction and performance cost of wastewater treatment installation;

- a pollution transport submodel that allows computation of the pollutant concentration in monitoring points for a discharge given at all sources.

The first and the second submodels were developed by experts on the basis of statistical and technological information. The third submodel

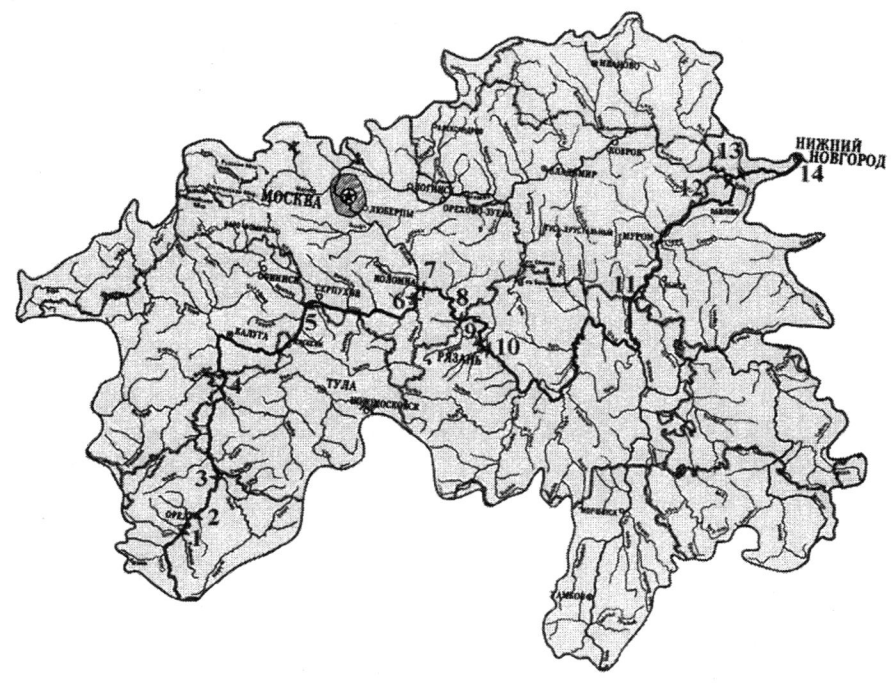

Figure 3.5. Map of the Oka River basin

was constructed by parameterization, i.e., approximation of the input–output dependencies of an original model. In this section, we discuss the development of the simplified integrated model; then we describe the DSS and application of the FGM/IDM technique in its framework. This section is based on the paper (Lotov, Bourmistrova, Bushenkov, Efremov, Buber, Brainin and Maksimov, 1999).

As it has been noted already, the most important form of simplified models is the linear one. The simplified linear description of dependence of output vectors upon input vectors has the form of a matrix. It is named the influence matrix. First, we consider possible ways of constructing influence matrices. Then, we describe our practical experience of constructing a simplified pollution transport model by parameterization of the system MIKE 11, which is a well-known system for modeling rivers and channels, produced by the Danish Hydraulics Institute (DHI).

Constructing influence matrices – general discussion. Let us start with the influence matrices for linear original models. Consider, for example, a stationary linear partial derivatives model that describes regional transport of a single pollutant from several pollution sources along a river. Assume that the pollution discharge per unit time is constant. Then, the model can be used to estimate the pollutant concentration at any point of the river.

If pollutant discharges are not known in advance, the point source method can be used. In the framework of the method, source functions can be constructed that provide concentrations resulting from individual sources with a unit rate of discharge. Since the pollutant transport model is linear, an individual source-related pollutant concentration at any point of the river equals the product of the source function value on its discharge. Due to the same linearity of the model, one has simply to sum up concentrations resulting from all sources to estimate the total concentration at that point. In other words, pollutant concentration at any point (for example, at a monitoring station) is a linear function of pollution discharges for all sources. It is given by its coefficients.

Let us consider several monitoring stations located along the river. The influence coefficients for all sources and monitoring stations provide the influence matrix. Let us consider the vector of pollutant concentrations at the monitoring stations. It is clear that the vector equals the product of the influence matrix on the discharge vector. It means that we can simply multiply the influence matrix by the discharge vector to compute the concentrations instead of solving the problem in partial derivatives.

Influence matrices can be constructed precisely in the linear case by using values of the source functions. In the non-linear case, an influence matrix approximates the discharge–concentration dependencies. The method for estimating an influence matrix may depend upon a particular scientific field. A universal approach may be based on the application of regression analysis of input–output dependencies, which can be obtained by simulation of non-linear models. Along with the approximation of input–output dependencies, simulation can provide their applicability ranges.

If there is no adequate mathematical model for a subsystem, an influence matrix can be constructed on the basis of statistical analysis of experimental or historical data. Sometimes, experts can provide both an influence matrix and its applicability range (as was done in the previous section).

A combination of influence matrices and other simplified descriptions as well as balance equations and constraints imposed on variables contributes to a simplified integrated model that describes the environmental system. Simplified integrated models are typically less precise than original models, but that fact is not of great importance since integrated models are used on the first stage of the decision process for screening of decision strategies. So, insufficient precision of the simplified models can be compensated on the stage of detailed analysis of the selected strategies.

Now let us describe our experience of constructing the influence matrices that describe transport of several pollutants along a river in the DSS under consideration. In contrast to the DSS described in the previous section, the coefficients of the influence matrices were constructed through simulation of the pollutant transport model of the system MIKE 11.

Constructing influence matrices for pollution transport. Constructing influence matrices was started with calibration of the hydrodynamic submodel (HD MIKE 11) of the system MIKE 11 to the stationary flow of the Oka River during the summer period with a minimal flow. There were 23 main tributaries considered in the model along with 33 conditional inflows that describe inflow from the neighboring land to the river. The inflow of water was given. The influence of the Volga River on the flow in the lower part of the Oka River was taken into consideration, too. It turned out that the geometric information and roughness coefficients used in the model were sufficient to determine the flow during the summer low-flow period. The hydrodynamic information on the river was used in the process of calibration of the advection–dispersion model (AD MIKE 11) of the system MIKE 11. AD MIKE 11 provided an opportunity to describe pollution transport in the river.

Multiple sources of pollution do exist in the river basin. They include natural sources, industrial wastewater discharge sources, municipal point and non-point wastewater discharge sources, agricultural non-point sources, large animal breeding enterprises, etc. In accordance with the information we managed to collect, six most important pollutants were considered in the model, namely concentrations of suspension, phosphates, nitrates, oil products, and ferrous combinations as well as biological oxygen demand. Both discharges and concentrations of pollutants were used to calibrate AD MIKE 11 for the summer low-flow period. Data on wastewater discharge from large cities and banks of the river were collected partially from state statistical institutions and partially they were obtained as a result of expert evaluation. Known pollutant concentrations in large tributaries were averaged on the basis of data for several years. The remaining discharge was spread among small rivers proportionally to their flow. In the process of calibration of the model, the decay constants of the advection–dispersion model were adjusted. Sometimes it was needed to correct data on wastewater discharge given by statistics.

The model AD MIKE 11 has an extremely important property: for a given water flow, a pollutant concentration at a monitoring station depends on the capacity of the pollutant sources in a linear way. One can prove this feature of the model theoretically, on the basis of its equations, or experimentally, by using simulation. This feature was used for constructing precise influence matrices for pollutants considered in the DSS.

The procedure of constructing the influence matrix looks as follows. In a particular river segment, three types of pollution sources are considered: sources located on the right and the left banks and sources located in the river up to the segment under consideration. The pollutant flow at the downstream end of the segment was assumed to be the sum of the flow from the upper segment and of the discharges from the banks multiplied by the related coefficients. The simulation experiment for a segment consists of four runs. In the first run, the pollutant flow at the upstream end of the segment (i.e., at the downstream end of the upper segment) was set equal the current value and the discharges from both banks of the river segment were set to be equal the background emission. In the second and the third runs, the pollutant flow at the upstream end was set to equal to the background flow. The discharge from the left-bank (or the right-bank) region was set to be equal to the current emission, while discharge from the region located at the opposite bank was equal to the background value. In the fourth run the capacities of all sources of pollution were set equal to the current level. Actually, the results of the three first runs were sufficient to estimate the coefficients of the linear model of the pollution in the segment, and the fourth run was used only to check the results. Though usually the additional pollutant flow computed in the fourth run was the same as the sum of the additional pollutant flows computed in the previous runs, sometimes mistakes in data were found. By this, it was experimentally proved that the pollution flow at the downstream end is related to the pollution sources in a linear way. This experiment had persuaded those people who did not trust the theoretical results received by the analysis of the equations of the simulation model.

Coefficients of the linear dependencies for all segments and all pollutants provided sufficient information for computing the influence matrices for the whole river. It made it possible to compute the pollutant concentrations at the downstream ends of the segments by multiplying the influence matrices by the discharge vectors. The influence matrices describe the pollutant transport by relating the decrement in the wastewater discharge to the decrement in concentrations of the pollutants at the downstream ends.

Constructing the influence matrices was carried out by Dr. A. Buber and N. Brainin from "Soyuzvodprojekt", Dr. A. Maksimov from the research institute VODGEO and Dr. R. Efremov from the Computing Center of Russian Academy of Sciences (CC RAS).

Other submodels. Unfortunately, available data on wastewater discharge happened to be very rough — it was attributed only to particular regions, but not to industries. So, in contrast to the DSS described in the previous section, we had to restrict to a regional pollution model, though we keep on hoping to receive related industrial data sooner or later. It is clear that the influence matrices constructed in the framework of the development of the current DSS can be easily combined with a multi-industrial/multi-services discharge model of the previous section.

The model of wastewater discharge treatment was based on a concept of wastewater purification technology. The concept provides an opportunity to include hundreds of possible technologies into consideration. Nevertheless, we had to restrict to a small database of discharge treatment technologies developed by Dr. A. Gotovtsev from the Institute for Water Problems of RAS using data given in (Henze and Oedegaard, 1995).

A decision variable of the model described an investment in a particular discharge treatment technology in a particular region at a particular river segment. The influence matrices were used to compute the resulting concentration of pollutants at monitoring stations for a given strategy, i.e., for given values of decision variables. The total and regional costs could be computed, too.

DSS description. The DSS included codes of the main technique used for decision screening, i.e., the FGM/IDM technique, as well as codes of auxiliary subsystems. Dr. L. Bourmistrova from the Computing Center of RAS has coded new time-efficient mathematical algorithms for the DSS (they are discussed in Chapter 6) and took care over the whole DSS. We describe the DSS on the basis of an example that helps to understand how the DSS works.

The following subsystems were included into the DSS for screening of water quality strategies:

1 a subsystem for visualization of the current pollution in the river,

2 a subsystem for specification of screening criteria and constraints imposed on the values of performance indicators,

3 a subsystem for the EPH approximation,

4 a subsystem for interactive display of decision maps and identification of a feasible goal,

5 a subsystem for computing the goal-related strategy,

6 a subsystem for visualization of the computed strategy.

The role of the subsystems is clear from their names. Only short comments are needed. Users receive information on the current pollution in the river in the form of diagrams. Figure 3.6 contains black and white copies of several diagrams provided by the second subsystem. The upper

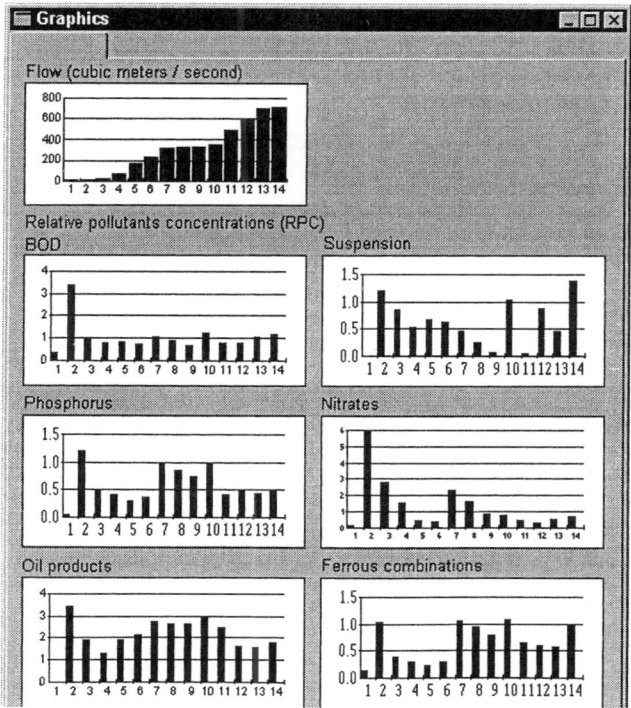

Figure 3.6. Current pollution concentration (PC) in monitoring stations. The upper black diagram shows the river flow during dry summer season

(black) column diagram provides data on the flow during dry summer season at 14 monitoring stations (in cubic meters per second). Six other column diagrams display pollution at the stations in relative pollutants concentrations (RPC) that are defined as the ratios of an actual concentration to maximum admissible concentration, which represents *a priori* environmental requirements. So, the required value of the RPC must be not greater than 1. Such diagrams are given as GIS-generated maps.

The information on the current situation is used in the process of formulation of a screening problem, i.e., in the process of specification of screening criteria and constraints imposed on the values of performance indicators. To satisfy different users with different interests, a large list of performance indicators is provided (part of it is given in Figure 3.7).

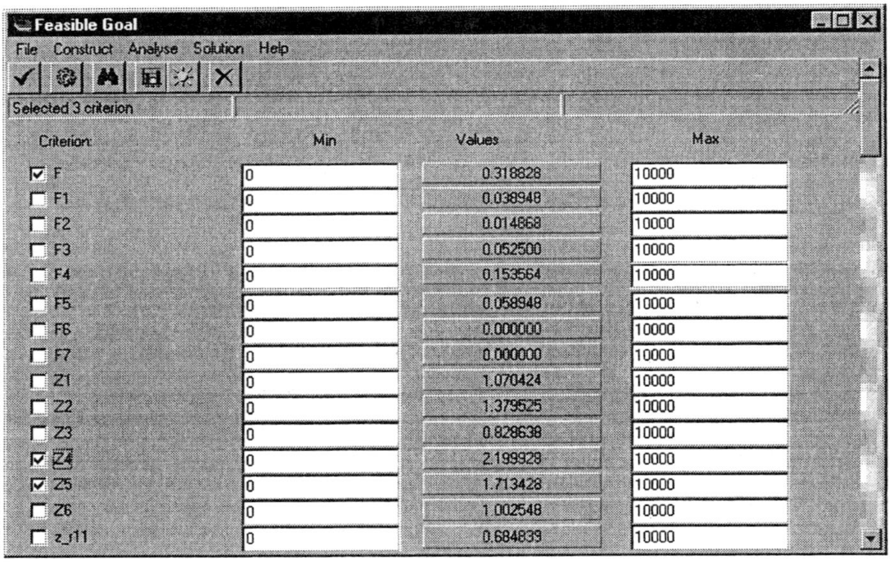

Figure 3.7. Part of the list of performance indicators

The list includes two kinds of indicators:

- environmental indicators — regional or maximal concentration of pollutants in a region or in the river (in RPC);

- economic indicators — investment in particular regions as well as the total cost of an investment project (in billions of US$).

Users can specify two to seven screening criteria directly in the list of performance indicators. Moreover, constraints on the value of any indicator may be imposed (one has to enter desired values into the left and right columns of the table). The central column of the table contains the performance indicators associated with a strategy that is the result of previous screening activities. At the very beginning, this column is empty.

Once again, we illustrate application of the DSS by an example, in the framework of which artificial data are used.

Example of DSS application. One can see in Figure 3.6 that the most drastic pollution problem is related to the upper part of the river (monitoring stations 2 and 3). This problem is related mainly to the low flow during the dry summer season. Another serious problem is related to pollution with oil products − it is too high in any segment of the river. In this example, we are going to deal with the latter problem.

Let us consider an example of exploring a possible conflict related to investment in wastewater treatment facilities in the Moscow region (*M*-region) and Nizhny Novgorod region (*NN*-region). Those regions are the economically most developed ones in the Oka River basin. The following five criteria were used in the study:

- maximal concentration of oil products at the monitoring stations located in the *M*-region (z_r45);

- maximal concentrations of oil products at the monitoring stations located in the *NN*-region (z_r75);

- total cost of the project (F);

- investment in the territory of the *M*-region ($F4$); and

- investment in the territory of the *NN*-region ($F7$).

The EPH was approximated for the five criteria listed above. Let us consider several decision maps. Figure 3.8 displays the feasible combinations of oil products concentration at monitoring stations located in the *M*-region (horizontal axis) and in the *NN*-region (vertical axis). The upper scroll-bar informs that the total cost is restricted by $233 million here. Constraints imposed on the values of $F4$ and $F7$ coincide with their maximal values and have no influence on the picture.

One can see that there is a conflict between values of oil pollution in these regions when the cost of the project is restricted. Pollution in the *M*-region can be decreased from the current pollution of 2.7 to a bit less than 1.5, that is, to the value that is minimal for this cost. However, if z_r45 is minimal, the pollution in the *NN*-region cannot be less than

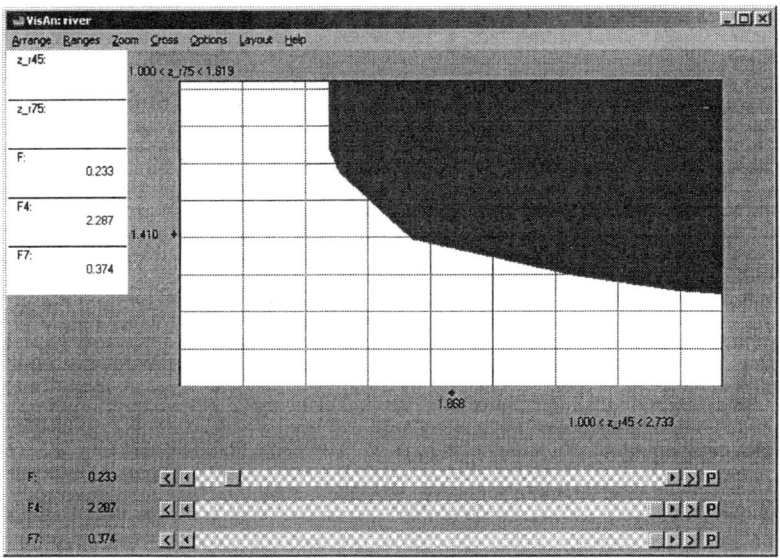

Figure 3.8. Feasible pollution concentrations at monitoring stations located in the *M*-region (horizontal axis) and in the *NN*-region (vertical axis) for the given cost of $233 million

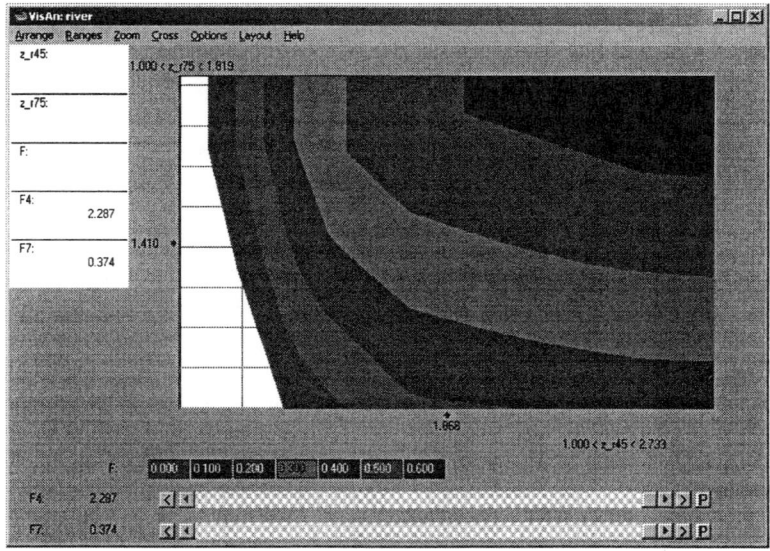

Figure 3.9. Black and white copy of the decision map for a restricted total cost

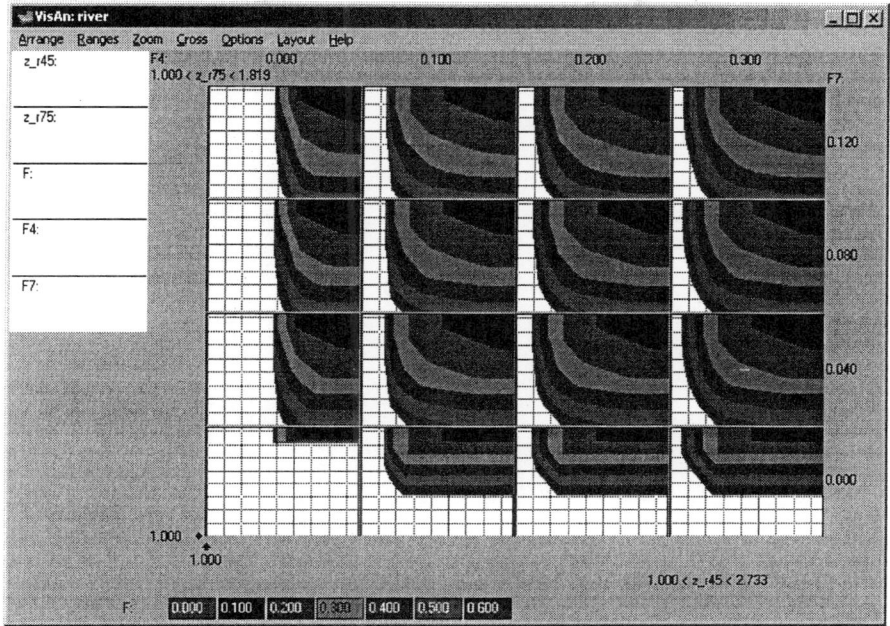

Figure 3.10. Black and white copy of a matrix of decision maps

about 1.65. Therefore, if \$233 million are applied in the interests of the M-region, the pollution in the NN-region can drop from the current 1.8 till about 1.65, but not less. In contrast, if \$233 million are applied in accordance to the interests of the NN-region, the pollution in it can be as low as 1.25. The frontier of the slice shows the criterion tradeoff between pollution in both regions in a clear form.

Figure 3.9 contains a decision map for several values of total cost (shading), while total cost is not greater than \$600 million. It is clear that the form of the tradeoff curve substantially depends on it. Total cost of \$100 million solves the pollution problem to some extent (compare pollution values related to the frontier of the related slice with those in the right upper corner that corresponds to zero cost). Additional \$100 millions are very effective, too (the next frontier). Proceeding to the next slices, we see that total cost of \$400 million could solve the problem for the NN-region if the investment is used according to the interests of the NN-region. At the same time, minimal pollution in the M-region for this cost is achieved while pollution in the NN-region is fairly high.

Now let us consider the influence of constraints imposed on regional investments, $F4$ and $F7$. The above decision map can be animated in accordance with the changes of $F4$ and $F7$. However, here we have to restrict to matrices of decision maps that provide snap-shots of the above animations. In the matrix of decision maps given in Figure 3.10, columns are related to constraints imposed on the values of $F4$ and rows are related to constraints imposed on the values of $F7$. These constraints are specified above the columns and to the right of the rows.

The decision map displayed in Figure 3.9 can be associated with the decision map located in the upper row of the extreme right column, which is related to such constraints imposed on $F4$ (not greater than $300 million) and on $F7$ (not greater than $120 million) that still do not influence the decision map.

By moving to the left in the same row, a user can see the influence of the constraint imposed on $F4$ while the $F7$-related constraint is fixed. By moving downward in the same column, the user obtains knowledge on the influence of the constraint imposed on $F7$ while the $F4$-related constraint is fixed. One can find various effects of such movement. For example, one can explore how a particular tradeoff curve, say, the frontier of the maximal slice, which is related to the total cost of $600 million, depends upon those constraints. One can see that the opportunities of water quality improvement are not broad for $F4 = F7 = 0$. Additional

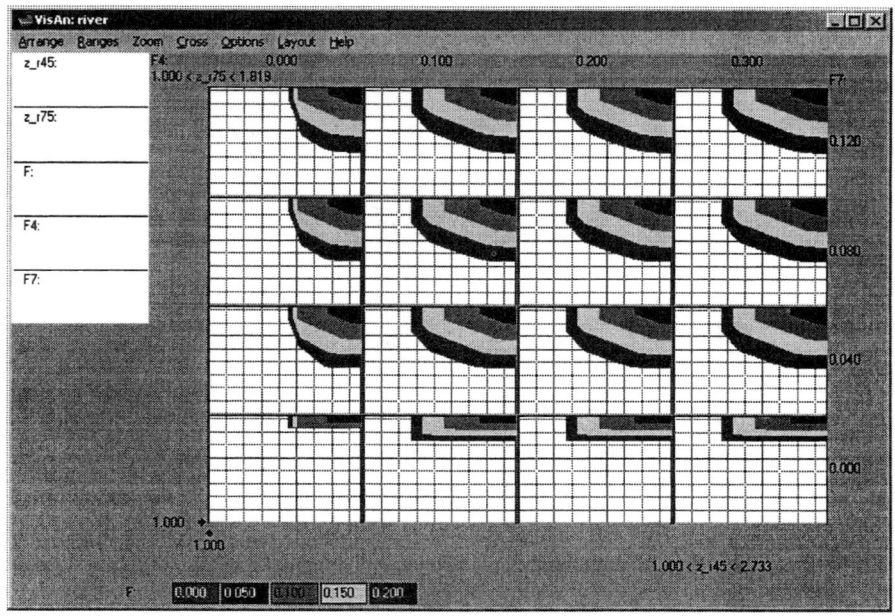

Figure 3.11. Matrix of decision maps for total cost restricted by $200 million

effects can be found on the color display. Once again, the number of columns and rows depends only on the quality of the display and may be regulated by the user.

Different matrices of decision maps can be displayed and explored, too. For example, let us consider a matrix of decision maps related to a substantial decrement of total cost to $200 million (see Figure 3.11). Slices are given for every $50 million. Though the cost is three times less, opportunities of water quality improvement still exist. Dependence of pollution on the total cost looks linear, except slices with zero regional investment.

Double click of a computer mouse on a decision map of the matrix results in the selection of the decision map for a detailed exploration. Let us suppose that a decision map was selected, which is related to $100 million investment in the M-region and $40 million investment in the NN-region. We split the range of total cost between zero and $200 million into nine shadings (Figure 3.12). One can see in detail how the increment in the cost influences the water quality.

A feasible goal is identified in the decision map by the cross. It is related to the total cost of $150 million. The pollution in the M-region is

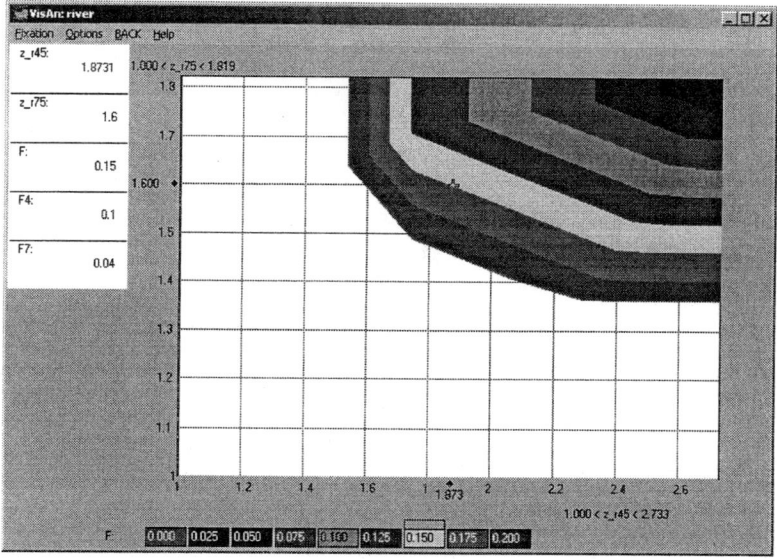

Figure 3.12. Black and white copy of the decision map related to $100 million investment in the M-region and $40 million investment in the NN-region. The goal is given by the cross.

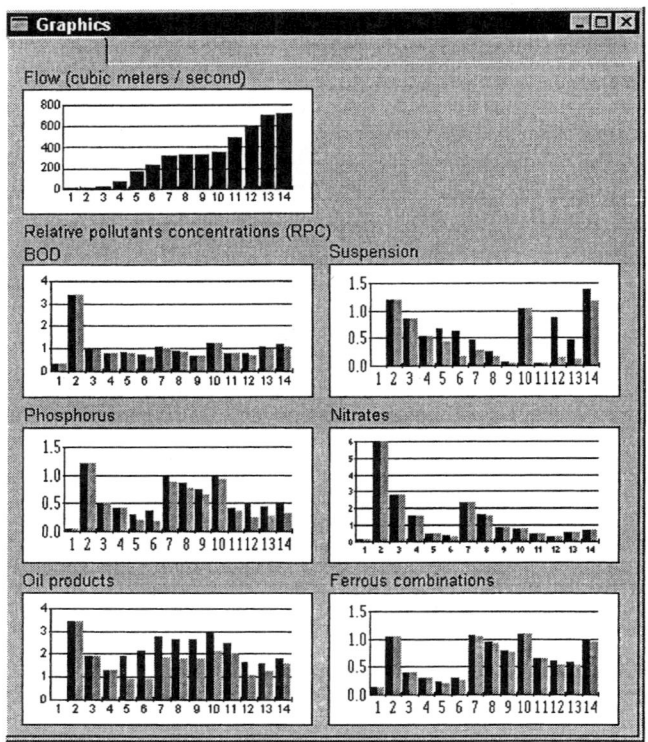

Figure 3.13. Pollution concentration that results from the strategy (gray shading) versus current concentration (black shading)

equal to 1.87, and in the *NN*-region it equals 1.6. The strategy associated with the identified goal is displayed in Figure 3.13.

Figure 3.13 contains the same column diagram that is provided in Figure 3.6. However, in addition to the current pollution concentrations, the new pollution concentrations (i.e., pollution concentrations resulting from the goal-associated strategy) are given; they are displayed in gray. One can see that the pollution with oil products is substantially less at the monitoring stations located in the Moscow region (stations 5, 6, etc.) and lower. Nevertheless, the problem of oil product pollution cannot be solved finally since investment is fairly low. It is important that the values of several other pollution indicators are improved, too. Spatial information on the selected water quality improvement plan can be given in a map (see Figure 3.14).

The map provided in Figure 3.14 displays a part of the river basin. It contains icons that help users receive diagrams with information on

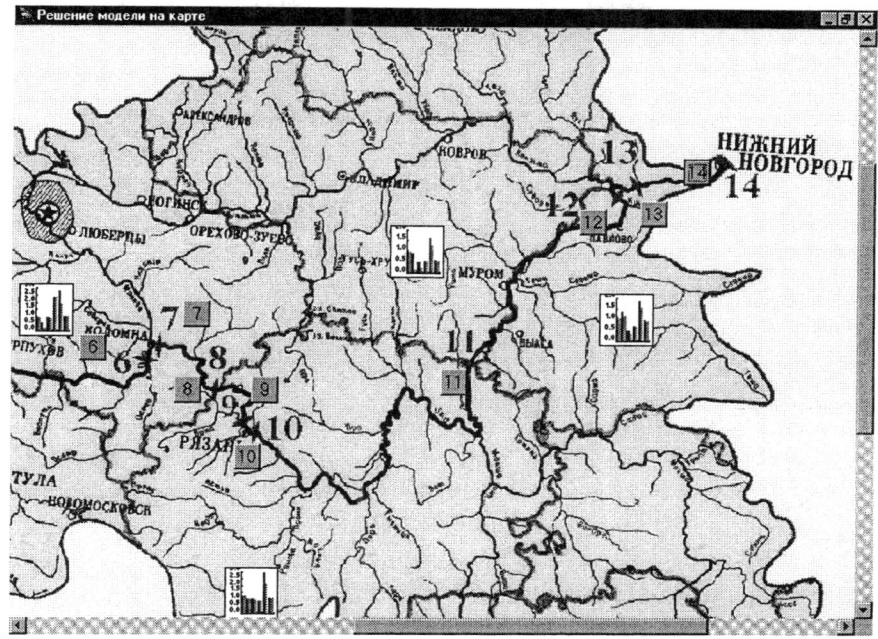

Figure 3.14. Black and white copy of the GIS-produced map. Icons are used to receive information on regional pollution levels, investments, pollutant discharge, etc.

regional values of discharge, pollution and investment. For example, these diagrams inform on the resulting investment distribution among the regions.

Since a simplified model is used for the screening stage, the selected strategies should be studied and refined in simulation experiments with MIKE 11 that provides an opportunity to explore them, in addition to the most important low-flow season, for other seasons as well. Robustness to changes in precipitation scenarios can be explored, too.

Evolutionary mode of DSS application. In the above subsection, the simplest straightforward mode of DSS application is described. Usually, users apply interactive (evolutionary) mode and several loops of strategy selection. The first loop may be associated with the display of a strategy in the geographic maps that provide a better understanding of its features. Often such a display results in the desire to change the specification of the screening problem. It means that the user returns to the second subsystem, specifies new decision criteria and imposes new constraints on the values of the performance indicators. Then, the user goes ahead and follows the screening procedure until the new strategy is generated and displayed.

Let us consider an example. Suppose that by exploring a strategy provided in the geographical map (say, given in Figure 3.14), the user has found that the pollution levels are too high at several important monitoring stations. Then, the user may decide to require ideal values (so-called first class water) at those important stations. By this, the user provides a new formulation of the screening problem. Exploration of decision maps may result in a new goal and a new goal-associated strategy. The strategy can be displayed in the geographical map. Information on the strategy can be used as a source of a new formulation of the screening problem.

Another loop of the procedure for strategy screening may be related to a modification of the integrated model used for the screening stage. As it has been said above, the generated strategy is studied and refined in simulation experiments on the basis of MIKE 11 for all seasons and precipitation scenarios. If the strategy happens not to be appropriate, the user may want to broaden the integrated model by combining the influence matrices for the low-flow season with influence matrices related to other seasons and (or) precipitation conditions. We have already described how the user can change the formulation of the screening problem by adding new (or more severe) constraints on pollution indicators. The option provided by the extended integrated model consists in an opportunity to formulate additional constraints on new pollution indicators of the extended model. Such loop of evolutionary exploration, however, has not been applied yet.

Real-life application of the DSS. As it was said already, the DSS described here was developed at the request of the Russian Federal Ministry for Natural Resources in the framework of the federal program "Revival of the Volga River". The DSS was developed in collaboration with engineers from the Engineering and Research Center on Water Management, Land Reclamation and Environment "Soyuzvodprojekt" headed by Dr. A. Buber. These engineers are the permanent users of the DSS.

The DSS turned to be a convenient and transparent tool for screening of water quality improvement strategies. It was planned that the DSS must be used in all river basins of Russia. However, a broad application of the DSS is inhibited by problems of data collection. Indeed, only a part of the data for the Oka River, namely the pollution transport model, is reliable. The wastewater discharge pattern and the wastewater treatment technologies are still only plausible. The problem of data collection is vital to real-life application in other river basins, too. Investment in data collection activities may solve the problem.

Summary. In the framework of the DSS described in this section, the models of the system MIKE 11 were applied as data preparation tools for the FGM/IDM technique. After selecting one or several strategies, the user can apply MIKE 11 once again for their simulation. Thus, the FGM/IDM technique can be considered as an additional system component that broadens the scope of decision support services provided by water quality simulation systems like MIKE 11.

The demo Web resource described in Chapter 1 proves that the above DSS can be easily implemented on the Internet. Federal and regional authorities could use it for negotiation preparation. However, it may be even more important that millions of ordinary Internet users could be able to apply such a Web resource individually to obtain information on the whole variety of possible strategies. Such opportunity is discussed in the Epilogue.

Chapter 4

REASONABLE GOALS METHOD AND ITS APPLICATIONS

The Reasonable Goals method (RGM) is introduced in this chapter, and several applications of the method are described. We consider a simplest form of the RGM here, which supports selecting of a small number of alternatives from given lists that contain a large, but finite, number of decision alternatives. Such lists may contain millions of alternatives. The RGM is based on representing decision alternatives in the form of criterion points and on approximating the convex hull (envelope) of a variety of points. To be precise, the EPH of the convex hull is approximated. Due to such enveloping, the IDM technique can be applied, but now the user studies proxy tradeoffs between the criteria. Application of the IDM technique for exploration of the Pareto frontier of the envelope and identifying a goal vector (so-called reasonable goal) on it are the main features of the RGM. Since the convex hull is explored instead of the variety of points itself, an identified goal may not be feasible, but only reasonable. As a result, several decision alternatives that are in line with the identified goal are selected.

Note that the identified goals can be considered as aspiration levels. In this case, the exploration of the Pareto frontier of the envelope can be understood as a process of developing the aspiration levels. It is clear that the RGM/IDM technique can be applied for visualization of a large relational database aimed at selecting of several preferred rows from the database. After an SQL-query results in a table of rows described by several attributes, the RGM/IDM technique provides visual display of the table. Such visualization can help to understand the relation among attributes for the given variety of items (rows) and to find some rows of the database that have interesting properties. In contrast to the usual "blindfold" selecting from a database, which is based on specification of a priori thresholds, the RGM/IDM technique informs on the frontiers of what is possible and on its form. So, the RGM/IDM technique can be considered as a specific graphic tool for data mining.

In this chapter, the main ideas of the RGM are introduced first on the basis of the example of real estate market. Then, two real-life applications of the technique (water quality planning in a small region of Russia and national energy planning in Israel) are described. In Section 4, application of the RGM/IDM technique in the framework of a DSS for location a health practice in Idaho, USA, is given. In Section 5, a Web application server based on the RGM/IDM technique is considered and its use in the framework of e-commerce for consumer support is discussed. Then, possible applications of the RGM/IDM technique in medical diagnostics, machinery design and financial engineering are outlined. Finally, the RGM/IDM technique is considered as a database visualization tool, as a graphic data mining tool and as a graphic tool for cost-benefit analysis with multiple costs and benefits.

1.   Introduction to the Reasonable Goals Method

We introduce the RGM using the example of real estate market. Each alternative (actually, a house) is described by several attributes such as price (thousands US$), age (years), lot size (acres), number of bedrooms, number of bathrooms, etc. The houses are given in a list, a part of which is provided in the table (see Figure 4.1).

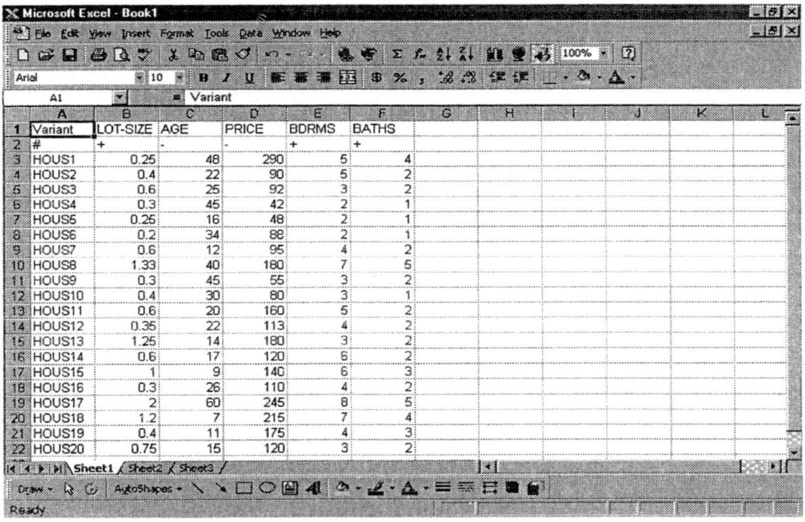

	A	B	C	D	E	F
1	Variant	LOT-SIZE	AGE	PRICE	BDRMS	BATHS
2	#	+	-	-	+	+
3	HOUS1	0.25	48	290	5	4
4	HOUS2	0.4	22	90	5	2
5	HOUS3	0.6	25	92	3	2
6	HOUS4	0.3	45	42	2	1
7	HOUS5	0.25	16	48	2	1
8	HOUS6	0.2	34	88	2	1
9	HOUS7	0.6	12	95	4	2
10	HOUS8	1.33	40	180	7	5
11	HOUS9	0.3	45	55	3	2
12	HOUS10	0.4	30	80	3	1
13	HOUS11	0.6	20	160	5	2
14	HOUS12	0.35	22	113	4	2
15	HOUS13	1.25	14	180	3	2
16	HOUS14	0.6	17	120	5	2
17	HOUS15	1	9	140	6	3
18	HOUS16	0.3	26	110	4	2
19	HOUS17	2	60	245	8	5
20	HOUS18	1.2	7	215	7	4
21	HOUS19	0.4	11	175	4	3
22	HOUS20	0.75	15	120	3	2

Figure 4.1.   Real estate database

We begin the description of the RGM with the case when only two attributes are used as choice criteria, namely price and age of a house. Though the use of the RGM is not advantageous in the case of two criteria, we use it anyway to explain the fundamentals of the technique.

RGM in the case of two criteria. In Figure 4.2, about two dozens of choice alternatives (houses) are depicted by points and crosses in the price and age plane. It is assumed that the user is interested in minimizing both price and age of a purchased house. The non-dominated alternatives are denoted by crosses and the dominated alternatives by points.

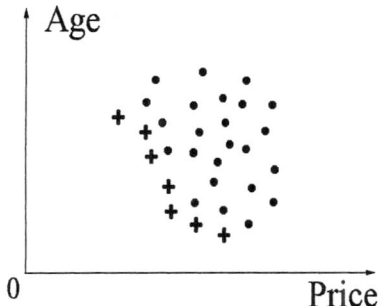

Figure 4.2. Real estate alternatives for two attributes. Pareto alternatives are depicted by crosses.

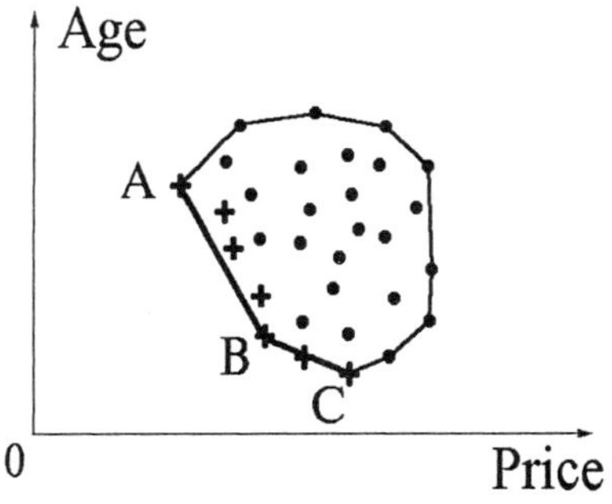

Figure 4.3. Convex hull of the variety of points (alternatives).

Since the number of decision criteria is two, Figure 4.2 provides full information about the variety of alternatives. However, one is unable to construct a similar graph for the case of three, four, five and more criteria, and so a different graphic technique must be used. In the RGM, the frontier of the envelope (convex hull of the variety of points) is displayed instead of points (see Figure 4.3). The convex hull also includes artificial alternatives, which help to explore the variety of feasible alternatives. The idea of using the convex hull of decision alternatives was introduced in (Raiffa, 1968).

Note that Figure 4.3 may be used to illustrate a shortcoming of a commonly used approach to express preferences − the criteria weighting procedures. Numerical weights may reflect, say, the relative importance of decision criteria. Weights help to construct a linear preference function and to apply optimization for screening the variety of alternatives. However, optimization with a linear function can only find those non-dominated points that are the vertices of the convex hull. In Figure 4.3, the non-dominated points A, B and C may be found in this way. In contrast, other non-dominated points (given by crosses, but not marked by letters) can not be found if a linear preference function is used. Consequently, some non-dominated alternatives that may be of interest to the user are not found.

There exist several other shortcomings of criteria weighting. One of them is revealed by the results of psychological experiments that prove that the identification of criterion weights is complicated even for experts, not to mention decision makers, and consequently the results of optimization are not reliable (Borcherding, Schmeer and Weber, 1993). Generally speaking, a linear preference function is often in contradiction with multiple criteria utility theory (see for details Keeney and Raiffa, 1976). For this reason we apply another technique based on direct visualization of the Pareto frontier of the envelope.

Since the user is interested in minimizing both price and age, the Pareto frontier of the envelope coincides with its lower left frontier (curve ABC in Figure 4.3). As usually, the Pareto frontier shows tradeoffs between two criteria — how much the decrement of age is related to the increment of price if points of the frontier are used. However, in contrast to the FGM, the Pareto frontier displays tradeoffs between the criteria for the convex hull in this case. Since the Pareto frontier of the convex hull roughly represents the Pareto frontier for a variety of the original decision alternatives, it can be considered as the proxy (averaged) tradeoff curve for the original alternatives.

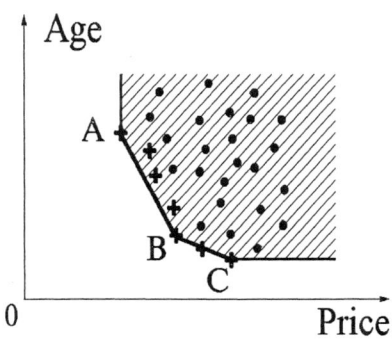

Figure 4.4. CEPH for two attributes.

In addition to the Pareto frontier of the convex hull (curve ABC), Figure 4.3 displays other frontiers. These frontiers are not needed since they only confuse the interpretation of the graph. As it was discussed in the previous chapters, such frontiers are especially harmful in the case of three and more criteria when the user has to compare multiple frontiers related to several values of the third criterion. In order to avoid the dominated frontiers, the EPH of the convex hull may be used, i.e., all dominated alternatives may be involved in the consideration as well. Once again, the EPH of the convex hull has the same Pareto frontier as the convex hull.

For example, the broadened variety in Figure 4.4 has the same Pareto frontier (curve ABC) as in Figure 4.3. The EPH of the convex hull of alternatives is denoted in this book by the Convex Edgeworth–Pareto Hull (CEPH) of the original alternatives. As in the case of convex models described in Chapters 1–3, using the CEPH does not influence a reasonable selection result, but its application makes the display simpler.

After exploration of a Pareto frontier is completed, a user has to identify a preferable combination of criterion values, which belongs to the Pareto frontier of the CEPH. This combination of criterion values is known as the reasonable goal (vector). In Figure 4.5 the identified reasonable goal is given by a circle. It is important to note that users do not need to be involved in complicated interactive procedures aimed at eliciting his/her preferences. Instead, a user has to identify the goal vector being supported by graphic display of the Pareto frontier of the envelope. Note that in contrast to other goal-based methods, the RGM restricts the identification of the goal to a proxy Pareto frontier. Due to this, the identified goal is close to the feasible points and is denoted as the reasonable goal.

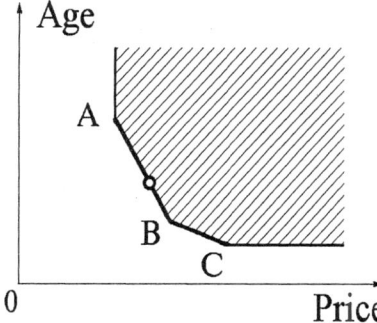 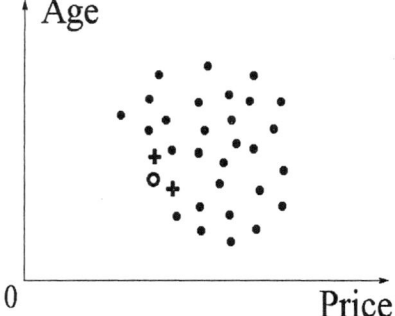

Figure 4.5. The identified reasonable goal is denoted by an empty point.

Figure 4.6. Points that are close to the goal are selected (depicted by plus sign).

Since the goal is identified in the graph, which displays a proxy Pareto frontier of the alternatives, the goal is not very likely to coincide with the feasible points. For this reason, a computer algorithm is used to select several feasible points, which are close to the identified goal (see Figure 4.6). The selected points (and the associated alternatives) are of interest to the user since they reflect both his/her subjective preferences expressed in the form of the goal and the objective situation represented by the variety of alternatives. Note that in Figure 4.6, selected alternatives do not coincide with the vertices of the convex hull. The selected alternatives may be studied using other tools. For example, spatial alternatives can be visualized on thematic maps. Such visualization of the selected alternatives is exemplified later in this chapter.

RGM in the case of more than two criteria. Now let us discuss the RGM in the case of more than two selection criteria. As in the bi-criterion case, the EPH of the convex hull (CEPH) of alternative-associated points is approximated. Visualization of the Pareto frontier of the CEPH is based now on the application of the IDM technique. Indeed, the problem of approximating the CEPH is not more complicated than the problem of approximating any other convex variety. Such a problem has already been discussed in short in previous chapters of the book and will be considered in detail in Part II. Visual on-line exploration of the CEPH is not more complicated as the exploration of the EPH in the framework of the FGM. Identification of a goal vector is based on the same ideas. The main difference between the application of the IDM technique in the FGM and RGM, as was exemplified by the bi-criterion case, consists in the fact that the IDM technique displays proxy (averaged) criterion tradeoffs in the framework of the RGM. Moreover, the goal identified by the user is not feasible. However, it is reasonable since it is located relatively close to the feasible points in criterion space.

The concept of reasonable goal was introduced in (Lotfi, Stewart and Zionts, 1992). A goal is considered to be reasonable if it is located close to the variety of points that represent decision alternatives. To support the choice of a reasonable goal, the method by Lotfi, Stewart, and Zionts lists alternatives selected by a user and provides statistical information to guide the selection process. In the RGM, the same meaning of a reasonable goal is used, but the information is provided in a different form − it is based on the graphic display of the frontier.

It is important that due to a graphic study of the Pareto frontier, the user identifies a reasonable goal consciously. Since the identified goal is usually non-feasible, but close to the feasible criterion points, it can be used for selecting a small number of non-dominated feasible criterion points that are in line with it. Of course, the user can change the goal (or even the specification of the criteria) several times during the decision-making process.

The RGM/IDM technique was introduced in (Gusev and Lotov, 1994) and implemented in several software tools. Visual Market 2 (VM/2) for Windows (Lotov and Bushenkov, 2000) turned out to be the most practical of them. It is used in all sections of this chapter, except Section 4, to illustrate the RGM/IDM technique for the case of more than two criteria.

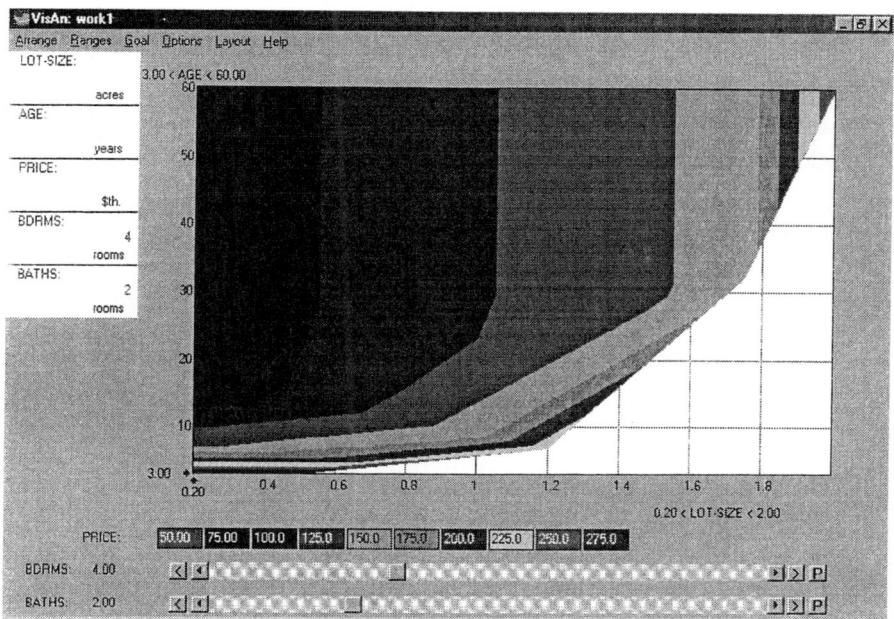

Figure 4.7. The black and white copy of a color decision map for all five selecting criteria

Let us consider an example of application of the RGM/IDM technique if all five criteria of the real estate list given in Figure 4.1 are used. The IDM technique displays the bi-criterion slices of the CEPH in the form of animated decision maps. The black and white copy of a color decision map is given in Figure 4.7. AGE and LOT-SIZE are given on coordinate axes, PRICE is given by color, Number of bedrooms and Number of bathrooms are given by scroll-bars. As usual in the IDM technique, the user can freely change the values given by sliders of scroll bars.

After studying various decision maps, the user identifies the goal. The information on the reasonable goal can be applied for developing a new domination (so-called meta-domination) between alternatives. A meta-domination is usually broader than Pareto domination, and so the number of non-dominated points for a meta-domination is less than for the original Pareto domination. Therefore, a meta-domination can be applied for selecting a small number of points from the variety of criterion points. A meta-domination results in points that are close to the identified goal in some sense. Several variants of the concept of meta-domination (and related proximity) can be used. Here we describe one of them that was introduced in (Gusev and Lotov, 1994) and is used in most of the studies and implemented in the Visual Market/2 software (Lotov and Bushenkov, 2000) and in the experimental Web application server (Lotov, Kistanov and Zaitsev, 2001).

The procedure of constructing the non-dominated alternatives for the meta-domination consists of three steps. At the first step, modified points are considered instead of original points. A modified point is defined in the following way: if a criterion value in the original point is better than the criterion value in the user-identified goal, the criterion value of the identified goal is substituted for the criterion value of the original point. It means that we consider the goal values as the aspiration levels and assume that the achievement of the aspiration levels is much more important, than "over-achievement" of them. At the second step of the procedure, the Pareto domination rule is applied to the modified points. As a result, non-dominated points are found among the modified points. Finally, those original alternatives are selected that gave rise to the non-dominated modified points.

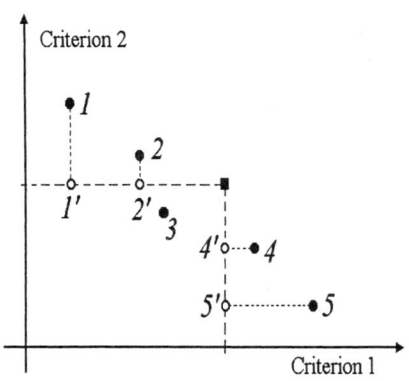

Figure 4.8. Selection procedure of the RGM (it is preferable to increase the criterion values).

The procedure is illustrated in Figure 4.8 for the case of two criteria, which are subject of maximization. The reasonable goal identified by the user is denoted by the filled square symbol. Original points are represented by the filled circles, modified points are represented by hollow circles. So, the feasible point 1 originates the modified point $1'$, etc. If all criterion values for an original point are not better than the goal values (for example, point 3), the modified point coincides with the original one. Then, the Pareto domination rule is applied to the modified points: non-dominated points are selected among them. Point $2'$ dominates point $1'$, and point $4'$ dominates point $5'$. So, three non-dominated modified points are selected: $2'$, 3, and $4'$. Finally, the original feasible points, which originated the non-dominated modified points, are selected. In Figure 4.8, these points are 2, 3 and 4. It is clear that alternatives selected through this procedure are non-dominated in the usual Pareto sense, too.

An example of the list of the decision alternatives according to a goal for the real estate problem is given in Figure 4.9. Four houses were selected. The goal values (aspiration levels) identified by the user

Visual Market / 2 - HOUSE [31]					
File Tree Filter Help					

Criteria

- ☑ BDRMS
- ☑ BATHS
- ☑ LOT-SIZE
- ☑ AGE
- ☑ PRICE
- ☐ SHOP
- ☐ SCHOOL

Current: Set2 [4]

	BDRMS	BATHS	LOT-SIZE	AGE	PRICE
Your Aspiration:	4	2	1.046	9.318788	150
1. HOUSE15	6	3	1	9	140
2. HOUSE13	3	2	1.25	14	180
3. HOUSE18	7	4	1.2	7	215
4. HOUSE29	4	2	1.75	32	135

Figure 4.9. The list of the alternatives that are in line with the goal

are given above the list of selected alternatives. One can see that any alternative is better for one or several criteria than the goal and worse for at least one criterion.

The steps of the RGM/IDM technique are given in Figure 4.10.

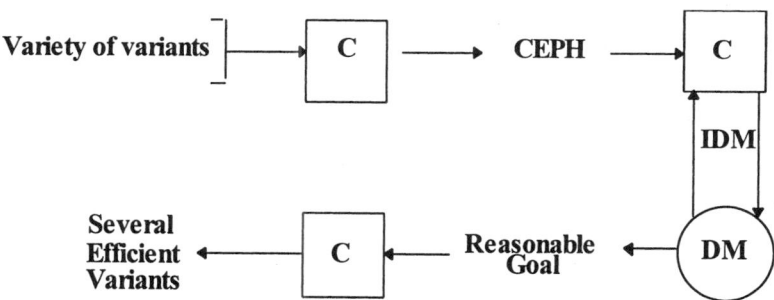

Figure 4.10. The steps of the RGM/IDM technique. Computer processing is denoted by C and decision maker by DM

Since animation of the decision maps can not be illustrated in a book, we propose to download the demonstration version of the VM/2 software from the Web page

http://www.ccas.ru/mmes/mmeda/soft/

and experiment with the animation of decision maps in the framework of the RGM/IDM technique.

Comparison with bi-criterion projections.   There is a well-known visualization method that is used in the case of multi-criteria (multi-attribute) selecting of alternatives from finite lists. The method consists in display of graphs containing all possible bi-criterion projections of feasible criterion points. First of all, it is important to stress that the RGM/IDM technique differs from what is used in the bi-criterion projection method — the IDM technique displays bi-criterion cross-sections (slices) of the CEPH of points in criterion space, but not bi-criterion projections of them. Decision maps differ from bi-criterion projections drastically, since decision maps display three-criterion information, and a matrix of decision maps helps to understand tradeoffs for five criteria.

Both the RGM/IDM technique and the bi-criterion projection method have advantages and disadvantages. The bi-criterion projection method is easy to apply since a bi-criterion projection of a finite number of points can be found easily. The method informs users on all bi-criterion non-dominated frontiers in a simple form, but it requires temporal exclusion of other criteria from the consideration (this is the sense of projection!). Therefore, the bi-criterion projection method does not inform the users on interaction between three or more criteria. Let us consider a simple three-criterion example given in Table 4.1.

Table 4.1. A simple three-criterion example (minimization of all criterion values is desirable).

Alternative	Criterion 1	Criterion 2	Criterion 3
1	0	1	1
2	1	0	1
3	1	1	0
4	0.2	0.2	0.8
5	0.2	0.8	0.2
6	0.8	0.2	0.2
7	0.4	0.4	0.4

One can easily check that all the alternatives are non-dominated. The 7-th alternative characterized by balanced criterion values may turn out to be the most preferable. However, it is located deep inside the variety of projected points at all the three possible bi-criterion projection graphs. One of such bi-criterion projection graphs is given in Figure 4.11. Other projection graphs look the same because of the symmetry of the problem.

So, in the case of a larger number of points, such balanced alternatives may be lost even for three criteria. Therefore, bi-criterion projection graphs, which provide excellent information on the conflict between two criteria, are not sufficient for identification of reasonable goals (or aspiration levels) even in the case of three criteria. It is desirable to supplement the study of projections with an exploration of decision maps.

However, application of the RGM/IDM technique is related to a complication — approximation of the convex hull of a large number of criterion points requires a software, which can not be coded by a user himself. For this reason, the user needs to download the VM/2 software from the Web cite mentioned above.

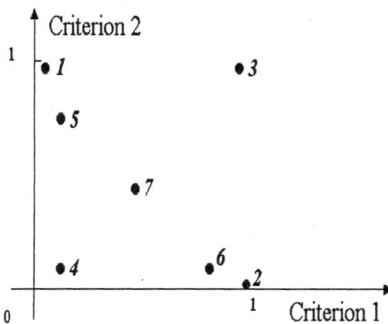

The RGM/IDM technique may be used for visual exploration of various databases and selection of a small number of preferable alternatives. Two following sections describe applications of the technique for selecting preferable decision alternatives in real-life problems.

Figure 4.11. One of the two-criterion projections for alternatives from Table 4.1.

2. RGM in water quality planning

This section is based on the paper (Bourmistrova, Efremov and Lotov, 2002). Let us consider a real-life application of the RGM/IDM technique for selecting a small number of decision alternatives in an environmental problem in the framework of a DSS for water quality planning in a small region. The DSS was adapted to a small region in the basin of the Oka River — the vicinity of the town of Kolomna, which is located at the point where the Moskva River meets the Oka River. Eight sources of wastewater discharge were specified in the vicinity of the town.

A large, but finite number of decision alternatives was prepared in the following way. Four technologies of the discharge treatment were considered. Therefore, five options of the discharge treatment were feasible at any wastewater discharge source — to implement one of four technologies or to do nothing at all. As the result, 390,625 decision alternatives were considered in total. Three pollutants were taken into account — nitrates, phosphates and biological oxygen demand (BOD). To measure pollutant concentration, conventional units were used. In the framework of these units, concentration equals 1 if some environmental requirements are satisfied precisely. The maximal (in the region) concentrations of these three pollutants were used as the screening criteria. The fourth criterion was the cost of the project measured in millions of US$.

To explore the influence of local pollution discharge, it was assumed that the pollution concentrations equal 1 in water coming from the reaches of the Oka and Moskva Rivers located up the river. This assumption is not true now, the pollution concentrations are much higher,

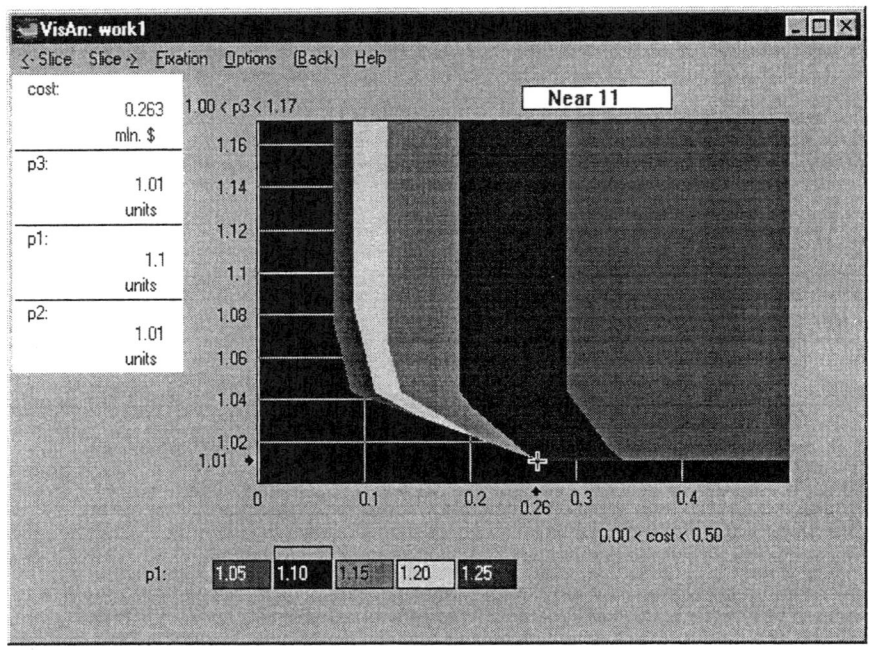

Figure 4.12. Decision map with the reasonable goal given by cross

but it was needed to develop local water improvement strategies. Therefore, the data used in the study are fairly artificial.

The software system MIKE 11 that has already been discussed in Chapter 3 was calibrated for the region and used for computing the pollution transport. The linear dependence of pollution concentrations on the pollution discharge in MIKE 11 made the estimation of all 390,625 decision alternatives a fairly easy task. Due to this, criterion values related to all decision alternatives were computed and the envelope of the associated points was approximated. Exploration of the variety of the alternatives was based on the Pareto frontier display of the CEPH by the IDM technique. Exploration of decision maps proved that it is reasonable to restrict cost to, say US$0.5 million and select phosphate concentration to be about 1.01. The related decision map is given in Figure 4.12. Cost is given in the horizontal axis, BOD ($p3$) is given in the vertical axis, concentration of nitrates ($p1$) is given by shading (color on display), and the value of phosphate concentration ($p2$) is given in the tablet.

One can see the cross in this decision map that is located at the point associated to $p3 = 1.01$, $p2 = 1.01$, $p1 = 1.10$, while cost is about

Goal point	0.2630	1.1000	1.0100	1.0100
Alternative	**Cost**	**p1**	**p2**	**p3**
C40431000	0.2630	1.1170	1.0010	1.0110
C40441000	0.2650	1.0980	1.0010	1.0110
C30441000	0.2560	1.0980	1.0010	1.0120

Figure 4.13. Decisions alternatives resulting from the goal point

US\$ 260 thousand. One can clearly see that about US\$70 thousand of additional cost is needed to decrease the value of $p1$ from 1.1 to 1.05 for the same values of $p2$ and $p3$. On the other hand, further increment of $p1$ to 1.15 or even 1.2 can not save much money — the associated Pareto frontiers are close to the point where the cross is located. So, a user may want to identify the reasonable goal at the position of the cross.

Several decision alternatives that are close to the identified goal are given in Figure 4.13. In the first line the goal point is given. The second row contains the names of the columns: the first column is the code name of the alternative; the second column is the cost; the third, fourth and fifth columns are concentrations of nitrates, phosphates and BOD after the project is completed. A digit of the code name (located in the first column) contains the numbers of technologies used at particular discharge sources. One can see that though the first alternative is formally efficient and its cost coincides with the goal cost, only an additional two thousand of US\$ are required to decrease the value of $p1$ for about 1%. The third alternative displays another opportunity — one can save US\$9 thousand (in comparison to the second alternative) while $p3$ increases only about 0.1%. Surely, the choice of a preferable alternative depends on the user's preferences. The selected decisions are visualized in the DSS using a GIS-generated geographic map. However, we do not display such a picture here.

The described DSS was developed at the request of the Russian Federal Ministry for Natural Resources in the framework of the Federal program "Revival of the Volga River".

3. RGM in screening of national energy production strategies

This section is based on the paper (Soloveitchik, Ben-Aderet, Grinman and Lotov, 2002) and is devoted to application of the RGM/IDM technique in the framework of long-term national energy planning. Since 2000, the RGM/IDM technique is permanently applied at the Ministry of National Infrastructures (former Ministry of Energy) of Israel for planning the capacity expansion of power generation systems taking several objectives (including environmental and social objectives) into account. The results are based on real data of Israel's electricity sector. The aim of the research consists in selecting a small number of alternative strategies that can be recommended for further detailed study and final choice.

The initial infinite variety of possible strategies was described in the framework of the computer software package CAPEX (see Levin, Tishler and Zahavi, 1985), which is a non-linear optimization software that can find the optimal strategy for various single optimization criteria. The optimization criterion can include the cost of fuel, operation and maintenance costs, capital investment, etc. Environmental indicators (like carbon dioxide or sulphur emission) can be included in the optimization criterion as well. In the study that is described here CAPEX was used for developing a relatively large, but finite list of strategies for subsequent screening. Such a list of strategies was developed by solving a large number of optimization problems with different criteria. Then, the RGM/IDM technique was applied. Several problems were considered during recent years. We describe here a study devoted mainly to pollution discharge reduction.

In this study the RGM/IDM technique was used for screening a list that included 24 national electricity production strategies that were developed with the help of CAPEX software. All the strategies were considered at the same time as the plans for reduction of air pollution, too. Therefore, five following choice criteria were used:

- percentage of CO_2 reduction (CO2_%) that is desirable to maximize;

- percentage of NO_x reduction (NOx_%) that is desirable to maximize;

- additional total abatement cost (NPV_D) that is desirable to minimize;

- marginal abatement cost (NPV_DC($)) that is desirable to minimize;

- average cost of electricity (AV.COST) that is desirable to minimize.

A decision map for five criteria is given in Figure 4.14. The map displays tradeoffs for CO2_% (horizontal axis), NPV_D (vertical axis) and NO_x reduction (shading) for given levels of marginal abatement cost (8.40 $/T) and of average cost of electricity ($2.57). The last two values are given by sliders of the scroll bars.

The Pareto frontier of the area that is given by the lightest shading (NOx = 0.16, i.e. $NO_x = 16\%$) proves that the minimal value of

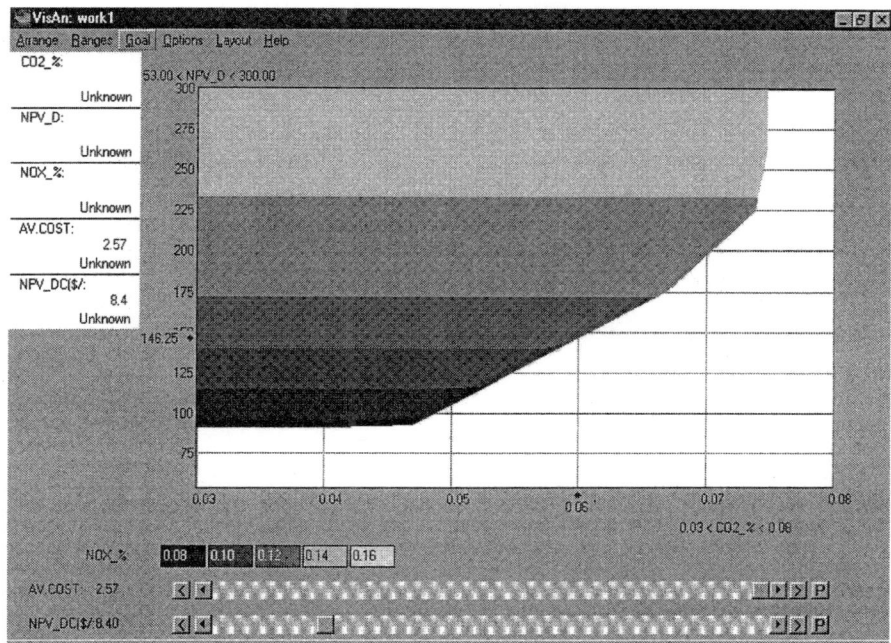

Figure 4.14. Decision map for five criteria

the additional total cost (NPV_D) for these values of NO_x reduction (16%), marginal abatement cost (8.40 \$/T) and average cost of electricity (\$2.57) equals about \$230 million. The minimal cost does not depend on CO_2 reduction while it is less than about 7.3%, but CO_2 reduction cannot be much larger than 7.3%.

The decrement in the value of NO_x reduction influences the picture. For NO_x reduction equal to 14% (a bit more dark shading), the minimal additional total cost is about \$175 million, and the CO_2-dependent growth of the cost at the Pareto frontier starts after CO_2 reduction is higher than 6.6%. If the NO_x reduction is 12%, the minimal additional total cost is about \$140 million, and the CO_2-dependent growth of the cost starts at about 5.7%. The Pareto frontiers between cost and CO_2 reduction for other values of NO_x reduction are clear as well. Note that for the minimal value of NO_x reduction in the picture, i.e., 8%, the minimal additional total cost is about \$90 million, and the CO_2-dependent growth of the cost starts at about 4.8%. The cost grows first in a linear way, and only after 6.7% the growth gets more steep. So, Figure 4.14 provides full information concerning relations among three criteria displayed in the decision map.

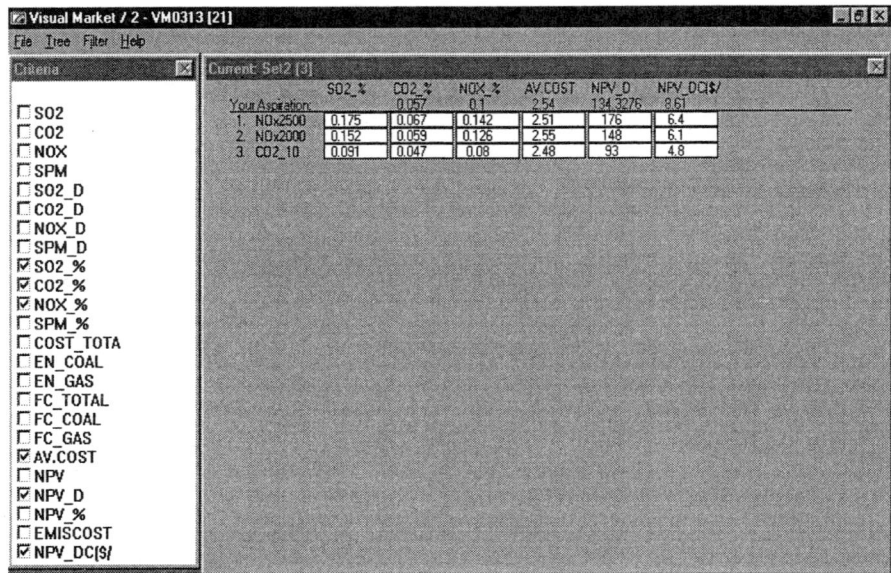

Figure 4.15. Selected alternatives

To explore the influence of two other criteria, i.e., of marginal abatement cost and average cost of electricity ($2.57), one needs to use animation. Since it is impossible to demonstrate the animation in a book or a paper, snap-shots of possible animation films must be given in the form of a matrix of decision maps. Such a map is provided in the paper (Soloveitchik, Ben-Aderet, Grinman and Lotov, 2002).

After examining various decision maps, the following goal was identified:

- CO_2 reduction equal to 5.7%;

- NO_x reduction equal to 10%

- additional total cost equal to $134 million;

- marginal abatement cost equal to 8.6 $/T;

- average cost of electricity equal to 2.54 c/kWh.

There are three selected strategies that are in line with the identified goal, and so they reflect the preference of the decision maker (see Figure 4.15). According to the goal values (aspiration levels), CO_2 reduction is between 4.7 and 6.7%, NO_x reduction is between 8.0 and 14.2%, additional total cost is $93 million and $176 million, marginal abatement

cost is between 4.8 and 6.4 \$/T, and average cost of electricity is between
\$2.48 and \$2.55. In addition, the values of SO_2 reduction are provided in
Figure 4.15. The decision maker has to study these strategies carefully
at the next step of the choice process. We do not discuss it here since
the IDM technique is not applied at this stage.

Another study was devoted to exploration of the influence of different
energy sources in electricity production. However, we do not describe it
here.

4. Selection of a location for rural health practice

The following section that is based on the paper (Jankowski, Lotov
and Gusev, 1999) describes the RGM/IDM technique application for the
problem of selecting a location for rural health practice in Idaho, U.S.A.
Many rural areas in the USA compete for physicians, thus creating choice
opportunities for those interested in practicing rural medicine. Yet, these
efforts coupled with various US federal programs have had mixed results
in attracting and retaining primary health care providers in rural lo-
calities. One possible cause for this is the lack of effective information
tools that would assist the prospective physicians in screening practice
location options and learning about tradeoffs involved. A prototype spatial DSS to aid health care practitioners in selecting a practice location
in Idaho was developed in (Jankowski and Ewart, 1996). This system,
however, is based on compensatory methods utilizing weights to represent physicians' priorities for decision criteria. An approach based on
the RGM/IDM technique allows a prospective physician to select preferable locations based on their closeness to the goal (a reasonable practice
location) without the difficulty of specifying criterion weights.

The DSS described here is comprised of two modules: the Visual Market (VM) software that implements the RGM/IDM technique, and a geographic data query and visualization module implemented in ArcView^{TM}
3.0. The modules supplement each other in providing decision support
functions. The prospective physician can learn quickly about the location of places offering practice opportunities, their physical and socio-
demographic characteristics, amenities offered by them, and relate this
information to the surrounding physical environment by viewing and
querying reference maps in ArcView. The information gained from spatial data query and visualization becomes useful in selecting a reasonable
goal for the health practice location. The goal selection, which is performed by the user in VM, results in a list of a few locations that are
close to the selected goal in the sense of satisfying all or some of the
criterion values set by the goal. These locations can be in turn displayed
and analyzed in ArcView. The process is interactive and iterative. Its

intended outcome is a better-informed decision on the part of prospective health care professionals about rural practice location selection.

Data representing health-care, social, economic, and environmental information were aggregated by 47 Primary Care Service Areas (PCSA) encompassing the entire state of Idaho. The attribute database describing the PCSA provided information for evaluation criteria. The criteria were grouped into professional and personal. Professional criteria included:

- need for physicians denoted by DOCS; this is a derived index measure, the higher the DOCS value the higher the need; DOCS can also assume negative values representing low demand for physicians or lack thereof,

- population in 1990 (POP90),

- percent of population receiving Medicare and Medicaid (MEDICARE),

- fertility rate (FERTILITY),

- loan repayment program (LOAN_REPA),

- number of hours per week on call (ON_CALL), etc.

  Personal criteria included:

- percentage of unemployed population (UNEMPLOYED),

- percentage of population below poverty level (POVERTY),

- percentage of population with college degree (POP_DEGREE), etc.

PCSAs are represented by primary cities for each area. Along with the list of PCSAs, their locations are given on the ArcView generated map of Idaho. Any PCSA is given by a point on the map. One can click a point and obtain the name of the PCSA along with associated data.

Let us consider five attributes for the location selection criteria:

- need for physicians (DOCS) − to be maximized;

- population in 1990 (POP90) − to be maximized;

- weekly number of hours on call (ON_CALL) − to be minimized;

- fertility rate (FERTILITY) − to be maximized;

- percentage below poverty level (POVERTY) − to be minimized.

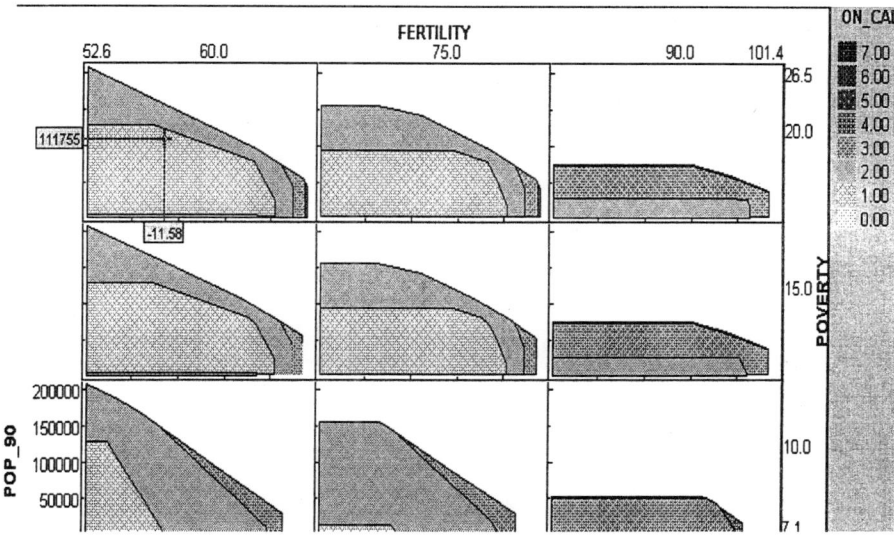

Figure 4.16. Matrix of decision maps for the problem of selecting a location for rural health practice

Matrix of decision maps for the problem with these five selection criteria is given in Figure 4.16.

Each of nine decision maps depicted in Figure 4.16 displays the trade-offs between DOCS, POP90, and ON_CALL. Values of the first two criteria are given on horizontal and vertical axes, and the increased intensity of gray-scale (upper right of the window in Figure 4.16) represents values of ON_CALL. Each column of the matrix is related to a certain requirement imposed on FERTILITY ("not less, than..."), and each row is related to a certain restriction imposed on POVERTY ("not greater, than..."). It is easy to see that class intervals constraining FERTILITY and POVERTY influence the feasible values of DOCS, POP90 and ON_CALL. As usually, a user can change class intervals creating a different set of decision maps corresponding to different ranges of requirements and constraints for fertility and poverty. The user can equally easily change the view of the decision space (decision maps) by choosing different criteria to be displayed along the axes of the matrix and represented by the sequence of gray-scale.

In the decision map, for which fertility rate is no less than 60 and percentage of population below poverty level is no greater than 20 (first row, first column in Figure 4.16), a reasonable goal was identified (DOCS = -11.58, POP90 = $111,755$, ON_CALL = 1). It means

DOCLOC2 close to goal							
	DOCS	POP_90	MEDICARE	FERTILITY	POVERTY	ON_CALL	POP_DEGRELOAN_REPA
Goal value	-11.58	111755		60.0	20.0	1.00	
St._Maries	-1.08	6400	24.9	65.4	16.3	0.92	9.5	0
Weiser	-0.91	7601	32.1	75.3	19.6	1.27	10.5	1
Nampa	-1.97	96404	21.2	81.6	15.5	1.29	10.8	1
Boise	-19.82	205775	14.4	64.7	8.8	1.64	21.4	0

Figure 4.17. Locations in Idaho in line with the identified goal

that the user who identified it, was not afraid of high competition (DOCS $= -11.58$), preferred to live in an urban area, but did not like to spend too much time on call. The identified goal is described by following criterion values: DOCS $= -11.58$, POP90 $= 111,755$, ON_CALL $= 1$, FERTILITY $= 60$, POVERTY $= 20$. After the goal was set, several related locations were selected: St.Maries, Weiser, Nampa and Boise. They are displayed in Figure 4.17.

Actually, Nampa is the only location which is close to the identified goal in the common sense of this word (note that the need for physicians is much higher in Nampa than it was set in the goal). Since the software does not know the user's preference, it displays three different locations that are not very near the goal, but may happen to be better locations from the user perspective than Nampa. Note that in Nampa the prospective physician has to spend 1.29 hours on call instead of one hour in the identified goal. In St.Maries one has to spend only 0.92 hours on-call per week. Perhaps, the user would agree to sacrifice the population level for this advantage? Only the user can decide it. Weiser was selected since it is a little bit larger than St.Maries. Finally, Boise, the capital of Idaho, was chosen since Nampa does not meet the population level of the goal. Perhaps this is important for the user? The user has to decide whether any of the selected places is attractive enough for the location of health care practice. It is important to note that all other places in the database were not in line with the identified goal, and so they were not selected.

To help the user to analyze the selected locations, a reference map is provided. In the map, the selected places are denoted by larger size square symbols. The user can click on a symbol and display the characteristics of the place. The additional information about the selected locations, including reference maps, photographs, quick-time movies, 3-D images, may be provided to support further analysis.

5. RGM on the Web: application server for e-commerce

This section shows how the RGM/IDM technique can be used in the framework of the Web application server that supports users in the process of selecting a small number of items from a large list of items. Such a server can be applied in e-commerce for supporting clients of various Web resources.

Business and individual consumers as well as users of various services often meet a problem: how to select a preferable good or service from a large variety of alternative options? Let us consider several examples.

- A family is looking for a house to buy. One can find many hundreds or even thousands of houses on sale on the Internet. How to select a small number of houses, which deserve special attention, i.e., it makes sense to have a look at them?

- Someone plans to buy a used car. How to select several cars from many thousands of them that can be found on the Internet?

- A firm develops a supply chain. A large list was prepared that contains possible suppliers from the whole world. How to select a small number of suppliers from the list that deserve special attention?

These examples are characterized by a large number of options. Traditionally, human advisers help to select the most deserving options. For example, support by a real estate agent is usually needed, though newspapers and special editions are full of information concerning real estate. Since Web services are not based on human advisers, computer tools must substitute for them.

Several approaches do exist. One of them is based on modeling of user preferences and further search for an option that could satisfy a user. In this case, however, a sophisticated problem of preferences identification does exist. The multi-attribute utility theory, which is a mature discipline now, has developed a lot of theory and algorithms aimed at solution of the problem (see Keeney and Raiffa, 1976). The theory proved that even under extremely restrictive assumptions, the preference identification is related to boring complicated interactions, in the process of which the user has to compare multiple pairs of feasible or artificial alternatives.

A different approach is based on the development of standard "preference patterns" and further classification of users. This idea seems to be more practical, but it restricts freedom of information and choice. It seems that it can be applied successfully in the case of beginners or users that are indolent. Moreover, such classification is not an easy task. For an experienced real-estate agent it is sufficient to look into the eyes of a client to classify him/her, but who will do it in the case of a computer network?

The RGM/IDM technique provides an effective alternative Web tool in this field. Such a tool is based on the fundamental feature of the IDM technique that consists in separating the phase of approximating of CEPH from the phase of Pareto frontier visualization and identification of the goal. Such a feature that has already been discussed in Chapter 1 makes it possible to apply the IDM technique in the framework of server–client structure. On the Web, such a structure can be applied by using the opportunities of Java. The server is used for approximation of the CEPH, and a Java applet provides visualization of the CEPH online. A demo application server of this kind has already been completed. It is located at

http://www.ccas.ru/mmes/mmeda/rgdb/index.htm

and can help to understand features of future customized Web application servers that will support easy selecting of preferable alternatives from various lists through the Internet using simple graphic interface of the RGM/IDM technique. The server was coded in C++ and Java by R. Efremov, A. Kistanov and A. Zaitsev. The structure of the Web application server is given in Figure 4.18. Development of the application server was partially supported by Fraunhofer Institute for Autonomous Intelligent Systems, Sankt Augustin, Germany.

To use the demo Web application server, a user has first to submit a list of decision alternatives prepared in a simple standard form. In

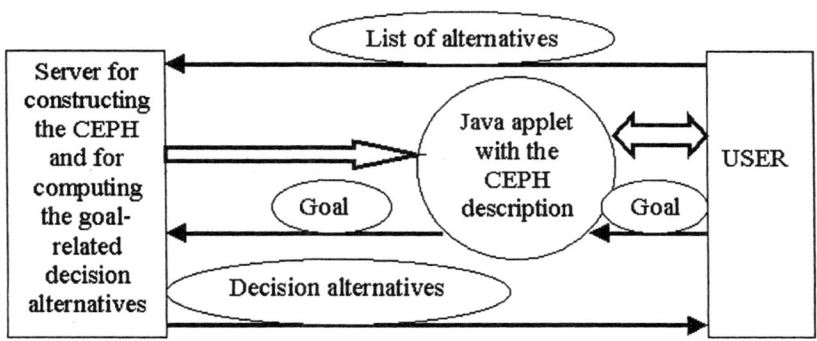

Figure 4.18. Scheme of the demo Web application server

the demo version of the application server, the user can type or provide it with the help of clipboard and browser. In customized servers different forms may be possible. Then, the server envelopes the variety of criterion points related to alternatives and approximates the CEPH. Approximating of the CEPH is carried out by a module coded in C++. The module is executed on the server. The visualization module was coded as a Java applet. The Java applet for visualization of the CEPH is then transmitted to the user's computer along with an approximation of the CEPH. Visualization is carried out at user's computer. The applet provides all opportunities of the IDM technique except matrices of decision maps. The user can study the decision maps in an interactive way, including animation of them. Then, he/she has to identify the reasonable goal that is transmitted to the application server. Several alternatives, which are in-line with the goal, are selected by the server and transmitted back to the user. Auxiliary operations of the demo application server were coded using CGI-scripts and Java middleware. To use the application server, one needs MS Internet Explorer. It is important that user interaction with the Java applet that implements the RGM/IDM technique is simple. Therefore, non-experts can master an RGM/IDM-based Web application server fairly easily.

Let us consider an example that is prepared by A. Zaitsev for the demo Web application server. The example is devoted to the Russian market of second-hand cars in the year 2000. The problem of selecting of several used cars from a list of 500 cars is considered. Each car is described by five parameters, which are used as the selection criteria (see Figure 4.19):

- *Year* − production year. Higher is better;

- *Price* − price in thousand $. Lower is better;

- *Run* − run in thousand of kilometers. Lower is better;

- *Power* − engine power in horse powers. Higher is better;

- *L/100* − fuel consumption in liters per 100 kilometer. Lower is better.

The black and white copy of one of the displays provided by the Java applet is given in Figure 4.20. The values of two criteria (*Year* and *L/100*) are given on scroll-bars located to the left from the decision map. Axes of the graph are related to *Run* and *Price*, and the color is associated with *Power*. As usual in the IDM technique, a user can change the values of two criteria located at scroll-bars by moving sliders

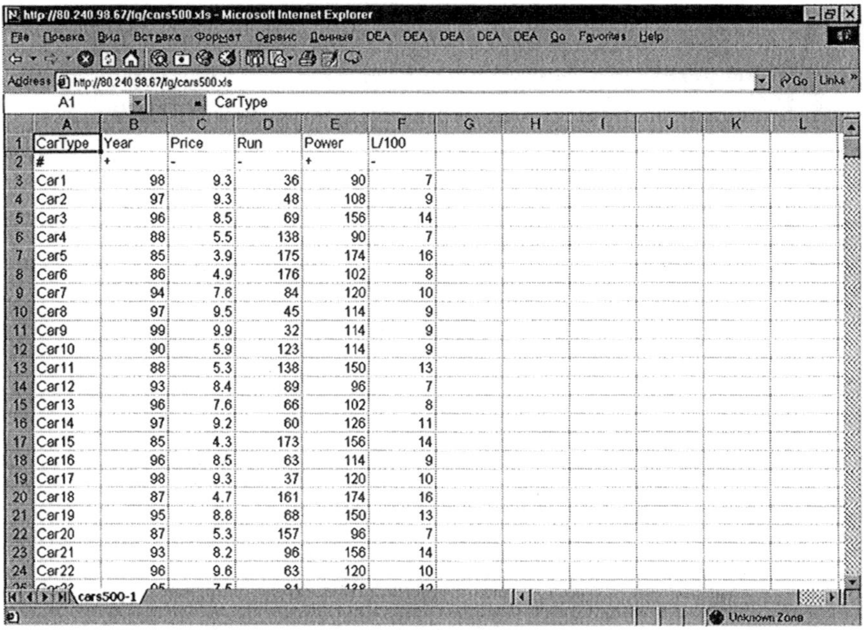

Figure 4.19. A part of the database of used cars

and watch the associated changes of the decision map. It is possible to use animation, too.

By moving the sliders the user can find that the value of *L/100* influences the color, i.e., *Power*, and increasing of the value of *Year* moves the frontier that describes the lowest value of *Price* upwards. In the figure, *Year* is not earlier than 1990, and *L/100* is maximal, i.e., 16 liters per 100 kilometers. The decision map reveals a very interesting feature of the Russian second-hand car market — *Price* depends on *Power* to a small extent (frontiers for different values of *Power* are very close). This feature can make a user think about an opportunity to buy a powerful car for just the same price (if he/she is not afraid of high cost of gasoline). Experienced buyers on the second-hand know and use this fact. However, it is amazing that one can find it easily using the Web application server.

Note that in addition to Pareto frontiers usually displayed in a decision map, a moving Pareto frontier is given in the figure. It is moving in accordance to the position of the slider of the scroll-bar associated with color and located under the decision map. The moving Pareto frontier

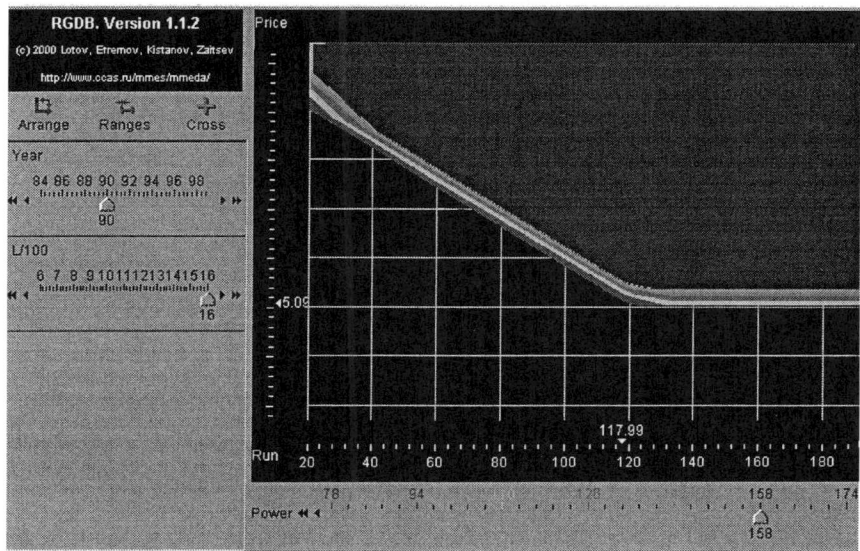

Figure 4.20. The black and white copy of a display provided by Java applet

helps to identify the goal. For example, the moving frontier is related to *Power* of 158 HP in the above figure.

Figure 4.21 shows the final decision map with the cross that is used to identify the goal position. Note that it is located at the moving frontier. The precise position of the goal is given in the window located to the left from the decision map. So, *Year* is 1990, *Price* is about 7 thousand US$, *Run* is about 80 thousand kilometers, *Power* is about 158 HP, and *L/100* is about 16 liters per 100 kilometers. The goal means that the user decided to buy a large powerful car for relatively small cost, and so he/she can be satisfied with a relatively old car with a high gasoline consumption.

The related cars from the list are found by server and provided to the user. The time delay depends mainly upon the quality of communication. One can see the selected cars in Figure 4.22.

One can see that the goal is not feasible – a car with such attributes does not exist in the database. This is the natural situation in the framework of the RGM. For this reason several cars, which are in line with the goal, are selected. As usually in the RGM/IDM technique, the user has to make a final choice using different computer support (say, photos provided by particular Web sites, etc.). However, one can see

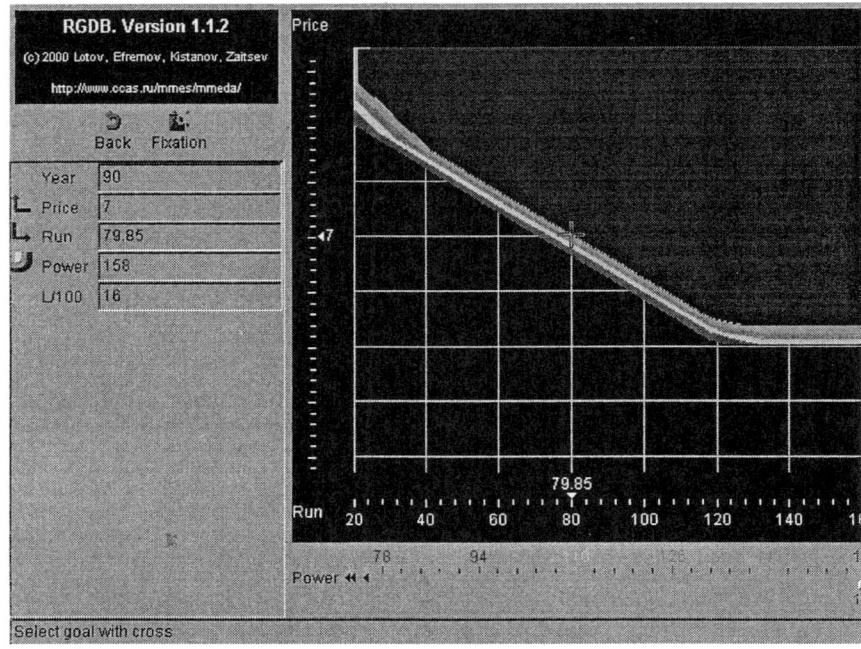

Figure 4.21. The final decision map with the cross

CarType	Year	Price	Run	Power	L/100
Your goal	*90*	*7.01*	*79.85*	*158*	*16*
Car454	93.0	7.0	85.0	162.0	14.0
Car267	94.0	7.2	81.0	138.0	12.0
Car385	95.0	7.1	79.0	114.0	9.0
Car338	93.0	6.9	84.0	90.0	7.0
Car261	95.0	7.5	76.0	144.0	12.0
Car178	93.0	7.6	84.0	150.0	13.0
Car119	94.0	8.1	78.0	156.0	14.0
Car77	95.0	8.3	70.0	174.0	16.0

Figure 4.22. The selected cars

that the first car given in the list of selected cars, is very close to the goal and may turn out to be a preferable one.

The provided example shows that the Web application server described in this section is a simple tool that can be used by any computer-literate person.

6. Several other applications

The RGM/IDM technique can be used in different fields including selecting from, say,

- investment projects described by such attributes as total discounted investment, discounted income, term of complete investment return, reliability;

- portfolio or assets allocations described by such attributes as dividend, income, variability, and uncertainty factors;

- business locations given in a Geographical Information System;

- possible designs of a machinery system.

The RGM/IDM technique has been experimentally used in the process of heart surgery for fast diagnostics of the state of the heart.

Let us consider several examples in short.

Application in the framework of machinery design. Let us give an example of applying the RGM/IDM technique for selecting a car gear transmission. In total, engineers have developed 990 alternative designs of the transmission. They are described by such attributes as strength of the transmission (to be maximized), distance between axes (to be minimized), width (to be minimized) and precision of transmission (to be maximized). All 990 alternative designs were initially given in a table. By using the RGM/IDM technique, they were transformed into the CEPH, which is given by three decision maps in Figure 4.23.

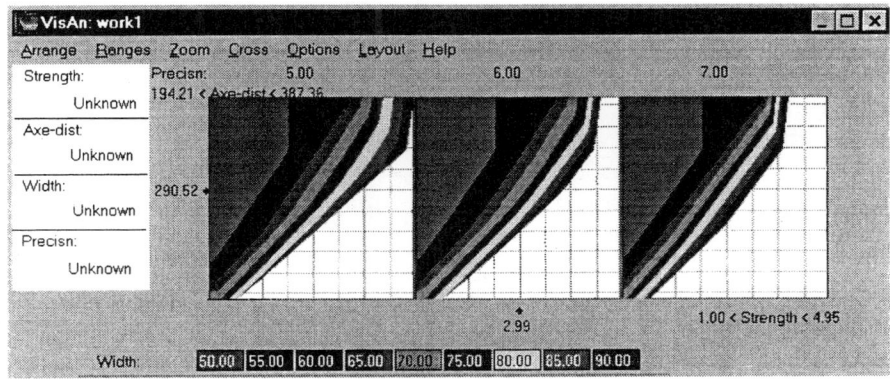

Figure 4.23. Decision maps obtained from the envelope of 990 alternative designs of car gear transmission

Any decision map is related to a certain value of precision. On a decision map, width is given in shading while axel-distance and strength are given on axes. One can easily understand the relation among three attributes for a decision map. The influence of precision is given by the differences between decision maps and can be easily understood, too.

Application in the framework of financial management. The RGM/IDM technique can be used as a DSS tool that can support a search for smart financial decisions while taking multiple risks into account. For example, in the framework of intraday trading, it can be used for:

- on-line display of the dynamics of a whole variety of stocks;

- supporting the technical analysis of a variety of stocks;

- supporting the selection of most appropriate stocks for immediate position opening;

- screening of stocks characteristics aimed at the search for ideas for the next day trade.

In general, the RGM/IDM technique can be used in

- fundamental analysis of marketed stocks aimed at the search for a most appropriate one for long-time investment;

- developing of winning portfolio;

- screening of possible alternatives of composite hedge strategies;

- evaluation and screening of the variety of futures and options under multiple risks;

- search for smart strategies in derivatives markets, etc.

Moreover, the RGM/IDM technique may be used as a part of an integrated DSS for financial consulting.

7.   RGM as a tool for database visualization, data mining and cost–benefit analysis

In the applications described above, the lists of alternative decision options are given in the form of a table, which is the essential element of relational databases. Such a table can be obtained from any relational database by using, say, SQL-query. Therefore, the RGM/IDM technique can be considered as a tool for visualization of relational databases. It is important that the RGM/IDM technique can be applied for visualization-based analysis of databases, which contain information that differs from decision alternatives. Say, databases can be analyzed that contain statistical data (medical, demographic, environmental, etc.), the results of experiments with technical or natural systems (say, data on a device performance, etc.) or simulation experiments with their models, etc. In the process of graphic analysis based on *database visualization*, interesting relations between the attributes can be discovered by the user and several items, which are responsible for these features, can be selected from the database.

It is important that the RGM/IDM technique can be used for visualization of *temporal databases*. Indeed, one can construct convex hulls of the data for all time-moments. Then it is possible to specify a decision map that describes the Pareto frontier for the initial time-moment and provide animation of the picture by changing the decision map in accordance with the monotonic automatic increment of time. By this, the user can obtain information on the *inter-temporal Pareto frontier* and try to discover inter-temporal properties of the data. This option can be especially important in financial management.

These examples show that the RGM/IDM technique can be used for database visualization aimed at the disclosure of information hidden in databases. For this reason, the RGM/IDM technique can be considered as a specific *data mining tool*. Data mining is a well-known approach to studying large volumes of data collected in databases. Statistical methods are usually used in data mining. These methods help to discover new knowledge concerning the data. Though the RGM/IDM technique does not use any statistical inference, it can support discovering of new information concerning the data collected in a relational database, too. However, the way the RGM/IDM technique supports data mining differs from the usual approach drastically — instead of informing on some laws, the RGM/IDM technique results in information on the unusual events given by the frontier of what is given in the database. Let us illustrate these words by a picture.

In Figure 4.24 a database is displayed that contains several houses on sale. The houses are given by the points in the "price–age" plane (as in the example given in Section 1). A regression of age on price is given by the dashed line. It proves that low-aged houses are more expensive than old houses. So, the law associated with houses on sale is found. It is important that the law is fairly stable, a perturbation of the price of any particular house can change the dashed line only to a small extent.

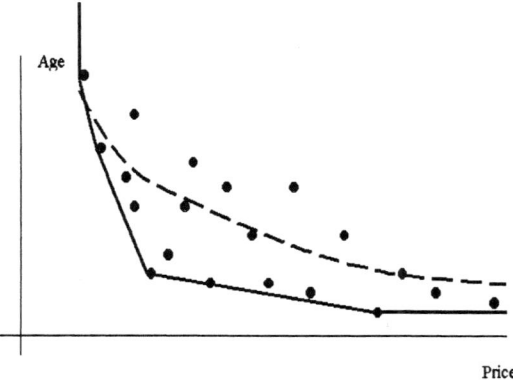

Figure 4.24. Comparison of IDM-based data mining with a usual data mining.

The solid line that is the frontier of the CEPH shows the momentary situation with the proxy tradeoffs between price and age. This is not a law since the frontier can change drastically after a point disappears or moves to a different location. However, such information is not less important since it can be effectively utilized by a person who managed to be the first to recognize it. For example, an Internet client of a firm that provides such visualization can use it to buy a house with a low age/price relation before other buyers will recognize the situation. The same opportunities can occur for an exchange trader who can try to use this information to be the first to buy a potentially profitable share. It is important that exploration of the frontier may help to find unusual (including erroneous) data in a database.

One of the possible applications can be associated with the cost–benefit analysis. The standard cost–benefit analysis is one of the well-known and widely applied tools of decision support. The RGM/IDM technique can broaden the scope of benefit–cost analysis to the case of multiple costs and benefits. The main advantage of the cost–benefit analysis consists in the explicit graphic display of an efficiency frontier for a single cost and a single benefit. Such a display helps the decision maker to identify a preferred combination of cost and benefit directly on the graph. However, the cost–benefit analysis is often not sufficient, especially in public problems, since several benefits and several costs may exist, say, separate benefits and costs for different population groups (Dorfman, 1996). So, a multiple criteria version of the cost–benefit analysis is needed. It is clear that the RGM/IDM technique can be used in the case of *several costs and benefits*. Section 2 of this chapter provides an example of such application — several benefits are given by decrements of several kinds of pollution. In the case of several regions (as it was studied in Chapter 3), several costs could be considered, too.

Chapter 5

NON-LINEAR SYSTEMS AND OTHER NEW DEVELOPMENTS

Chapter 5 is devoted to new developments in the field of Pareto frontier visualization, which is based on approximating the feasible criterion set (FCS), its EPH and related sets such as the convex hull of the FCS or the EPH. First we describe application of Feasible Goals Method (FGM) in the case of non-linear models, for which the FCS and EPH are usually non-convex. Two different approaches that implement the FGM in the case of non-linear systems are described in the next two sections of this chapter.

Then, Section 3 outlines the RGM/IDM technique for non-linear systems with an infinite number of feasible decision alternatives. Section 4 is devoted to application of the RGM/IDM technique for visualization of decision problems under risk. Finally, Section 5 describes a new IDM-based technique for the analysis of performance of a system of homogeneous production units, Data Envelopment Analysis (DEA).

1.   Feasible Goals Method for non-linear models

In both linear and non-linear cases, the FGM is based on two separate steps:

- approximation of the FCS (or its EPH); and

- visualization of the approximation and identification of a feasible goal vector on it.

In the linear case characterized by convex varieties of feasible criterion vectors, the polyhedral approximation is used. In the non-linear case, the FCS and its EPH are usually non-convex, and so different methods must be developed. Such methods are described here. They are based on simulation of a large number of random alternatives and special processing of the output. Due to simulation, one does not need to assume that models that describe the variety of feasible decisions and the relation between decision and criterion vectors are given by mathematical formulae – they can be described by complicated computational algorithms. To approximate an FCS or its EPH for such complicated systems, the software developed for the non-linear case can be interfaced with any simulation module. In particular, such simulation modules can be based on finite element methods (FEM), finite difference methods (FDM) or their combination. Moreover, the FGM can be implemented in the framework of parallel computing and meta-computing, that is, parallel computing using large uncertain numbers of networked PCs and workstations.

In multi-criteria problems, simulation of a random feasible decision results in a criterion vector associated with the decision. Therefore, one can develop a list of outputs related to feasible random inputs and provide it to the user (see, for example, Statnikov and Matusov, 1995). Instead, we use such outputs as a source of information for approximation and visualization of FCSs or EPHs. In particular, approximating of an FCS applies filtering of such outputs (constructing an *approximation base*) and subsequent covering of the FCS by a collection of boxes located at the criterion points given by the approximation base. Such a collection of boxes can be displayed on-line by its bi-criterion slices.

Two approaches to application of the FGM in the non-linear case were proposed. The first one, introduced in (Kamenev and Kondrat'ev, 1992), is based on approximating of an FCS and subsequent on-line visualization of its slices. The second one, introduced in (Lotov, Kamenev and Berezkin, 2002), is based on approximating of an EPH and application of the IDM technique for Pareto frontier visualization. In contrast to the convex case, these approaches are fairly different, and so they are considered separately. This section is devoted to the first approach. The second one is considered in the next section.

Approximation and visualization of FCSs. In the framework of the first approach, which is based on approximation and visualization of FCSs for non-linear models (Feasible Goals method for Non-Linear models, FGNL method), an FCS is approximated by a relatively small number of boxes. Centers of the boxes are located at the points of the FCS. Such points (the so-called approximation base) are obtained through simulation and filtering of random feasible decision alternatives. It is important that simulation of random feasible decisions helps to evaluate the quality of the approximation, too. Though an approximating

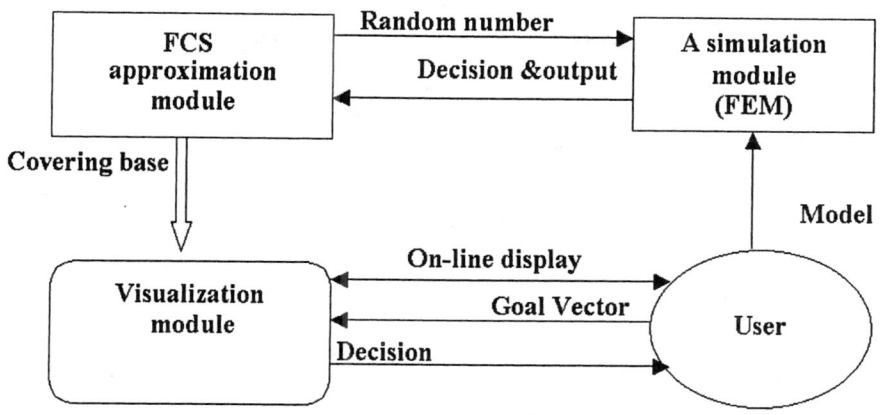

Figure 5.1. The scheme of the FGNL method

process may take a fairly long time (days and even weeks for complicated models), exploration of the FCS and identification of the feasible goal are carried out on-line. Visualization is based on the same ideas as in the convex case — collections of bi-criterion slices are displayed. Different slices are given in different color or shading in accordance to the value of a third criterion. The user can explore any number of such pictures on-line and identify a preferred goal vector directly on display. Then, the related vector of the covering base and the associated feasible decision are provided to the user. The scheme of the FGNL method is given in Figure 5.1. Mathematical description of the FGNL method is given in Section 4 of Chapter 6.

Two software packages were coded that are based on the FGNL method. The first one implements the method in MS Excel environment (see Berezkin, Kamenev and Lotov, 2000). Application of the FGNL method in such an environment is convenient for several reasons. First, the FGNL software was implemented in MS Excel environment in the simple form of the so-called add-in tool. Secondly, MS Excel provides a simple tool for model formulation. Finally, multiple economic and managerial models have already been coded in MS Excel, and now they can be studied using the FGNL method.

The second, more powerful software package was coded in C++ programming language (Berezkin, 2002). It requires professional programmers to make the software interact with a simulation module of a particular mathematical model. Here, we start illustrating the FGNL method by an example of application of the first software package.

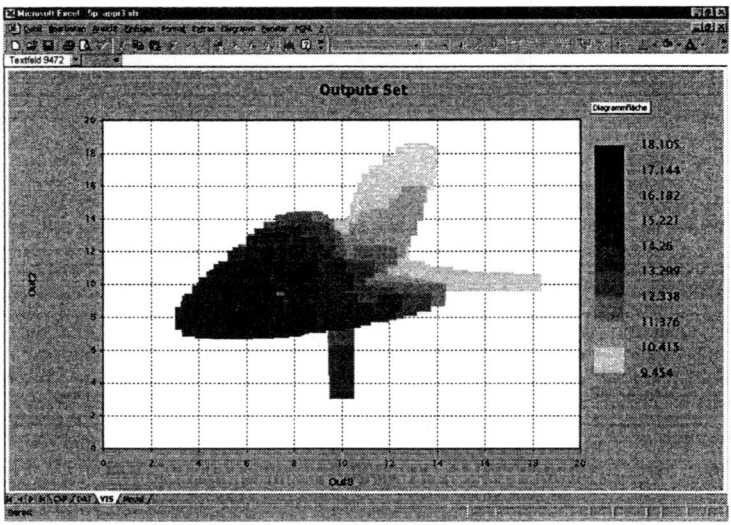

Figure 5.2. Three-criterion picture (collection of slices of the FSCS). The picture is the black and white copy of a color display.

Illustrative example. As an example of a non-linear multi-criteria optimization problem, we here handle a decision problem with five criteria and two decision variables. The problem is to locate a pollution monitoring station in some region. Decision variables are its coordinates. The goal is to locate the station in a place where the forecasted pollution level is maximal. As it usually happens, five experts have given five different forecasts for the pollution level. The forecasts are described by five functions. The decision maker has to identify the location by balancing between the five possible pollution levels.

Fairly arbitrary functions were taken to describe pollution forecasts. Mathematically they are given in Appendix 5.A. Several pictures related to the example problem are given in Figures 5.2–5.6, where black and white copies of the display are provided. The criteria are denoted by *Out1*, *Out2*, *Out3*, *Out4* and *Out5*.

Figure 5.2 displays a collection of bi-criterion slices of the non-convex output set for criteria *Out1*, *Out2* and *Out3*, while the values of *Out4* and *Out5* are free. Differently colored (here differently shaded) slices are related to different values of the value of *Out1*. The values of the criteria *Out2* and *Out3* are given on the axes. The relation among the range of the values of *Out1* and the color (shading) is given in the diagram located to the right from the picture. All the criteria have approximately the same range − their values are between 0 and 20. In Figure 5.2 the

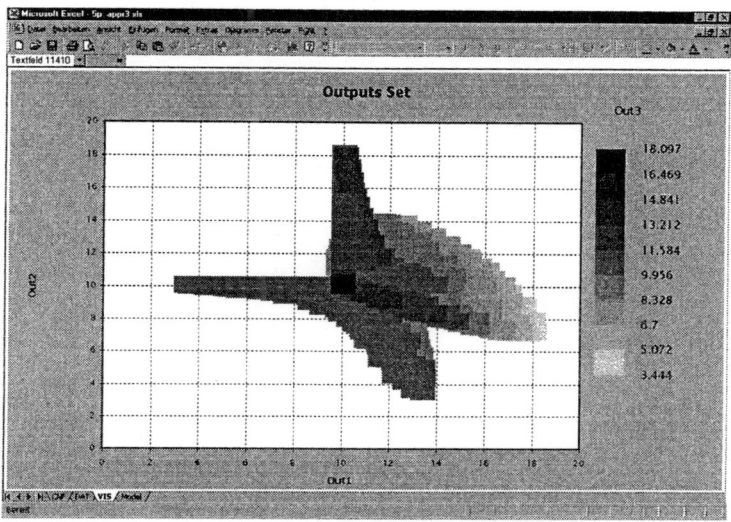

Figure 5.3. The value of *Out3* is provided in color scale, and the values of *Out1* and *Out2* are given on axes (black and white copy of a color display).

range of *Out1* was squeezed: the minimal value of the selected range is about 9.5. Due to this, a more precise relation between these three criteria is provided.

One can see that the maximal values of *Out1* (given in black shading — more than 17.1) are related to relatively small values of *Out2* and *Out3*. In contrast, maximal values of *Out2* and *Out3* (more than 18) are related to medium values of *Out1* (gray shading in the figure corresponding to values from 9.5 to 10.5). At the same time, the points with maximal values of *Out2* and *Out3* do not coincide — for the maximal value of *Out2* the value of *Out3* is about 14, and for the maximal value of *Out3* the value of *Out2* equals 10. If needed one can find more information about the relations among these three criteria. Pictures that display the FCS from different points of view (say, *Out2* may be given in color) may be requested, too. In addition, it is possible to squeeze ranges of one or several criteria and obtain a more detailed picture.

Let us have a look at different pictures related to the problem under consideration. In Figure 5.3 the relation between the above three criteria are displayed from a different point of view — the value of *Out3* is provided in the color scale, and the values of *Out1* and *Out2* are given on the axes. One can see that the large values of *Out2* (more than 18) are related to medium values of *Out1* and *Out3* (about 10).

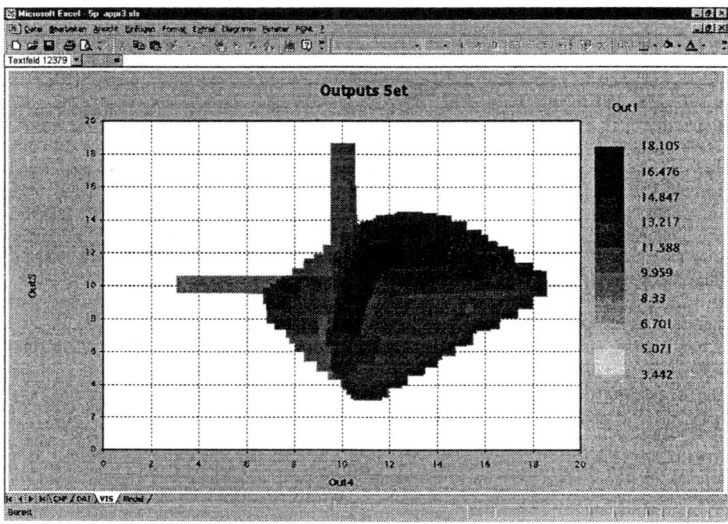

Figure 5.4. Values of *Out1* are given in color, values of *Out4* and *Out5* are given on the axes (black and white copy of a color display).

Relations among different three criteria are given in Figure 5.4 that is devoted to values of *Out1*, *Out4* and *Out5* (values of *Out1* are given in color, values of *Out4* and *Out5* are given on the axes), while *Out2* and *Out3* are free. Figure 5.4 differs from Figure 5.2 a lot. One can see that the maximal values of *Out1* (more than 16.5) are related to large values of *Out4* and *Out5*.

Now let us impose a restriction on the value of *Out3*, which is free in Figure 5.4: now it must be not less than 9.9. The resulting image (Figure 5.5) contains only a part of Figure 5.4. In particular, large values of *Out1* and *Out4* disappear in it. However, large values of *Out5* still remain in the picture. In Figure 5.6 the same criteria are displayed, but now the value of *Out3* is not less than 10.04. Note that a small change in its value has transformed the picture drastically!

Many other displays can be studied on-line to obtain full understanding of the situation, but we have to stop because of the restricted volume of the book. Once again, to find a preferred feasible decision, it is sufficient to identify a preferable criterion point (goal vector) in the picture with the computer mouse. Immediately, a point from the covering base, which neighborhood contains the identified goal vector, is found and displayed. The related decision can be displayed, if needed.

Several real-life applications of the FGNL method have taken place. One of them, its application in the framework of conceptual design of

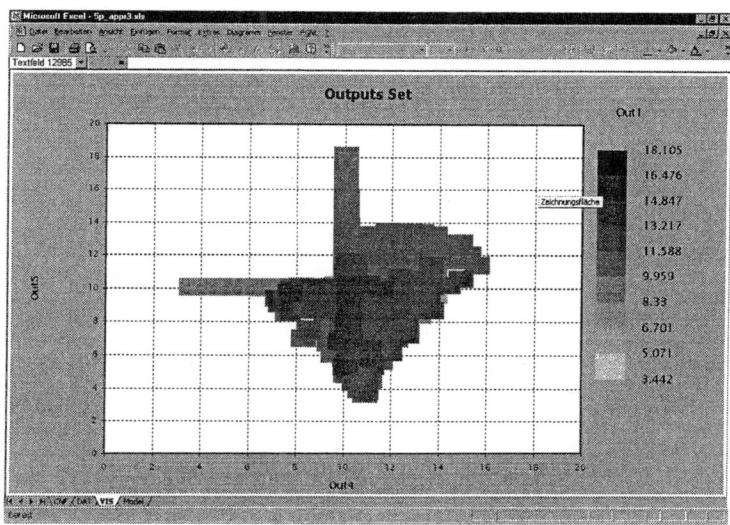

Figure 5.5. Restriction on the value of *Out3* was imposed: *Out3* is not less than 9.9 (black and white copy of a color display).

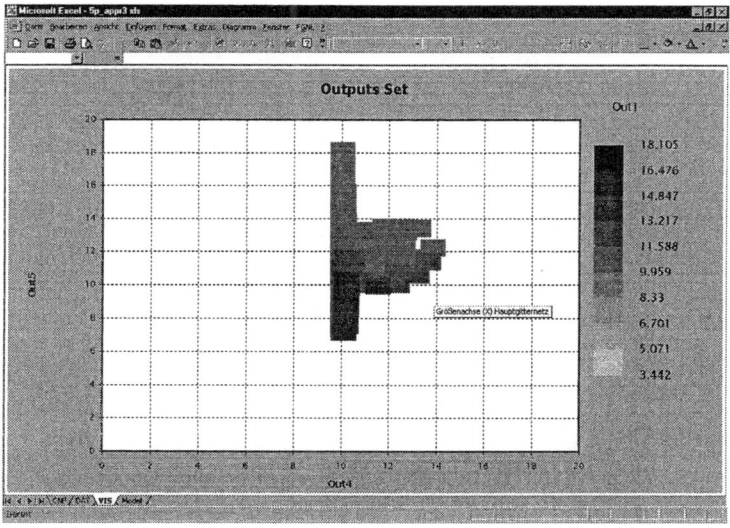

Figure 5.6. New restriction on the value of *Out3* was imposed: *Out3* is not less than 10.04 (black and white copy of a color display).

a future aircraft, is described here. Another application, devoted to the search for reasonable strategies of economic reform in Russia in 1992, is described in (Petrov, Pospelov and Shananin, 1999) and (Lotov, Bushenkov and Kamenev, 1999).

FGNL in conceptual design of aircrafts. Conceptual design is the starting phase of a design process. Decisions that are made at this phase can define the quality of a technical system. Multiple conflicting goals related to various performance characteristics are considered usually at the conceptual design phase. In the framework of aircraft design, goals in such fields as technical perfection, economic efficiency, environmental acceptability, investment availability, etc. may be taken into account. It is needed to find a balanced combination of particular goals that is feasible, i.e., it can be used as a starting point in subsequent phases of the design process. Mathematical models used at the conceptual design phase contain decision variables and parameters, whose influence on performance characteristics must be analyzed and the values of which must be selected.

The conceptual design phase has several important features that distinguish it from the other phases of the design process and make it especially complicated. In addition to conflicting goals, it is desirable to consider all possible alternative designs, at least roughly and in an aggregated form. Fortunately, time restrictions are usually not too strict at the conceptual design phase. This feature distinguishes the conceptual design phase from other phases, say, from the phase of engineering design. For this reason, time-consuming mathematical methods and software can be applied for supporting the designers at this phase.

Computer simulation is often used in the framework of computer support of the conceptual design. Indeed, it is possible to simulate models of the designed objects, say, future airplanes, and study important characteristics of flight scenarios. It is important that simulation can be easily interfaced with modern multimedia tools such as virtual reality, etc. However, simulation by itself cannot provide a tool for resolution of conflict among multiple goals − it must be combined with methods that can help to select a balance among conflicting requirements. The FGNL method can provide the needed information in a graphic form and support possible discussions on the preferable balanced goal vector.

An experimental application of the FGNL method at the conceptual design phase started in the beginning of the 1990s. An integrated model of a future aircraft was used. It was a non-linear simulation model, which was able to simulate the flight of the aircraft depending on its construction parameters and to estimate by this the flight characteristics.

The model was fairly rough, and the data were likely only. So, the study was aimed at demonstrating the features of the method.

Four construction parameters of the aircraft were considered:

- draft of the engine per weight;

- frontal resistance;

- inductive resistance;

- elevating force.

The following flight characteristics were considered:

1 maneuverability at the earth surface;

2 maximal vertical acceleration at a given height;

3 rotation speed at a given height;

4 elevation time, i.e., minimal time of elevation to a given height;

5 acceleration time, i.e., minimal time of acceleration to a given speed.

It was found that the first two characteristics are in line with the third one. So, only the three last characteristics were studied graphically (see Figure 5.7). The values of elevation time (*TimeH*), acceleration time (*TimeV*) are given on axes (it was needed to decrease their values). The intervals of the value of rotation speed at a given height (increment of its value was desirable) are given by shading (by colors on the display). Relative units of measurements were used.

One can see that the minimal value of elevation time is about 11, the minimal value of acceleration time is close to zero (roughness of the model is responsible for it), and the maximal value of rotation speed is about 0.77. However, these goals cannot be met simultaneously: designs with a great rotation speed (not less than 0.72) are related to minimal values of elevation time (between 11 and 12), but only to medium values of acceleration time (about 20). A minor decrement of rotation speed (values between 0.68 and 0.72, given by a more dense shading) allows reducing acceleration time only a bit. To improve the value of acceleration time drastically, one needs to reduce the rotation speed to 0.63 and to increase the elevation time to about 17. Such information is extremely valuable for designers. It is clear that one could discover other properties of the variety of possible designs that we do not discuss here.

As usually in the FGM, it is possible to identify a preferable feasible goal vector. Then, the computer immediately informs the user about the

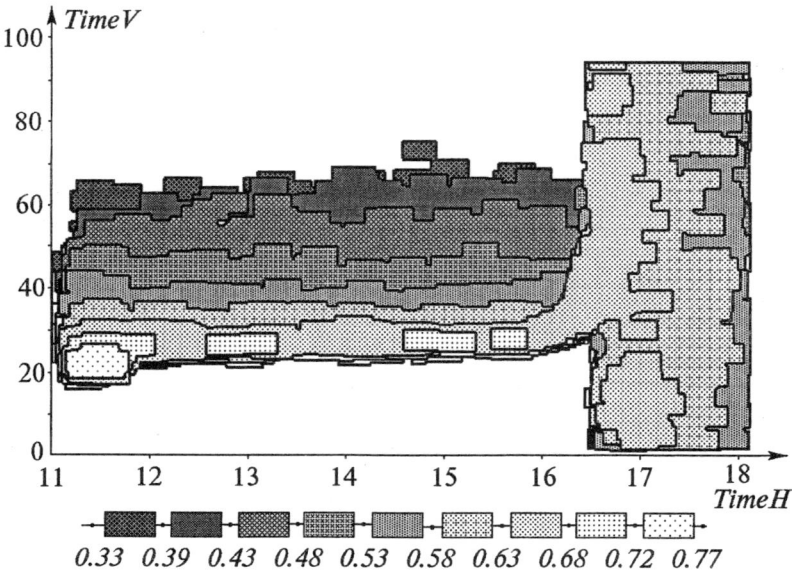

Figure 5.7. The variety of feasible values of three characteristics: elevation time (*TimeH*), acceleration time (*TimeV*) and rotation speed at a given height. Relative units of measurements were used. The values of the first two characteristics are given on the axes (it was needed to decrease their values), while the intervals of the value of the rotation speed (increment of its value was desirable) are given by shading. The picture is the black and white copy of a color display.

related design. However, there is an additional opportunity provided by the FGNL method in the case of a small number of design parameters taken into account in the model. The user can impose restrictions on the characteristic values and display the variety of design parameters, which output satisfies these restrictions. It is possible to explore such a variety of design parameters using the same graphic tools of the FGNL method.

Let us consider an example display of a squeezed variety of design parameters, which satisfies certain requirements imposed on characteristics (Figure 5.8). Three design parameters are considered: draft of the engine, frontal resistance of the aircraft, and elevating force. One can see two different clusters in the figure. It is possible to display any number of pictures of this kind to inform the user on possible parameters of the aircraft that satisfies his/her requirements.

The user may want to study the influence of restrictions imposed on feasible parameters of the aircraft in detail. To do so, one needs to

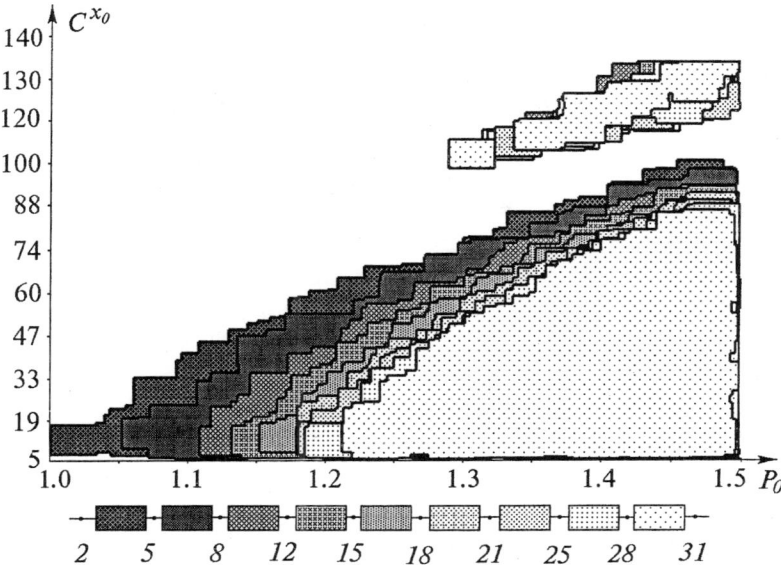

Figure 5.8. Squeezed variety of feasible values of draft of the engine (P), of frontal resistance of the aircraft (C^{x_0}) and of elevating force. Draft of the engine (P) and frontal resistance of the aircraft (C^{x_0}) are given on the axes, elevating force is given by shading (black and white copy of a color display).

approximate the FCS for criteria that include both characteristics and parameters. This is an important opportunity in the framework of design procedures, since constraints imposed on the design parameters are known only approximately. The same approximation technique provides an opportunity for the user to consider the influence of small data variations on the feasible goals.

Integration of the FGNL method with optimization tools: fine tuning. As one can see, the FGNL method results in a fairly rough approximation of the FCS. For this reason, the balanced feasible goal vector and the associated design are found fast and easily, but fairly roughly. The roughness of the FGNL method must be compensated by a further fine tuning of the feasible goal vector and the associated design. This problem can be solved by exploration of a feasible variety of the design space with outputs close to the identified goal vector. The output of such a variety can be explored by approximating of the FCS related to this variety only. Then, a new, more precise goal vector can be identified, etc. Such procedure of fine tuning of the feasible goal vector is actually an FGNL-based graphic method for sequential improvement of the goal and the associated design.

Another approach to the fine tuning of the feasible goal vector can be based on combination of the FGNL method with methods that apply the single-criterion optimization. It can help to find the preferable design in a simple way. First, the FCS is explored. Then, the designer expresses his preferences in the form of the feasible goal. Finally, the identified goal is refined using a single-criterion optimization. For example, this idea has already been implemented in the framework of experimental combination of the FGNL method with the optimization technique NIMBUS (see Miettinen, Berezkin, Kamenev, and Lotov, 2000; Miettinen, Lotov, Kamenev and Berezkin, 2003).

2.   FGM/IDM technique for non-linear models

On first glance, the FGM/IDM technique for non-linear models does not differ very much from the FGNL method − it applies approximation and visualization of an EPH instead of an FCS. However, as we have said already, approximation procedures for an EPH in the non-linear case cannot be based solely on the ideas used in the methods for FCS approximation. It turned out that simulation of random alternatives is not sufficient for constructing a reasonable approximation of an EPH in the case of a relatively large number of decision variables. So, different techniques must be involved, too. We apply a combination of single-criterion optimization with random simulation and filtering. This section proves that such an idea can solve the problem. Note that single-criterion optimization plays an important role in the convex case, and it is natural to use it in the non-linear case as well.

A combination of optimization with random simulation and filtering is the main idea of the FGM/IDM technique described in this section. To preserve the merits of the FGNL method (such as an opportunity for application of complicated computational modules instead of mathematical formulae), we use simulation-based optimization in addition to random simulation and filtering. The idea of the method was introduced in (Lotov, Kamenev and Berezkin, 2002).

Approximation and visualization of EPH.   Recall that an EPH is an FCS broadened by all dominated points (see Chapter 1). For this reason, one can easily understand that an approximation of an EPH can be constructed as a collection of domination cones located at the points of an FCS. However, in contrast to the case of approximating of an FCS, now it is desirable to obtain such approximating figures that are close to its Pareto frontier. To satisfy such a requirement, the single-

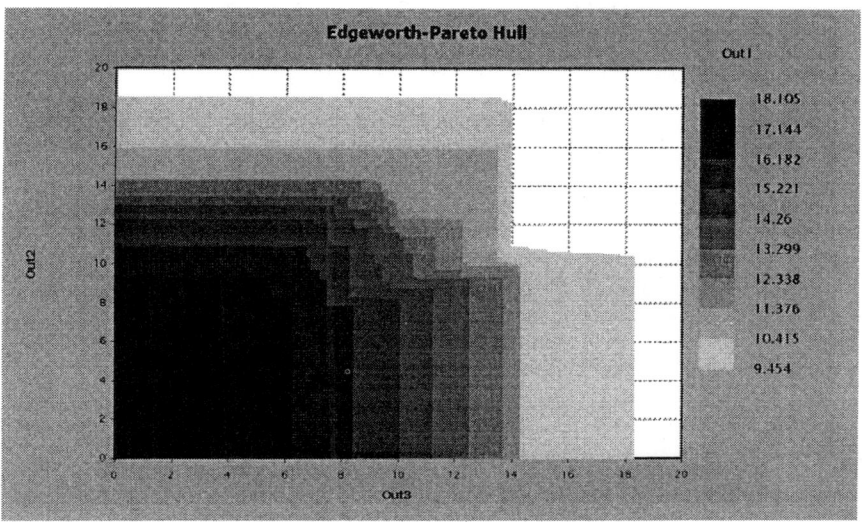

Figure 5.9. Slices of Edgeworth–Pareto Hull in the form of decision map.

criterion optimization is used in the process of EPH approximating for correcting the criterion points that are the result of simulation of random feasible decisions. Such a correction results in the shift of the criterion points in the direction of the non-dominated frontier. As the result, a collection of cones is constructed, each cone of which is located at a point of the FCS close to the Pareto frontier. The evaluation of the quality of an approximation is provided, too. Details of the method are given in Section 5 of the next chapter. As usual in the non-linear case, the approximation process may take a fairly long time, but visualization and identification of the feasible goal are carried out on-line. Visualization of the EPH helps to concentrate on the Pareto frontier and to forget about other frontiers of a bi-criterion slice. Let us consider the EPH for the example model (selecting of location for a monitoring station) that was discussed in the previous section.

A collection of slices of the EPH (decision map) is given in Figure 5.9. This collection of slices is related to the same set of criteria as in Figure 5.2. One can see that the tradeoffs among criteria can be understood in Figure 5.9 easier than in Figure 5.2. For example, one can immediately realize that the Pareto frontier of the largest (white) slice, which corresponds to relatively small values (9.5–10.4) of *Out1*, has two separated zones of the Pareto frontier. One zone is very narrow; the criterion *Out2* is close to its maximal value (about 18.2). Another zone is relatively broad. It starts at the maximal value of *Out3* (about 18.2); the value of *Out2* can be increased a bit if the value of *Out3* decreases drastically — from 18.2 to 14.0.

By comparing this slice with the next one (where the value of *Out1* is between 10.4 and 11.4), we see that the increment of *Out1* simplifies the Pareto frontier — it has now only two non-dominated points that can be related to two zones of the white slice. One can see how the increment of *Out1* influences the values of other criteria. The next slice with *Out1* between 11.4 and 12.3 has a totally different Pareto frontier. It provides an opportunity to exchange values of *Out2* and *Out3* in a broad range.

Properties of other slices could be discussed as well. Dependence of the Pareto frontier on the value of *Out1* is clear, too. Pictures of this kind could be animated, that is, they could be changed in accordance with an automatic change of the fourth or fifth criterion. However, such an option was not provided in the framework of the first FGM/IDM software that was coded as a part of the add-in for MS Excel software (Berezkin, Kamenev and Lotov, 2000), which has already been considered in the previous section.

The new powerful software package coded in C++ programming language helps to explore complicated technical systems including those that are given by computational modules (Berezkin, 2002). One example of this kind is provided by application of the software in the framework of the study of secondary steel cooling equipment used in the process of continuous steel casting. Here we outline the study in short; a detailed description is given in (Berezkin, Kamenev, Lotov and Miettinen, 2003).

Multi-criteria study of steel cooling process.   The technology of continuous steel casting was developed and implemented in the 20th century.   Though it has wide recognition and is used universally, its development continues.   In particular, the technology is studied and improved with the help of mathematical models of the cooling process. The process of steel casting is illustrated in Figure 5.10.

Molten steel is poured down from the tundish into a water cooled mold, where the strand of steel obtains a solid shell. After the mold exit (point z_1), the strand is supported by rollers and cooled down by water sprays.   In this spray cooling region (the so-called secondary cooling region) more heat is extracted, so that the solidification is completed. After the water sprays region (point z_2), the strand is cooled down by radiation. Then, the strand is straightened at the unbending point z_4, and then in the cutting point z_5 it is cut up.

The secondary cooling plays a major role in the steel casting process, since the intensity of the water sprays affects highly the solidification

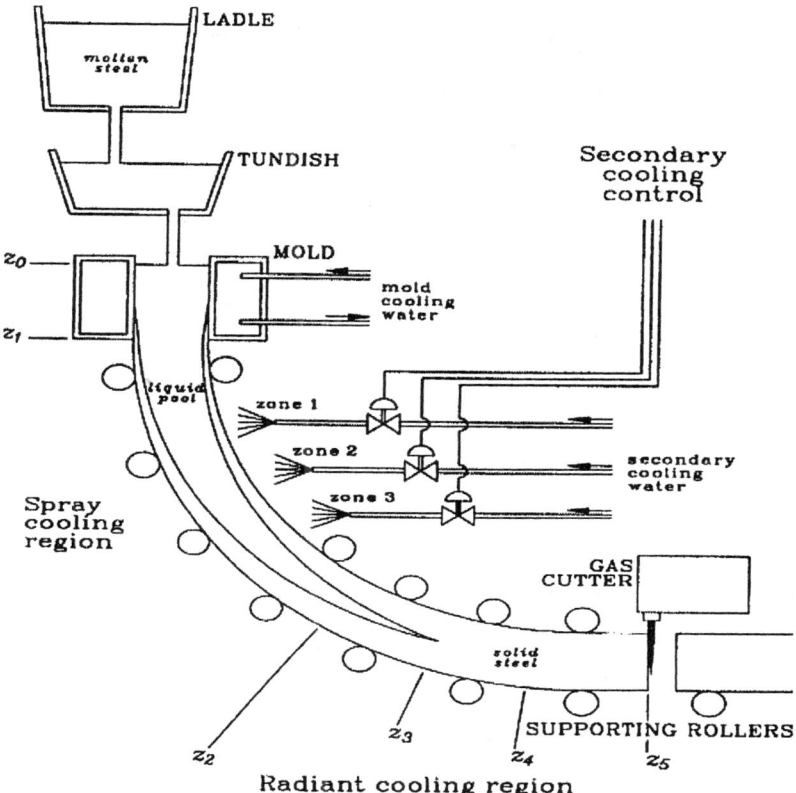

Figure 5.10. Steel cooling in the process of continuous steel casting, adapted from (Miettinen, Mäkelä and Männikkö, 1998)

rate. First of all, overcooling can lead to the formation of cracks. Moreover, there must be a smooth transition of the surface temperature as the steel passes through in the secondary cooling region. In addition, undercooling of the strand in the process of secondary cooling can result in a liquid pool that is too long. These technological requirements result in constraints that must be imposed on the secondary cooling process. Fortunately, it is quite easy to adjust the intensity.

It is clear that a mathematical model of the cooling process must be non-linear and fairly complicated, especially if one takes into account the fact that the position of the solidification front is not known a priori and depends upon the secondary cooling. Moreover, the solidification front is not sharp, between the solid and the liquid phases there exists a mushy zone, which is considered to be partially solid and partially liquid. A mathematical model that takes these features into account was developed at the University of Jyväskylä, Finland (Laitinen and Neittaanmäki, 1988). The model was implemented in the form of a computational module based on application of the FEM/FDM techniques and used for specification of parameters of the secondary cooling process.

The task of selecting the intensities of the water sprays was formulated in the form of a multi-criteria optimization problem in (Miettinen, Mäkelä and Männikkö, 1998). In total 325 control variables were considered that described the intensity of water sprays from different sprayers in the secondary cooling zone. The technological constraints imposed on the steel cooling process include constraints on the surface temperature of the strand, on the variation of the surface temperature along the strand, on the temperature after the point z_3 and at the point z_5. Originally, a single optimization criterion J_1 was considered that described deviation from the desired surface temperature of the steel strand. However, it turned out that this criterion is of minor importance since the constraints imposed on the steel cooling process cannot be satisfied simultaneously. For this reason, a multi-criteria problem of minimization of constraints violation was formulated in (Miettinen, Mäkelä and Männikkö, 1998), where four additional criteria J_2 to J_5 were considered that are the penalty criteria introduced to describe the violations of constraints. Namely,

- criterion J_2 is related to the surface temperature;

- criterion J_3 is related to variation of the surface temperature along the strand;

- criterion J_4 is related to the temperature after point z_3; and

- criterion J_5 is related to the temperature at point z_5.

Criteria from J_2 to J_5 were considered in the study that applied the FGM/IDM technique (Berezkin, Kamenev, Lotov and Miettinen, 2003). The EPH was approximated for the non-linear model of the steel cooling process (Laitinen and Neittaanmäki, 1988), which contains 325 decision variables and four criteria. Let us provide an example of decision maps considered in this study (see Figure 5.11).

In Figure 5.11, the values of J_4 are given on the horizontal axis and the values of J_5 are given on the vertical axis. Values of J_3 are given by colors (shadings in the black-and-white copy given in the book) that correspond to value intervals of J_3. The range of the criterion J_3, which values change from 0 to 0.05, is split into five shading intervals. The

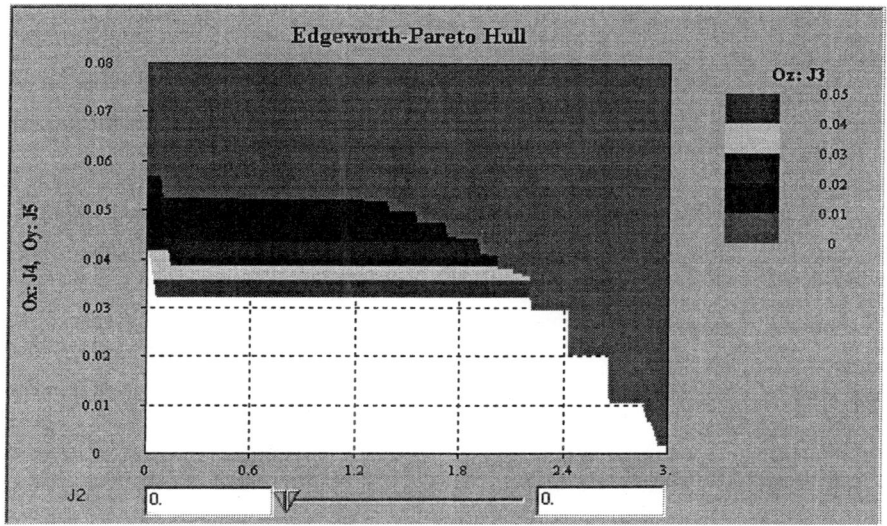

Figure 5.11. A decision map for the steel cooling problem

correspondence between the shading and the value interval is given in the palette located to the right of the decision map.

The slices for smaller values of J_3 are superimposed over the slices for higher values. Range of J_2 can be specified by two sliders of the scroll-bar located under the decision map. It turned out that positions of the sliders practically did not influence the decision map. Therefore, zero values for both sliders were specified in Figure 5.11. It means that the value of J_2 equals zero.

The decision map given in the figure proves that the ideal criterion values cannot be achieved simultaneously, and so there is a conflict among criteria. Look at the large grey area given by the slice that is related to the interval of the criterion J_3 with minimal values (between zero and 0.01). It is clear that the criteria J_4 and J_5 cannot achieve their minimal (zero) value simultaneously for these values of J_3. Zero value of J_4 can be achieved only while the value of J_5 is about 0.06. At the same time, the smallest value of J_5 (it is close to zero) can be achieved while the value of J_4 is about 3.0. The tradeoff curve between J_4 and J_5 is given in a clear way for the interval of J_3 with minimal values.

Figure 5.11 proves that the situation is not getting much better while one turns to larger values of J_3. The decision map informs on how the value of J_3 influences the feasible combinations of J_4 and J_5. Though larger values of J_3 provide new feasible criterion points (we can see those points that are not covered by the slices with smaller values of J_3), the increment of slices is fairly small. The minimal value of J_5 for J_4 being about zero is decreased a bit, but not substantially. Say, it is close to 0.032 for the maximal value interval of J_3 in the figure (J_3 is between 0.04 and 0.05). Larger values of J_3 do not help either (additional figures that are not given in the paper prove it). Note by the way that the slices of the decision map are not convex, and so the problem under study is not convex, either.

This and other figures that are not given here (see Berezkin, Kamenev, Lotov and Miettinen, 2003) prove that the control of the intensity of the water sprays cannot solve the conflict between the constraints imposed on the process. The user (designer in this case) has to find a balance between violations of the constraints or propose some changes in the steel casting equipment that will solve the conflict.

3.   RGM/IDM technique for non-linear models

In this section a form of the RGM/IDM technique is described that can be used in the general case of non-linear models. In contrast to the version of the RGM/IDM technique considered in Chapter 4, the technique supports selection of a small number of decision alternatives from the varieties with an infinite number of feasible decision alternatives.

The scheme of the RGM/IDM technique for the general non-linear case consists of six steps (see Figure 5.12):

1 Approximation of the EPH for a non-linear model in the form of the collection of cones with apexes located at criterion points that belong to the approximation base;

2 Approximation of the convex hull of the EPH (CEPH);

3 On-line exploration of the CEPH by the user who applies the IDM technique; then, the user identifies the reasonable goal;

4 Automatic selection of a small number of criterion points from the approximation base, which are in line with the identified goal;

5 Display of the selected criterion points and associated decision alternatives.

Let us consider the scheme step by step. The first step, i.e., approximation of the EPH, has already been discussed in the previous section; mathematical details of the approximation method are given in Chapter 6. It is important to stress that a problem with an infinite number of decision alternatives is converted by this into a problem with a finite

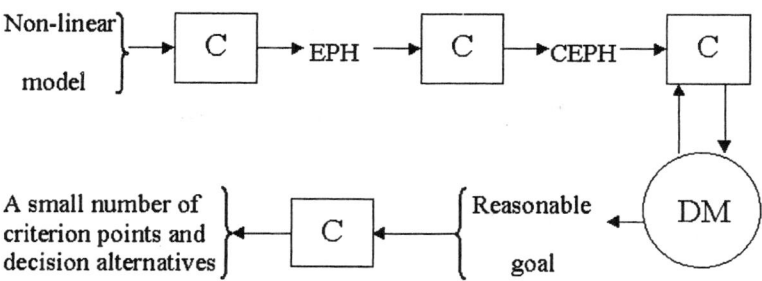

Figure 5.12. The steps of RGM/IDM technique. Computer processing is denoted by **C** and decision maker by **DM**.

number of alternatives, which outputs are included in the approximation base. The second step, that is, approximation of the convex hull of the EPH, which is given by a finite number of criterion points of the approximation base, coincides with approximating the CEPH for the approximation base. Such problems have already been considered in Chapter 4.

At the third step, the user is involved in the process of selection. He/she explores the CEPH using the IDM technique and identifies the reasonable goal. This procedure coincides with the identification of the reasonable goal described in Chapter 4, and so we do not need to pay additional attention to it here. Then, points from the approximation base are selected that are close to the identified goal in some sense. Finally, decision alternatives, which are associated with the selected criterion points, can be easily found by computer, since they are stored together with points of the approximation base.

As an illustration of the RGM/IDM technique in the general nonlinear case, we provide the convex hull of the EPH for the problem of steel cooling considered in the previous section (see Figure 5.13).

In this figure, the same view is provided as in Figure 5.11. As earlier, the criterion J_2 is located at the scroll-bar. The criteria J_4 and J_5 are given on the axes, and the criterion J_3 is given by shading. Now the decision map informs about a proxy Pareto frontier and can be understood more easily. However, the goal identified on it is usually not feasible. Therefore, as it usually happens in the RGM, several feasible decision alternatives are provided to the user.

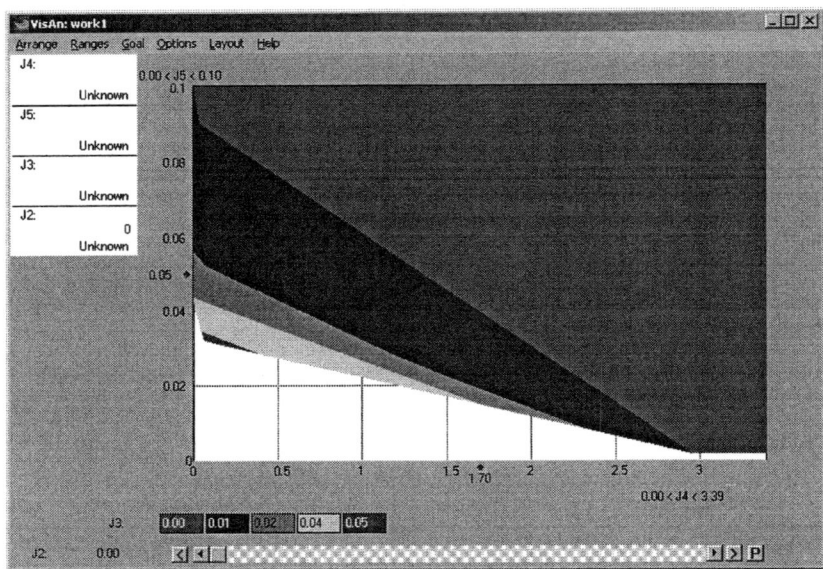

Figure 5.13. CEPH for the steel cooling problem

4.   Visualization of risk-related decision problems

Visualization of risk-related decision problems can play an important role in searching for reasonable decision alternatives. It has been recognized that risk management can not be restricted to a minimization of the mathematical expectation of possible losses: other risk indicators must be taken into account, too. In particular, Howard Raiffa (Raiffa, 1968) criticized application of single-criterion optimization methods, which take into account the expectation only. The book (Haimes, 1998) contains examples of a misleading role of expectation.

Often it is erroneously assumed that, though expectation is not sufficient, one can solve this problem by taking into account only two criteria — expectation and variance. Reasoning and the examples given in the book (Raiffa, 1968) prove that these two values are not sufficient for decision making under risk, either. Raiffa proposes to construct a value function instead. Maximization of the value function can result in identification of the "best" decision. The value function must be estimated on the basis of subjective comparison of lotteries. By this, subjective appreciation of risk is evaluated and applied in the process of selecting the best alternative. This approach, which is brilliant from the theoretical point of view, has a feature that prevents its real-life application

– comparison of lotteries turned out to be too complicated for human beings (Larichev, 1984; Larichev, 1992).

Another approach to decision making under risk can be based on exploration of a larger number of performance characteristics. The multicriteria decision support methods provide the techniques that can deal with them. The methodology of risk management based on multi-criteria techniques was elaborated in (Haimes, 1998). Here, we apply the same concepts that are aimed at converting the problems of decision making under risk into multi-criteria decision problems. However, we apply a different multi-criteria method, namely the RGM/IDM technique. We show that both versions of the RGM/IDM technique, which have been considered in this book, can be used for visualization of decision problems under risk. Once again, these versions of the RGM/IDM technique are related to two different descriptions of the decision situation:

- a large, but finite list of decision alternatives does exist, for which several performance characteristics can be evaluated; and

- an infinite variety of possible decision alternatives is described by a mathematical model that describes the consequences of feasible alternatives.

The case of a finite number of decision alternatives. Let us consider a typical problem of decision making under risk that is related to the management of risks of natural disasters (such as forest fires, sea tides, mountain or river floods, etc.). Let us assume that there exists a list of possible alternative actions (as constructing of fire-breaks, dams, levees and dikes) that are aimed at the decrement of losses resulting from particular disasters. We assume that the consequences of an alternative are given in probabilistic form. To be precise, the random value (stochastic variable) π_i is considered that describes losses that still remain after the i-th decision alternative has been implemented. It is assumed that the values of the cumulative probability functions

$$F_i(v) = \mathbf{P}(\pi_i < v), \ i = 1, 2, ..., N,$$

may be estimated somehow (using mathematical models, subjective expert knowledge or empirical data) at least for several values of v. Moreover, some moments of the cumulative probability functions (as their expectation and variance) may be estimated, too. Furthermore, we assume that the user considers this information to be sufficient for risk management (say, he/she has specified the values of v). By this, the problem of decision making under risk is converted into a decision problem with a finite number of decision alternatives given by a finite number of attributes. Then, the RGM/IDM technique can be applied for visualization of decision problems of this kind.

Table 5.1. Database of decision alternatives

NAME	Expectation of losses	P_h – Probability of high losses	P_c – Probability of catastrophic losses
Alt1	4	10	0
Alt2	3.3	25	1
Alt3	2.6	20	0
Alt4	1.1	35	1
Alt5	3	20	0.5
Alt6	2	25	1
Alt7	3.1	15	0
Alt8	3.2	30	2
Alt9	1.5	40	1
Alt10	1	30	1
Alt11	3	20	0.25
Alt12	1	30	5
Alt13	2	40	3
Alt14	0.96	25	1
Alt15	3.5	20	0.03
Alt16	3.2	15	1
Alt17	2.6	25	0.5
Alt18	1.7	30	2
Alt19	1.1	28	1.5
Alt20	2.3	25	0.5
Alt21	3.5	16	0.01
Alt22	4.1	10	0
Alt23	3.8	15	1
Alt24	3.3	20	5
Alt25	2.2	25	0

Let us consider an example. For the sake of simplicity, we restrict ourselves to three attributes. We assume that Table 5.1 has already been constructed that contains values of these attributes for any of $N = 25$ alternative decisions of, say, a dam, a dike, etc. The following three attributes are used:

- expectation of "losses" that include (along with losses resulting from the disaster) the constructing and maintenance cost, (in million US$);

- probability of high losses (in percent); and

- probability of catastrophic losses (in 0.01 of percent).

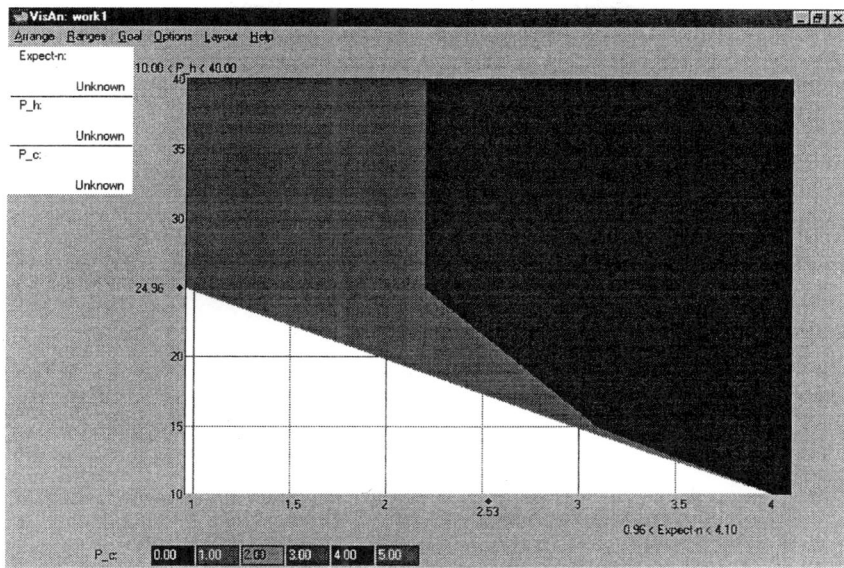

Figure 5.14. Decision map that describes the database given in Table 5.1

It is assumed that the meaning of high and catastrophic losses was identified by experts. Though the number of alternatives in the table was restricted to 25, it is clear from the results described in Chapter 4 that the RGM/IDM technique makes it possible to consider any finite number of alternatives. These three attributes are used as the selection criteria. It is desirable to minimize the values of all three criteria.

We apply the RGM/IDM technique to explore the list of alternatives. Figure 5.14 contains a black and white copy of a color decision map related to the problem. The values of expectation are given on the horizontal axis, and values of probability of high losses, denoted by P_h, are given on the vertical axis. Pareto frontiers among expectation and probability of high losses for several values of probability of catastrophic losses (denoted by P_c) are given. The relation among the shading and the values of P_c is given in the palette located under the picture. Black shading is related to the slice, that is, the combinations of values of expectation and probability of high losses that belong to the convex hull for $P\_c = 0$. The slice related to $P\_c = 1$ is displayed by a lighter shading. The slices related to the values of P_c that are higher than 1 are not seen in Figure 5.14 since they are covered by the slices for $P\_c = 1$. So, Figure 5.14 proves that it makes no sense to have P_c more than 0.01%.

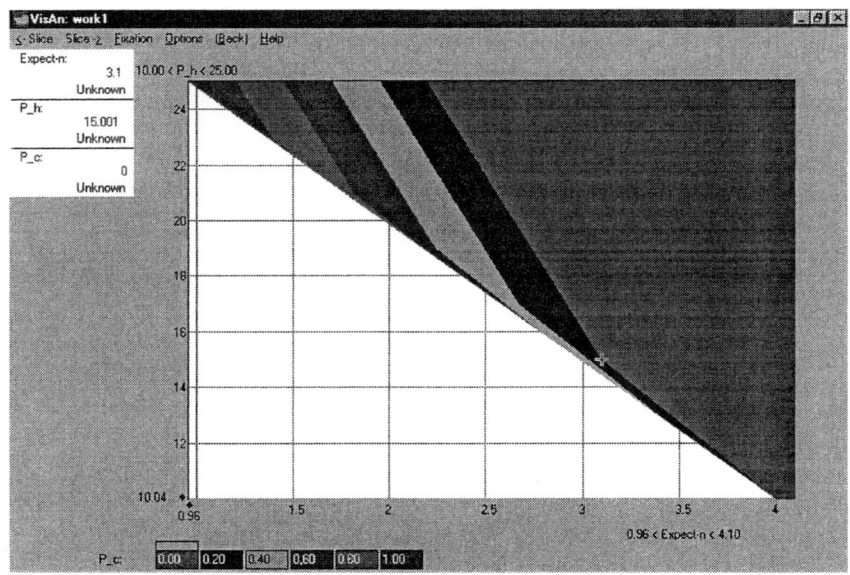

Figure 5.15. Detailed decision map.

One can see in Figure 5.14 that the Pareto frontier for $P\_c = 1$ is a straight line. It is possible to decrease the expectation of losses from about 4 million to about 1 million by increasing the probability of high losses from 10% to 25%. In contrast, the Pareto frontier for $P\_c = 0$ is kinked. So, if the range of expectation is between 3 and 4 million, both frontiers are extremely close. If expectation of losses is less than 3 million, it is necessary to pay by substantially larger increment of P_h for a decrement of expectation of losses. It is important that expectation of losses can not be less than about 2.3 million for $P\_c = 0$.

Let us squeeze the maximal value of the range of P_h to 25% and of P_c to 1. The related picture is given in Figure 5.15. Note that the value of P_c changes with the step equal to 0.2. One can easily note one important feature of the alternatives under study — the fact that if P_h is less than about 15%, all Pareto frontiers are very close. It means that an increment of P_c from zero to 1 results in an extremely small increment of expectation of losses for these values of P_h. Further, Pareto frontiers for low values of P_h display linear dependence of P_h on expectation of losses. It means that the user has to make a difficult tradeoff among P_h and expectation. For this reason, the point marked in Figure 5.15 by the cross may be preferable — it may be identified as the reasonable goal. The precise position of the cross is given in the tablet: expected losses equal to 3.1 million, $P\_h = 15\%$, $P\_c = 0$. The computer finds the alternatives that are close to the goal − these are alternatives 1 and 7 from the above Table 5.1.

It is worth noting that the user informed by Figures 5.14 and 5.15 about the criterion tradeoffs among different risk indicators can identify a preferable goal consciously, that is, using this knowledge. Subjective preferences and experience of the user play an important role in the selection of the goal. Using decision maps, the user can inform other stakeholders and general public of the reasons for his/her decision. In turn, people interested in the problem can use the same decision maps to criticize him/her.

The case of an infinite number of decision alternatives. In the case of an infinite variety of feasible decision alternatives, such a variety must be described by a mathematical model. Then, the RGM/IDM technique for non-linear models described in the previous section can be applied. Since it reduces the infinite choice problem to a problem with a large, but finite number of decision alternatives, its exploration has the same form as has already been described in this section.

Visualization in robust analysis. Robust analysis is an approach to decision support in the case of radical uncertainty about the future (see Rosenhead and Mingers, 2001). Say, probability of different futures or other information of this kind is not assumed to be given. In the framework of robust analysis, a system of possible alternative decisions is formulated, which are evaluated against possible futures. One needs to find such a balanced decision that is relatively good for all possible futures.

Let us assume that a single qualitative performance indicator is used. In this case, a numerical table can be developed, which rows represent the alternative decisions, and columns are associated with possible futures. The columns of the table, that is, performance indicator values related to particular futures, can be used as the selection criteria. Let us assume that the number of futures does not exceed seven, and all of them have already been specified by the user as the selection criteria. (If the number of futures is larger than seven, constraints can be imposed on the performance indicator values for several futures; alternatively, several futures can be combined into one criterion in some way).

The next steps of the decision support procedure coincide with the steps of the RGM/IDM technique in the process of supporting the multi-criteria choice from a finite number of decision alternatives: the user explores the Pareto frontier of the envelope, identifies a balanced reasonable goal, obtains a small number of decision alternatives and selects the most preferable one. It is clear that such an approach to robust decision making can be especially effective in the case of a large number of decision alternatives.

An example of real-life application of this approach is related to private money management during the time before Russian financial default in 1998. Graphic exploration of decision maps related to the envelope of different strategies evaluated against three possible futures (normal development of the national financial market; devaluation of ruble about 50%; total collapse of the national financial market) helped to find a decision, most reasonable from the subjective viewpoint. This decision proved to be the best one during the default.

In the case of multiple performance indicators, multiple tables of indicator values can be formulated.

5.   Visualization in Data Envelopment Analysis

A new approach to the analysis of performance of a system of homogeneous production units, Data Envelopment Analysis (DEA), was introduced in the end of the 1970s (Charnes, Cooper and Rhodes, 1978). It is based on the comparison of input/output values of the homogeneous production units. DEA has found a wide range of applications (see, for example, Thanassoulis, 2000). The DEA studies are mostly centered on three main topics:

1 deriving an integrated measure of the efficiency of each production unit,

2 identifying of target levels of input and output for a particular production unit;

3 selecting of role-model units of good operating practice.

In this section we describe application of IDM-inspired ideas in the framework of DEA. This visualization-based approach to DEA was proposed in (Lotov, 2003). It provides a graphic support of the search for preferable target levels of input/output for a particular unit. The approach is based on approximation and visualization of the *production possibility set* (PPS) for the whole system of units, which is given by the envelope (convex hull) of the points associated with the production units. Visual analysis of the position of a particular unit in the production possibility set helps the decision maker, who makes decisions on improvement of performance of the unit, to specify the target levels consciously and to find the role-model units.

The problem of identification of the target input/output levels for a production unit has traditionally been considered in DEA on the basis of a proportional adjustment of outputs (or inputs) of the unit under consideration. Such an adjustment defines a ray in the input–output space. The point where the ray crosses the frontier of the PPS was recommended for the target input/output level. Such an approach, however, does not take into account the objective opportunities of a production unit to modify its inputs or outputs. For example, it may turn out that the unit can decrease application of production funds, but not of the labor force (or vice versa). Also, it may happen that a rise of output of some products is not reasonable because of the absence of demand, etc.

Starting with papers (Golany, 1988; Thanassoulis and Dyson, 1992), the issue of target input/output levels is considered as a multi-criteria decision problem. The relation between DEA and multi-criteria decision making was studied in several papers (see, for example, Stewart, 1996). Here, we propose to apply the IDM-related visualization in DEA. Such an approach provides the decision maker with graphic information on the position of a production unit in the system of units and gives an opportunity to identify the target levels of input and output directly. To be precise, visualization of the PPS is used for informing the decision maker and providing him/her with the option to set the target level. Mathematical description of the approach is given in Appendix 5.B. In what follows, an informal description of the approach is given.

There exist several definitions of the PPS. One of them coincides with the definition of the EPH of the convex hull of the variety of points associated with a finite number of decision alternatives. As was shown in Chapter 4, approximating of the EPH used in the RGM/IDM technique makes it possible to display the Pareto frontier to the decision maker in the form of a large number of bi-criterion slices of EPH, to give him/her an opportunity to identify the goal point on the computer display and to select by this several alternatives. The same idea can be used in DEA for informing the decision maker on the position of the production unit, thus supporting him/her in the process of identification of the target input/output levels and selecting the role-model production units. We show in this section how one can implement such an idea. From the very beginning, we want to stress that the visualization procedure used in DEA has important features that distinguish it from RGM-related visualization.

Let us describe the procedure, in the framework of which visualization of a PPS helps to develop target input/output values for a production unit. It is assumed that the information on performance of a group of production units has already been prepared. First, a decision maker has to specify a list of inputs and outputs of interest (the number of them must not exceed seven–eight). Then, the PPS approximation is carried out automatically. Before visualization starts, the decision maker has to specify a production unit of interest. If the unit is not located at the frontier of the PPS (the problem of identification of the target values has a sense only in this case), it is needed to improve its performance. The decision maker is informed about the location of the specified unit in the PPS. To be precise, bi-criterion slices of the PPS that pass through the point associated to the unit under consideration are displayed. All the slices are displayed simultaneously in small windows, which are displayed in the form of a matrix. Rows and columns of the matrix are associated with the inputs or outputs. The production unit under study is displayed by a point at all slices. Due to it, the decision maker obtains an integrated view of the position of the unit in the PPS. Note that the unit-associated point can be close to the frontier of the PPS for some inputs and outputs, while for other pairs of inputs and outputs the unit can be fairly distant from the frontier. In addition, as has already been said, values of some inputs/outputs can be changed, while it is impossible or too expensive for other inputs/outputs.

The decision maker has to specify one of the slices (that is, to specify a combination of two inputs/outputs), values of which can be used for improvement of the position of the unit. He/she receives a zoomed display of the slice related to the specified inputs/outputs. Then, the decision maker has to move the point in the plane of the specified inputs/outputs into a new position at the slice (the computer mouse is used) into the direction of the frontier. The values of other inputs/outputs do not change in the process of such a movement, but frontiers of other slices of the PPS change simultaneously with the movement of the point. After experimenting with various trial movements, the decision maker can set a new location of the unit-related point in the plane of two specified inputs/outputs. The new location of the point is considered as the current target. If the target does not belong to the frontier, the decision maker has to select another slice (that is, specify new inputs/outputs) and move the current target into a new position, etc. After a position at the frontier of the PPS is selected, it is considered as the final target.

Then, the procedure turns to selecting role-model production units, which can be used as a standard of good operating practice for the unit under consideration. Usually in DEA, such units are proposed to be the role-model units that give rise to the hyperplane, which describes the frontier of the PPS in the vicinity of the target input/output. Real-life applications of DEA prove that such units may be located far off the target. For this reason, these units cannot claim to be selected as role-model units. We use another approach, which is applied in the RGM/IDM technique and is based on the search for non-dominated production units that are close to the target.

An experimental variant of the software based on the described approach was implemented in MS Excel environment. It was used for exploration of the relative efficiency of the capital funds and labor force in a group of 28 regions of the European part of Russia. The experimental software and the example were developed by A. Biryukov, a diploma student of Lomonosov Moscow State University.

Data on the regions of the European part of Russia for the year 1997 were taken from the official statistical source (Goskomstat of Russian Federation). Then, several regions with relatively large population and industrial production (like Moscow and Sankt Petersburg) were excluded from the list. Regions were described by five attributes:

1 Industrial production (billions of rubles of the year 1997),

2 Agricultural production (billions of rubles),

3 Production of constructing industry (billions of rubles),

4 Labor force (hundred thousands of people);

5 Funds (billions of rubles).

The production was considered as outputs, while labor force and funds as inputs. Part of the list of regions with their attributes is given in Figure 5.16.

In accordance with our approach, the decision maker has first to identify a production unit of interest. It turned out that most of the regions under consideration were located at the frontier of the PPS. Several other regions were located fairly close to the frontier. In our example study, we considered a region for which the related point was located sufficiently deep inside of the PPS. It is Chuvash Republic. In the original study, we used all five inputs/outputs, but here, for the sake of simplicity, we consider only four of them − two outputs (industrial production and agricultural production) and both inputs (labor force and funds).

To identify the target, the decision maker can move the unit-related point in the display of the DEA software (its black and white copy is given in Figure 5.17).

The small window (Object name) informs that the object of the study is Chuvash Republic. The tablet in the lower left corner informs on the current values of inputs/outputs for the unit, the new values and

Region	Industrial production	Agricultural production	Construc ting	Labor	Funds		
Karelia	61.63	7.731	6.75	3.71	89.51		
Komi	106.68	10.114	19.86	5.46	188.2		
Arkhangelsk	111.02	16.928	11.96	6.37	157.9		
Murmansk	122.76	4.201	10.08	4.72	163.8		
Novgorod	39.91	12.962	5.17	3.33	60.1		
Pskov	23.29	18.315	4.19	3.27	65.17		
Bryansk	40.56	29.122	10.23	5.9	96.19		
Vladimir	88.45	23.574	8.98	7.11	105.6		
Ivanovo	53.33	14.638	5.31	5.17	78.61		
Kaluga	43.6	22.061	9.44	4.97	85.07		
Kostroma	35.33	16.558	5.48	3.52	73.19		
Oryol	35.59	21.137	7.01	4.04	70.49		
Ryazan	96.99	28.278	12.57	5.79	102		
Smolensk	62.88	24.203	6.43	4.75	108		
Tver'	76.06	27.933	11.86	6.97	121.8		
Tula	121.49	34.624	12.85	8.15	155.2		
Yaroslavl'	129.61	21.166	14.2	6.52	130.2		

Figure 5.16. Part of the list of regions of the European part of Russia

whether an attribute is input or output. In Figure 5.17, the current and new values coincide since the movement of the point has not been carried out. The right-hand side of the display shows the slices of the PPS that pass through the point related to inputs/outputs of Chuvash Republic. Let us consider them more carefully.

The first row displays the slices that are related to the first output, that is, industrial production. The left slice shows possible values of industrial production and agricultural production (grey color in the black-and-white copy) that belong to the PPS if the values of both inputs are fixed at the current values for Chuvash Republic. The dark point at the slice describes the original position of Chuvash Republic in the PPS. Since it is desirable to increase both outputs, the upper right frontier is displayed.

The central slice describes possible values of industrial production and labor for the PPS. Since, for the sake of efficiency, it is desirable to increase industrial production and decrease labor application, the upper left frontier is shown. The right slice describes possible values of industrial production and funds for the PPS. Once again, the upper left frontier of the PPS is displayed for this output and this input.

The second row is related to the second output. Two slices are given that display the PPS frontier for agricultural production and both inputs

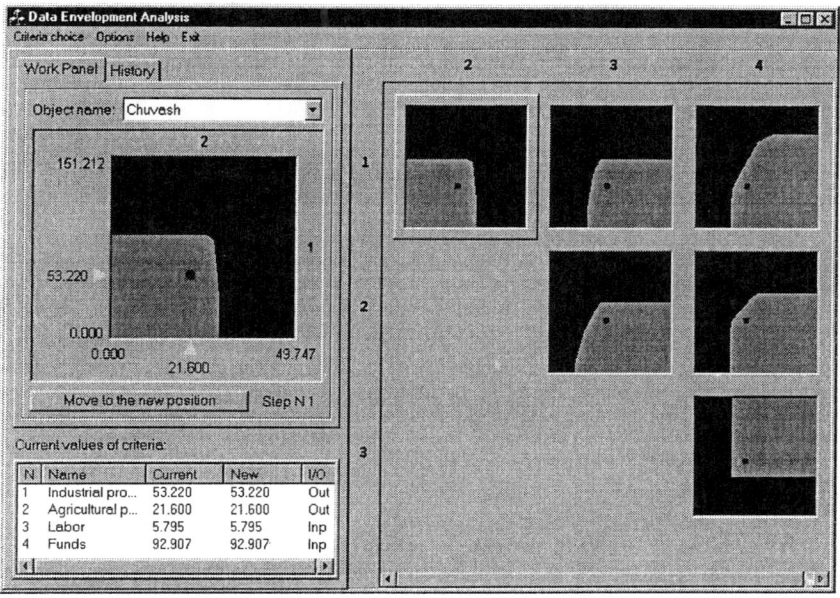

Figure 5.17. A white and black copy of the initial display of the DEA software: the current target values coincide with the original values. The slice for industrial production and agricultural production is selected for detailed study and identification of a current target.

(the slice for the pair provided by agricultural production and industrial production, which has already been given in the first row, is not repeated). The slice given in the third row displays possible values of both inputs. For this reason, the lower left frontier is displayed.

The dark points describe the current position of the unit in the PPS. To propose an improved position of the unit, the decision maker has first to select a pair of inputs/outputs, that is, one of the slices. In our example study, the slice related to both outputs was selected. The enlarged copy of this slice is given in the left-hand side window of the display. The decision maker has to identify new improved values for the selected outputs. To do it, he/she has to move the point given at the enlarged slice to a new position.

Figure 5.18 displays the situation after the movement has already been completed. In this figure, one can see the old point given by the cross and the new target given by the dark point in the enlarged slice. The dark point is located now at the frontier of the slice. It means that the decision maker has decided to identify the final target value in one step. Usually, such a result requires several movements displayed in

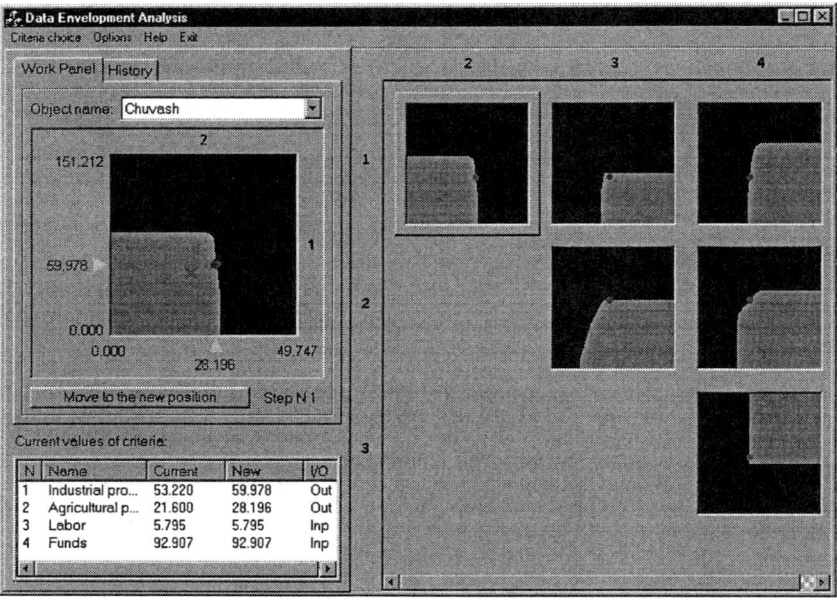

Figure 5.18. The result of movement of the point to a new position. The original
position is given by a cross in the left-hand side window

different slices, but we avoid them here for the sake of simplicity. The
new coordinates associated with both productions have changed, and
their old and new values can be compared in the tablet. Though the
values of inputs did not change, the slices have changed in such a way
that now the points approximately belong to the frontiers.

Since the process of target identification is completed, we can start the
search for the units (regions) that are close to the target input/output
given in the tablet in Figure 5.18, that is

industrial production=60.0, agricultural production=28.2, labor=5.8,
funds=92.9.

Several regions were selected by the software to be in line with this
target. However, only one of them turned out to be sufficiently close
to the target in the sense of production resources and output structure.
Therefore, it can be considered as a pretender for the role-model unit.
It is Ryazan' region, which is characterized by the numbers

industrial production=97.0, agricultural production=28.3, labor=5.8,
funds=102.0.

One can see that Ryazan' region has the same values as the target input/output, except its funds are about 10% greater and the industrial production is about 60% greater. Comparing with Chuvash Republic, original values for which are given in Figure 5.17, we can find that Ryazan' region can be used as the role-model unit.

APPENDIX 5.A: Mathematical formulation of the example model

The example multi-criteria non-linear problem studied in Section 1 has the following form. It is a decision problem with two decision variables and five criteria. The values of decision variables x_1 and x_2 satisfy the box constraints

$$x_1 \in [-4.9; 3.2], \quad x_2 \in [-3.5; 6].$$

The criteria are defined on the basis of the so-called peak function, very well known by optimization professionals,

$$f(x_1, x_2) = u_1(x_1, x_2) + u_2(x_1, x_2) + u_3(x_1, x_2) + 10,$$

where

$$u_1(x_1, x_2) = 3 (1 - x_1)^2 \, e^{-x_1^2 - (x_2 + 1)^2},$$

$$u_2(x_1, x_2) = -10 \left(\frac{1}{4} x_1 - x_1^3 - x_2^5 \right) e^{-x_1^2 - x_2^2},$$

$$u_3(x_1, x_2) = -\frac{1}{3} e^{-(x_1 + 1)^2 - x_2^2}.$$

The criterion functions are generated as follows:

$$f_i(x_1, x_2) = f(x_1 + \alpha_i, \ x_2 + \beta_i), \ i = 1, ..., 5.$$

The following values of the displacement values α_i and β_i were used:

$$
\begin{aligned}
\alpha_1 &= 0, & \beta_1 &= 0; \\
\alpha_2 &= -1.2, & \beta_2 &= -1.5; \\
\alpha_3 &= 0.3, & \beta_3 &= -3; \\
\alpha_4 &= -1, & \beta_4 &= 0.5; \\
\alpha_5 &= -0.5, & \beta_5 &= -1.7.
\end{aligned}
$$

The model exemplifies a problem related to the location of a monitoring station, while five different pollution scenarios were provided by experts.

APPENDIX 5.B: Mathematical description of the DEA/IDM technique

Let all of N production units use the same p production resources (inputs) and produce the same q products (outputs). In this case the space of inputs/outputs $\mathbb{R}^{p+q}$ can be considered. Any production unit is associated with a point in this space. Let the j-th unit be associated with the point

$$(x^j, y^j) = (x_1^j, x_2^j, \dots, x_p^j, y_1^j, y_2^j, \dots, y_q^j).$$

Then, the system of N production units is described by N points

$$\left\{ \left(x^1, y^1\right), \left(x^2, y^2\right), ..., \left(x^N, y^N\right) \right\}. \tag{5.B.1}$$

The production possibility set can be defined as the convex hull (envelope) of these points (the so-called empirical production possibility set)

$$T_E = \left\{ (x, y) : x = \sum_{j=1}^{N} \lambda_j x^j, \ y = \sum_{j=1}^{N} \lambda_j y^j, \ \lambda_j \geq 0, \ \sum_{j=1}^{N} \lambda_j = 1 \right\} \tag{5.B.2}$$

or, as a broader set,

$$T = \left\{ (x, y) : x \geq \sum_{j=1}^{N} \lambda_j x^j, \ y \leq \sum_{j=1}^{N} \lambda_j y^j, \ \lambda_j \geq 0, \ \sum_{j=1}^{N} \lambda_j = 1 \right\} \tag{5.B.3}$$

that includes all points of the space $\mathbb{R}^{p+q}$, which are characterized by greater inputs and (or) smaller outputs. Usually in DEA, the role of the production possibility set is restricted to illustration of the main concepts of DEA in the case of two inputs or outputs. In this chapter, the set T is approximated and visualized in the form of a collection of its two-dimensional slices.

II

COMPUTATIONAL METHODS OF THE IDM TECHNIQUE

Chapter 6

COMPUTATIONAL METHODS FOR STATIC MODELS

Computational methods developed in the framework of the IDM technique are introduced in Part II of the book. This chapter concentrates on methods for static models. Methods for dynamic models and distributed systems are provided in the next chapter. The mathematical basis of the computational methods is described here in a simplified way. Complicated mathematical issues of the theory of the methods are considered in Part III. However, in contrast to the previous chapters, certain mathematical background is required from the reader of this part of the book.

Let us start with the general mathematical statement of the IDM technique. We consider problems with a finite number of decision variables, namely n. It is assumed that the decision variables constitute a vector x that belongs to the linear decision space $\mathbb{R}^n$. Let the variety of feasible decision vectors be denoted by $X \subset \mathbb{R}^n$. Let us assume that criterion vectors z are composed of m coordinates (partial criteria) and are the elements of the linear criterion space $\mathbb{R}^m$. Criterion vectors z are related to decision vectors x by *objective functions*, which are given by the mapping

$$f : \mathbb{R}^n \to \mathbb{R}^m.$$

Then, the variety of feasible criterion vectors Z (known in the MCDM theory as the feasible criterion set, FCS) is defined as $f(X)$, i.e.,

$$Z = \{z \in \mathbb{R}^m : z = f(x),\ x \in X\}.$$

Let us assume that the user is interested in decreasing the criterion values. In this case a criterion vector z dominates (is better than) a criterion vector z', if and only if $z \leq z'$ and $z \neq z'$. This kind of domination is usually denoted as *Pareto domination*. The non-dominated criterion vector z is a feasible criterion vector, for which then does not exist another feasible vector z' that dominates it, that is,

$$\{z' \in Z : z' \leq z, \ z' \neq z\} = \varnothing.$$

The *Pareto frontier* $P(Z)$ of the variety Z is defined then as the variety of its non-dominated feasible vectors, that is

$$P(Z) = \{z \in Z : \{z' \in Z : z' \leq z, \ z' \neq z\} = \varnothing\}.$$

Other terms such as non-dominated frontier or Pareto, Pareto-efficient or Pareto-optimal set are used to denote the variety $P(Z)$, too.

If the user is interested in the Pareto frontier only, it may be reasonable to approximate and visualize another variety that is simpler than the variety Z but has the same Pareto frontier. The *Edgeworth–Pareto Hull* (EPH) of the variety Z is the broadest of such varieties. It is denoted by Z^* and defined as the variety of $\mathbb{R}^m$ that along with the vectors from Z contains non-feasible criterion vectors dominated by the vectors of Z. Since the vectors dominated by any vector z from $\mathbb{R}^m$ are given by the non-negative cone $z + R_+^m$, where R_+^m is the non-negative cone of $\mathbb{R}^m$, the EPH is given by

$$Z^* = \{z^* \in \mathbb{R}^m : z^* = z_1 + z_2, \ z_1 \in Z, \ z_2 \in R_+^m\},$$

or, in short,

$$Z^* = Z + R_+^m.$$

It is important that the non-dominated frontiers of the varieties Z and Z^* coincide, but the dominated frontiers vanish in Z^*. Therefore, the frontiers of the EPH have a simpler form and can be understood easier.

The IDM technique consists in approximating the variety Z^* (or, sometimes, the variety Z) using simple figures such as polytopes, balls, boxes and cones, and in its further display using various collections of its bi-criterion slices (cross-sections). To define the notion of the bi-criterion slice of the variety Z^*, we consider any pair of components of the criterion vector, say, u and v. Let us denote the vector of the remaining criteria by y. To define a slice of the variety Z^* in the plane of the criteria (u, v), one has first to specify the coordinates of the vector y. Let us assume that the values of y have been specified. Then, a bi-criterion slice $G(Z^*, y)$ of the variety Z^* for the given vector y is defined as

$$G(Z^*, y) = \{(u, v) : (u, v, y) \in Z^*\}.$$

A bi-criterion slice of Z is defined in the same way.

The EPH has several important properties that make its visualization more convenient than visualization of the variety Z. All these properties can be proven easily using the definition of the EPH. The most important property that has already been mentioned is that

PROPERTY 6.1 *Pareto frontiers of the varieties Z and Z^* coincide.*

It is important that the dominated frontiers of Z do not remain as the frontiers of the EPH (we do not speak about dominated criterion points of the so-called weak-efficient frontier). For this reason, only two direct lines that are parallel to axes are seen in addition to the Pareto frontier in a bi-criterion slice of the EPH.

An extremely important feature of the EPH is related to the monotonic dependence of its bi-criterion slices on coordinates of the vector y' used in the definition of a slice. Let us consider any vector y', for which the slice $G(Z^*, y')$ is not empty. It is possible to prove that the slice $G(Z^*, y') \subset G(Z^*, y'')$ in the case if $y'' \geq y'$, that is, y'' is not better, than y'. This property holds for bi-criterion slices that differ only by one coordinate of the vector y, too. Due to this feature, it is convenient for the user if such collections of slices of the variety Z^* are displayed, for which the value of only one coordinate of the vector y' changes. Such collections of slices are known as decision maps. This property of bi-criterion slices results in the following important property of decision maps:

PROPERTY 6.2 *The frontiers of the slices of a decision map never intersect (though they can touch each other sometimes).*

Due to such a feature, decision maps are used as the main form of visualization of the variety Z^* throughout the book.

Another important property of the EPH is related to the goals identified with the help of decision maps. Let us recall that, along with the feasible criterion vectors, the EPH includes non-feasible dominated vectors. So, it may happen that the user identifies a non-feasible vector z^* as a goal. However, it is not of great importance since the EPH has the following important property:

PROPERTY 6.3 *For any vector z^* that belongs to Z^*, there exists another vector z^{**} that belongs to Z and dominates z^*.*

It means that in the case of a non-feasible goal vector z^* one can find a better feasible vector z^{**} and an associated decision alternative. Therefore, the non-feasible criterion vectors that are included in the EPH and displayed by decision maps for the sake of simplicity, do not hinder to identification of the goals.

Let us describe how decision maps can be used for identification of the goal vector z^*. Exploration of multiple decision maps including their animation assists the user in developing an idea of the preferable feasible combination of the criterion values, i.e., an idea of the feasible goal vector. To identify a feasible goal directly on the decision map, the user has first to select a convenient decision map and a slice on it (by this the values y^* of all criteria except two are fixed). Then, the identification of a goal vector is reduced to fixation of the values (u^*, v^*) of two criteria given on the axes. It can be done by a click of the computer mouse. By this the goal vector $z^* = (u^*, v^*, y^*)$ is identified.

It is important to note that the problem of reconstructing a decision alternative that results in the identified criterion vector is not correctly posed − a minor disturbance of the goal may result in the existence of an infinite number of decisions or, in contrast, the decision may not exist at all. This fact is very important in the case of the IDM technique since the EPH is constructed approximately, and so the identified goal vector belongs to an approximate frontier. To solve this problem, we regard the identified goal z^* as the "reference point" (Wierzbicki, 1981), that is, the point that represents the user's preferences and can be used in the process of correctly posed procedures for selecting a neighboring non-dominated criterion vector and the associated decision. In the convex case, the concept of a "reference point" results in the following optimization task that can be used for computing an efficient decision alternative:

$$
\text{minimize} \quad \left\{ \max_{i=1,2,\ldots,m} (z_i - z_i^*) + \sum_{i=1}^{m} \varepsilon_i(z_i - z_i^*) \right\}
$$
$$
\text{subject to} \quad z = f(x), \ x \in X,
$$

where $\varepsilon_1, \ldots, \varepsilon_m$ are small positive parameters. Since the identified goal is close to the Pareto frontier, the solution of the task is an efficient decision, which output is close to the identified goal. Application of the concept of a reference point is especially useful in the linear case. If the models and/or objective functions are non-linear, different methods may be used, but in any case they result in decisions, which outputs are close to the identified goal.

Now let us consider a table that contains N rows and several columns. Any row of the table is associated with a decision alternative and any of the columns is associated with an attribute. Such tables may be constructed using a mathematical model of a decision problem under consideration. Alternatively, it may be based on other data as evidence, experimental results or expert estimations.

Let us assume that the user has specified m numerical attributes to be the decision criteria. Then, each row can be associated to a point of the m-dimensional linear criterion space $\mathbb{R}^m$. Criterion values for the row number j are described by the point z^j, which coordinates are $z_1^j, z_2^j, \ldots, z_m^j$. Since N rows are considered, we have N criterion points $z^1, z^2, \ldots, z^N$. Application of the IDM technique in this case is based on enveloping the variety of points, i.e., on constructing their convex hull defined as

$$\text{conv}\left\{z^1, z^2, \ldots, z^N\right\} \equiv \left\{z \in R^m : z = \sum_{k=1}^N \mu_k z^k, \ \mu_k \geq 0, \ \sum_{k=1}^N \mu_k = 1\right\}.$$

Since the set $Z_C = \text{conv}\left\{z^1, z^2, \ldots, z^N\right\}$ is convex, methods for approximation of convex bodies can be used for approximation of the Z_C and Z_C^*. A bi-criterion slice of Z_C^* passing through its point z^* is defined in the same way as for the feasible criterion set. Therefore, decision maps are defined in the same way and the IDM technique can be applied immediately. The specific features of selecting the decision alternatives that are close to the goal have already been discussed in Chapter 4.

Chapter 6 is devoted to approximation of the sets Z and Z^*. We start with approximating the set Z and only then turn to approximating Z^*. Three basic cases are considered. In the first case, we assume that X is a convex polyhedral set and the objective functions are linear. Such a situation is met in the case of linear models. Note that the set X may be not bounded. Constructing the set Z for such a system is based on the convolution of linear inequality systems proposed by J.B. Fourier as early as in the 19th century. This approach is described in Section 1.

In the second case considered in Section 2, the set Z is assumed to be convex and bounded. Polytopes are used to approximate it in this case. The computational methods are based on evaluating the support function of the set Z. Methods of this kind do not depend upon the way the support function is evaluated, and so they can be adapted to more general decision spaces.

Section 3 is devoted to methods for approximating the set Z^* for the first two cases. These methods are based on convolution or evaluating the support function.

In Section 4, the set Z is not assumed to be convex. This feature may result from the non-linearity of the objective functions and/or from the non-convexity of the set X. It is assumed that both sets X and Z are bounded. Collections of boxes are used for approximating the set Z. Approximation methods are based on simulation of random feasible decisions.

Approximating the set Z in the non-convex case is considered in Section 5. Collections of the non-negative cones with apexes located in feasible criterion points are used as the approximating set. In addition to simulation of random feasible decisions, local single-criterion optimization is used for obtaining the apexes of the approximating cones.

1.   Methods based on convolution of linear inequalities

In this section, computational methods for convolution of linear inequality systems are introduced, and then their application for constructing the set $Z = f(X)$ is described.

Formulation of the problem.   Once again, it is assumed that the set X is a polyhedral set in $\mathbb{R}^n$ and the mapping $f : \mathbb{R}^n \to \mathbb{R}^m$ is linear. The set $f(X)$ is a polyhedral set in this case. Let us assume that the set X is specified by a linear inequality system

$$X = \{x \in \mathbb{R}^n : Hx \leq h\} \tag{6.1}$$

where H is a given matrix and h is a given vector. The set X may be not bounded. The linear objective functions are given by mapping $f : \mathbb{R}^n \to \mathbb{R}^m$, which is assumed to be given by the matrix F,

$$f(x) = Fx. \tag{6.2}$$

We are going to construct the set

$$f(X) = \{z \in \mathbb{R}^m : z = Fx, \ Hx \leq h\} \tag{6.3}$$

in the form

$$Q = \{z \in \mathbb{R}^m : Dz \leq d\}, \tag{6.4}$$

where D is a matrix and d is a vector to be computed.

Constructing the set $f(X)$ by projecting polyhedral sets.   The idea to apply methods for projecting of polyhedral sets for constructing the set $f(X)$ was introduced in (Lotov, 1973b; Lotov, 1975a).

DEFINITION 6.4 *The (orthogonal) projection of the set $M \subset \mathbb{R}^p \times \mathbb{R}^q$ onto $\mathbb{R}^q$ is defined as the set*

$$M_w = \{w \in \mathbb{R}^q : \quad \exists v : (v, w) \in M\}.$$

To represent the set $f(X)$ in the form of a projection, let us consider the graph Y of the mapping (6.2) defined on the set (6.1), i.e.,

$$Y = \{(x, z) \in \mathbb{R}^n \times \mathbb{R}^m : \quad z = Fx, \ Hx \leq h\}. \tag{6.5}$$

Note that the set $f(X)$ given by (6.3) may be represented as

$$f(X) = \{z \in \mathbb{R}^m : \exists x \in \mathbb{R}^n : z = Fx, \ Hx \leq h\}.$$

Therefore,

$$f(X) = \{z \in \mathbb{R}^m : \exists x : (x, z) \in Y\}.$$

That is, the set $f(X)$ is the projection of the set Y onto $\mathbb{R}^m$. Because the set Y is specified by equations of the system (6.1–6.2), the task of constructing the set $f(X)$ reduces to constructing the projection of the polyhedral set Y from $\mathbb{R}^n \times \mathbb{R}^m$ onto $\mathbb{R}^m$. Recall that this projection must be constructed in the form (6.4).

Note that if $m = n$ and the matrix F is of a complete rank, one can easily find the set $f(X)$ by expressing the vector x in terms of z as

$$x = F^{-1}y$$

and by inserting this expression into the inequality system (6.1). Therefore,

$$f(X) = \{y \in \mathbb{R}^m : HF^{-1}y \leq h\}.$$

That is, the vector x may be excluded from (6.5) by an inversion of the matrix F.

However, when $m < n$, which is most often the case in applied problems of constructing the set $f(X)$, it is no longer possible to find this set in this simple way. The available equalities permit the elimination of only some of the coordinates of x from (6.5). In this case, the coordinates are expressed in terms of the remaining variables and can be inserted into the inequalities. To eliminate the remaining variables and, thus, to construct the desired projection of Y, one can use the methods for the convolution of systems of linear inequalities proposed by J.B. Fourier (Fourier, 1826).

The convolution methods help to construct the projection M_w of a convex polyhedral set $M \subset \mathbb{R}^p \times \mathbb{R}^q$ in the form (6.4), that is,

$$M_w = \{w \in \mathbb{R}^q : Dw \leq d\}. \tag{6.6}$$

To be precise they find the matrix D and the vector d of the description of M_w. Since the convolution methods are not widely known, we find it necessary to describe them briefly.

Convolution of linear inequality systems. It is assumed that the convex polyhedral set M is specified as

$$M = \{(v, w) \in \mathbb{R}^p \times \mathbb{R}^q : Av + Bw \leq c\} \tag{6.7}$$

where A, B are pre-specified matrices, and c is a pre-specified vector. It is needed to find a matrix D and a vector d of the description of the projection M_w in the form (6.6).

J.B. Fourier proposed the convolution method for eliminating the vector v from the finite system of linear inequalities (6.7) in a way that results in the constructing of the projection (6.6) of the set (6.7). The method starts with the elimination of the first coordinate of the vector v, that is, with constructing the projection of the set M onto $\mathbb{R}^{p-1} \times \mathbb{R}^q$. Then the process continues until all components of the vector v are eliminated and the required projection is constructed. So, in the framework of the Fourier convolution method, every next projection is found from the previous one by the elimination of one coordinate of the vector v.

Eliminating one component of the vector v is carried out by summation of pairs of inequalities. Fourier method resembles to some degree the method proposed by Gauss for solving systems of linear equations. However, Fourier method is a bit more sophisticated.

Fourier method for $p = 1$

In this case the system (6.7) may be recast as

$$a_i v + \langle b_i, w \rangle \leq c_i, \quad i = 1, 2, ..., N, \qquad (6.8)$$

where a_i are numbers and b_i are vectors. We break up all inequalities (and their numbers $i = 1, 2, ..., N$) into three groups, Π_+, Π_- and Π_0, in correspondence to the sign of the coefficient by the variable v. The system describing the projection includes:

- all inequalities from (6.8), which have zero coefficients at the variable v (that is, the inequalities that belong to Π_0);

- all possible linear combinations of pairs of inequalities with opposite signs of the coefficients, i.e., all inequalities

$$\langle a_j b_i - a_i b_j, w \rangle \leq a_j c_i - a_i c_j,$$

where $i \in \Pi_-$, $j \in \Pi_+$.

When excluding a greater number of variables by Fourier method, the procedure remains the same: on a current elimination step, the system, which is the result of elimination of the previous coordinate, is taken as the starting system.

Convolution of particular systems shows that number of inequalities resulting from the consecutive elimination of variables by Fourier method grows extremely fast. For this reason, the array of their coefficients soon overfills the memory of any existing computer. Fortunately, there is an opportunity to improve the situation considerably. It is based on the fact that many of the resultant inequalities are redundant, i.e., they follow from other inequalities of the system. Therefore, their elimination from the system does not affect the solution set. The ability to identify and eliminate the redundant inequalities is crucial for successful application of the convolution methods.

To show how a part of the redundant inequalities can be eliminated by convolution methods, we use the results of the theory of linear inequalities, in particular, Farkas Lemma and its corollaries.

Farkas Lemma and its corollaries. Farkas Lemma is a fundamental result of the theory of linear inequalities. Among other things, it provides the basis for the duality theory of linear programming. Farkas Lemma has two formulations: a geometrical formulation and an algebraic one. First, we introduce several concepts needed to formulate Farkas Lemma.

DEFINITION 6.5 *A polyhedral cone in $\mathbb{R}^N$ is the set*

$$K = \{b \in \mathbb{R}^N : b = A\lambda, \ \lambda \geq 0\},$$

where A is a matrix, which rows are provided by a collection of vectors $\{a_1, a_2, ..., a_N\}$.

DEFINITION 6.6 *The conjugate cone of a cone $K \subset \mathbb{R}^N$ is the set*

$$K^+ = \{u \in \mathbb{R}^N : \langle u, b \rangle \geq 0, \ b \in K\}.$$

For a polyhedral cone K specified by a matrix A, its conjugate takes the form
$$K^+ = \{u : uA \geq 0\}.$$

In turn, the conjugate of the cone K^+ may be defined as

$$K^{++} = \{w \in \mathbb{R}^N : \langle u, w \rangle \geq 0, \ u \in K^+\}.$$

LEMMA 6.7 (**Farkas**, GEOMETRIC FORMULATION) *For a polyhedral cone $K \subset \mathbb{R}^N$, it holds $K^{++} = K$.*

LEMMA 6.8 (**Farkas**, ALGEBRIC FORMULATION) *Let there be given a matrix A. For a vector b, there exists a vector $\lambda \geq 0$, such as $b = A\lambda$ if and only if for any solution of the system $uA \geq 0$ it holds that $\langle u, b \rangle \geq 0$.*

An obvious corollary of Farkas Lemma, convenient for an analysis of the convolution methods, is the lemma by Ky Fan (Ky Fan, 1956).

LEMMA 6.9 (***Ky Fan***)   *The system $Av \leq b$ has a solution if and only if for any solution of the system*

$$uA = 0, \ u \geq 0 \tag{6.9}$$

it holds that

$$\langle u, b \rangle \geq 0.$$

By the way, A.D. Aleksandrov (Aleksandrov, 1950) seems to be the first one to prove the above lemma and use this result.

Let U be a finite matrix whose rows are vectors, which generate a polyhedral solution cone for the system (6.9). Then, the lemma may be reformulated as follows.

LEMMA 6.10 (***Ky Fan***, ALTERNATIVE FORMULATION)   *The system $Av \leq b$ is compatible if and only if it holds that $Ub \geq 0$.*

Ky Fan Lemma can be used to describe the projection M_w of the polyhedral set (6.7) in the following way. Consider the system

$$Av \leq b,$$

where $b = c - Bw$. By the definition of a projection and according to Ky Fan Lemma, the system is compatible for a given w, that is, $w \in M_w$ if and only if

$$U(c - Bw) \geq 0.$$

Therefore,

$$M_w = \{w \in \mathbb{R}^q : UBw \leq Uc\}. \tag{6.10}$$

In accordance with (6.9), the elements of the matrix U are non-negative. Therefore, the elements of the matrix D and of the vector d can be derived upon multiplying by a positive number and summing up the rows of the original inequality system (6.7). This is what happens in Fourier method.

Note that there exist an infinite number of matrices U that can generate the solution cone for (6.9). However, matrices with a minimal number of rows do exist. Such matrices, denoted by U_f, are called the fundamental matrices of solutions for (6.9). Examples show that, for $p > 1$, the Fourier method does not usually result in such a system (6.10) that can be related to a fundamental matrix of solutions, that is, such that

$$D = D_f = U_f B, \quad d = d_f = U_f c.$$

Important ideas for constructing the matrix U_f were developed in (Motzkin, Raiffa, Thompson and Thrall, 1953), however, the ideas were suggested in a rather vague form and without a proper proof. In (Burger, 1956) the ideas were reformulated and the necessary proofs were provided. Using these ideas, S.N. Chernikov (Chernikov, 1965a; Chernikov, 1965b; Chernikov, 1968) invented several methods for the convolution of inequality systems, which make it possible to find D_f and d_f. The gist of these methods is as follows.

Chernikov methods. The methods for convolution of linear inequality systems are based on the result provided in (Chernikov, 1968, Theorem 5.1). Here, we describe the same ideas in a somewhat simplified form, which relies on the generally known concept of the affine dependence of vectors.

DEFINITION 6.11 *The vectors a_i, $i = 1, 2, ..., r$, are called* affine-dependent *if there exists a non-zero vector $\lambda \in \mathbb{R}^r$ such that*

$$\sum_{i=1}^{r} \lambda_i a_i = 0, \quad \sum_{i=1}^{r} \lambda_i = 0, \quad \exists \lambda_i \neq 0, \quad i = 1, 2, ..., r.$$

To apply this definition for the description of Chernikov convolution methods, we need the notion of extreme vectors of a cone. An extreme vector of a cone is one that cannot be represented as a linear combination with positive coefficients of two non-collinear vectors from the cone. Let us restrict ourselves to the pointed cones, i.e. the cones that do not contain subspaces of non-zero dimension. It is known that the minimal subset of vectors, which generate a pointed cone, coincides with the subset of its extreme vectors. The solution cone for (6.9) is a pointed cone because it belongs to the non-negative orthant. Therefore, to obtain the minimal description of the cone (6.9), it is needed to find the subset of its extreme vectors.

The necessarily and sufficient condition for a solution of (6.9) to be an extreme vector of the solution cone is given by the following lemma formulated and proved by O.L. Chernykh.

LEMMA 6.12 (*Chernykh*) *The vector u is an extreme vector of the solution cone for the system (6.9) if and only if the set of rows of the matrix A, corresponding to the positive coordinates of u, is affine-independent.*

Since the author of the lemma did not publish its proof, we have provided it in (Lotov, Bushenkov and Kamenev, 1999, pages 220–222).

The lemma shows that the extreme vectors of the solution cone for the system (6.9) can be identified by testing for affine independence those rows of the matrix A that enter into a linear combination with positive coefficients.

S.N. Chernikov (Chernikov, 1968) proposed to solve the problem of computing the matrix D_f and the vector d_f by the convolution methods in the following way. Note that any inequality of the system (6.7) can be associated with its number in the system. Let us denote by I_i the set of inequalities of the original system (6.7) that have entered into a linear combination upon the formation of the i-th inequality of the system (6.6). This set of numbers is *the index* of the i-th inequality of the system (6.6).

The following rule for identifying redundant inequalities in the system (6.6) can be used: if

$$I_i \subset I_j, \tag{6.11}$$

then the rows of the matrix A with numbers from I_j are affine-dependent. Therefore, the j-th inequality is redundant.

The rule (6.11) can be derived from the results given in the book (Chernikov, 1968). On the other hand, it follows immediately from Chernykh Lemma. The rule (6.11) helps to derive a new method based on Fourier convolution. The new method consists in testing the rule (6.11) at any step of elimination of a component of vector v. The new method is known as the *fundamental convolution method* (Chernikov, 1968). We do not describe the method in detail. Instead, we consider another method that is used in our software for constructing the projections of polyhedral sets. The method also was proposed by Chernikov and is known as the *reduced fundamental convolution method* (Chernikov, 1968). The idea of this method is as follows.

It is easy to show that a matrix of p columns cannot contain more than $p + 1$ affine independent rows. Therefore, the number of elements in the index of the j-th inequality of the system (6.6) denoted by $|I_j|$ can be used for a preliminary test of its redundancy: if

$$|I_j| > p + 1,$$

then the j-th inequality is redundant. Hence, the condition (6.11), which requires a comparison of the indices of pairs of inequalities, should preferably be preceded by testing the number of elements in its index. This is the main idea of the reduced fundamental convolution method.

Reduced fundamental convolution method. An elimination step of the reduced fundamental convolution method can be described as follows. Suppose we have eliminated the variables v from the system (6.7) and derived a system of the type (6.6). In doing so, for each i-th inequality of the system (6.6), we retain its index I_i. Suppose that the further elimination of one more variable from the system (6.6) is required. As with Fourier method, we break up all inequalities into three groups, Π_+, Π_- and Π_0. Now, we include all inequalities from Π_0 in the new system (index is not changed). We also include in the new system the linear combinations of pairs of inequalities, one of which is taken from Π_+, and the other, from Π_-, however, in contrast to Fourier method, we limit ourselves to those combinations whose indices satisfy a condition defined below. Let us consider two inequalities $i \in \Pi_+$ and $j \in \Pi_-$ from the system (6.6):

$$b_i^1 w_1 + \sum_{k=2}^{q} b_i^k w_k \le c_i,$$

$$-b_j^1 w_1 + \sum_{k=2}^{q} b_j^k w_k \le c_j.$$

Here both b_i^1 and b_j^1 are positive. We include the new inequality

$$\sum_{k=2}^{q} (b_j^1 b_i^k + b_i^1 b_j^k) w_k \le b_j^1 c_i + b_i^1 c_j$$

with the index $(I_i \cup I_j)$ in the new system if and only if:

1 $|I_i \cup I_j| \le p + 2$;

2 there is no k-th inequality of the system (6.6), such that $I_k \subseteq I_i \cup I_j$.

This completes the steps of the method. Note that the first condition is given in a simplified form; actually, it is possible to make it more effective (Chernikov, 1968).

Discussing convolution methods. Note that, although the number of resultant inequalities is much smaller in the methods proposed by Chernikov than it is in Fourier method, eliminating the variables may still result in an exponential growth of the number of inequalities. This fact poses difficulties in using convolution methods in real-life problems, even in the case where the number of decision variables is relatively small.

Let us consider a typical example. In a practical decision problem (not described here) it was necessary to eliminate six variables from a system that contains 36 original inequalities. Two alternative variants of input data were available, which differed solely in several coefficients of the matrix A. With the first data set, the number of inequalities exceeded 500 after the elimination of the third variable, and the computation was terminated. With the second data set, it was possible to eliminate all six variables, and the number of the inequalities did not exceed 500. This example shows that the efficiency of convolution methods depends substantially on the values of elements of the matrix A. So, it seems to be reasonable to change the elements of the matrix a bit to decrease the number of inequalities. The influence of modification of elements of the matrix A on the solution set of the system is studied in Chapter 9. It is interesting that the results of Chapter 9 can also be used for estimation of the influence of rounding errors on the stability of the Fourier convolution (Lotov, 1986).

The experience gained in using convolution methods in real-life problems shows that these methods can guarantee the final result only for systems of a very small dimension (say, with 10 variables and 20 inequalities). At the same time, there are practical tasks where convolution methods are effective because the elimination of variables results in minor growth of the number of inequalities. The order in which variables are eliminated also has a strong influence on the progress of a convolution process. This implies that the desired result can sometimes be obtained through a change in the elimination order.

Let us stress that application of the reduced fundamental convolution method solves the problem of redundant inequalities only partially. Even the minimal matrix U_f usually generates redundant inequalities in the description of the set M_w. Indeed, the matrix U_f describes the edges of the solution cone for (6.9). This cone depends on the matrix A only. Since the matrix U_f must include rows that help to describe the projection of the set M for *any* B and c, a great number of redundant constraints does arise in constructing a projection for specific B and c.

To extend the applicability of convolution methods, it has proved necessary to invoke methods for identification of redundant inequalities in linear systems. A mathematical statement of the problem is as follows. In a finite system of linear inequalities

$$\langle a_i, x \rangle \leq b_i, \ i = 1, 2, ..., N,$$

where $a_i \in \mathbb{R}^p$ are the given coefficient vectors, and b_i are the given numbers, it is necessary to identify a subsystem that has the same solution set, but that does not contain redundant inequalities. The problem of eliminating redundant inequalities in linear systems has been known for a long time (see, for example, Karwan, Lotfi, Telgen and Zionts, 1983). Most methods are based either on iterations of the prime or dual problems of linear programming or on some heuristic techniques. We have also developed such methods and used them to construct sets of the type (6.3) in practical tasks.

We have found that there is no way of finding a universal method for elimination of the redundant constraints that is effective in most of the problems we have studied. Some methods identify practically all redundant inequalities, but they require an enormous amount of time; others operate quickly, but they do not give any information about how many redundant inequalities still remain. An original convolution-based method was proposed in (Bushenkov and Lotov, 1980a) that can study the redundancy of inequalities in the case of parametric right-hand sides.

To make the convolution methods more effective, algorithms were proposed that, along with redundant inequalities, can exclude inequalities that are "quasi-redundant" (Lotov and Ognivtsev, 1980). To be precise, an inequality is quasi-redundant, if its exclusion from the inequality system results in a minor change of its solution set. The papers (Bushenkov and Lotov, 1980b; Bushenkov and Lotov, 1982) give more details on application of such techniques.

Speaking about possible application of convolution methods for constructing the set $Z = f(X)$, one can state that each problem calls for separate experimenting using various orders for the elimination of variables, as well as various methods for the elimination of redundant and quasi-redundant inequalities. Therefore application of convolution methods cannot be completely automatic, and the need arises for intervention by a human operator. For convenience, a package of programs called POTENTIAL was developed (Bushenkov, 1982; Bushenkov and Lotov, 1984) in the beginning of the 1980s. It offered a way to solve several applied problems. Using the POTENTIAL package, the following record result was obtained (take into account the volume of computer memory in the beginning of the 1980s). A model was studied that consisted of three blocks, each with 12 decision variables and 18 equalities and inequalities. The coordinating problem had four additional variables and 17 global constraints. The set $f(X)$ for seven criteria that was constructed included 180 inequalities. Moreover, the set $f(X)$ for a finite-difference approximation of controlled boundary problem in partial derivatives was constructed by using the POTENTIAL package (Lotov and Stolyarova, 1984).

Modern hardware provides new opportunities for implementation of convolution methods. For example, these methods were applied recently in a dynamic control problem (Bushenkov and Smirnov, 1992; Bushenkov and Smirnov, 1997).

A major advantage of convolution methods is the opportunity to construct the sets $f(X)$ for systems with large numbers of criteria. Another advantage consists in the opportunity to construct the sets $f(X)$ for non-bounded sets X. However, it is clear that convolution methods have multiple disadvantages. First of all, it is obvious that one cannot apply these methods to real-life models that involve hundreds of decision variables. Another important disadvantage of convolution methods consists in the following: in cases where the convolution process cannot be carried out to its completion, the effort fails to yield even an approximate result.

Note that the elimination of almost redundant inequalities from a system describing a projection leads to a fundamentally new situation; the task of constructing the precise projection is replaced by the task of approximating the projection. It is logical to ask, if one is ready to end up with an approximation, why he should strive for the precise projection rather than construct an approximation from the very beginning in an optimal way (e.g., by seeking an approximating polytope, which has minimal number of vertices or faces for a specified accuracy). Such an approach is used in the Projection Estimate Refinement (PER) method proposed in (Bushenkov, 1981) and described in detail in (Bushenkov, 1985). The PER method helped to overcome the principal drawback of convolution methods, that is, the absence of an approximate estimate of the projection in cases where it is impossible to complete the task.

Projection Estimate Refinement method. In the PER method, projection M_w of the polytope $M \subset \mathbb{R}^{p+q}$ is approximated by a sequence of polytopes $P^0, P^1, ..., P^k, ...$, that tend to M_w. The sequence of polytopes is constructed in an iterative way, beginning from P^0, upon solving auxiliary convolution and optimization problems.

Prior to the $(k+1)$-th iteration of the PER method, a polytope P^k must be specified, such that
  (a) vertices of P^k belong to the frontier of M_w; and
  (b) P^k is specified in two forms simultaneously, namely:
    (b$_1$) as the convex hull of vertices $\{w^{(1)}, w^{(2)}, ..., w^{(r)}\}$;
    (b$_2$) as the solution set of the system of linear inequalities

$$P^k = \{w \in \mathbb{R}^q : \langle c_j, w \rangle \le d_j, \; j = 1, 2, ..., N\} \qquad (6.12)$$

where $c_j \in \mathbb{R}^q$ and $d_j \in \mathbb{R}^1$.

Transition to P^{k+1} consists of the following steps. For faces of P^k, which are specified by vectors c_j and numbers d_j, the following N linear optimization tasks are solved:

$$\left. \begin{array}{c} \text{maximize } \langle c_j, w \rangle \\ \text{subject to } (v, w) \in M \end{array} \right\}, \; j = 1, 2, ..., N. \qquad (6.13)$$

From the solutions $(v_{(j)}, w_{(j)})$ of (6.13), a solution with maximal distance of the point $w_{(j*)}$ from P^k is selected. Then we put

$$P^{k+1} = \text{conv}\{w^*, P^k\},$$

where $w^* = w_{(j*)}$ and construct the polytope P^{k+1} in the form (6.12). This completes the $(k+1)$-th iteration of the method.

The algorithm for constructing the polytope P^{k+1} in the form (6.12) that was proposed in the framework of the PER method deserves a special discussion. The problem consists in constructing the convex hull of a polytope, which is given as the solution set of the linear inequality system (6.12), and of a point w^*. The convex hull must be constructed in the form of the solution set of a linear inequality system. To be precise, a linear inequality system must be constructed, the solution set of which coincides with the above convex hull. The method has the following steps:

1 Inequalities of the system (6.12) violated by the point w^* are detected;

2 Vertices $w^{(i)}$, $i = 1, 2, ..., L$, of the polytope P^k are detected, which belong to the faces related to the violated inequalities;

3 The cone $K \subset \mathbb{R}^q$, specified by edges $w^{(i)} - w^*$, $i = 1, 2, ..., L$, is constructed in the form of the solution set for a system of linear inequalities (which is equivalent to constructing the edges of the conjugate cone K^+);

4 In the inequality system (6.12), the inequalities violated by the point w^* are eliminated; new inequalities are included, which were found at Step 3.

To construct a conjugate cone, the method applies the methods for the convolution of linear inequality systems. Using our experience in the convolution of systems of linear inequalities and the appropriate computer software, it was possible to use the PER method as a powerful tool for approximating the projections of polyhedral sets. Note that in problems related to approximating the set $f(X)$ the conjugate cone is constructed in the criterion space $\mathbb{R}^m$, but not in the original decision space $\mathbb{R}^n$. Since the number of criteria m is relatively small, it is possible to approximate the set $f(X)$ for problems with a large number of decision variables.

On the other hand, it is clear that the basic ideas of the PER method can be used for invention of methods for solution of a more general problem — the problem of iterative approximation of bounded convex sets by polytopes. The only requirement that exists is that one must be able to solve optimization problems with linear criteria for the approximated body. Such methods could be used for approximation of the set $f(X)$ in the convex case, too. A class of methods for polyhedral approximation of convex bounded bodies is described in Section 2. These methods use the following ideas of the PER method:

(a) approximating polytopes are constructed iteratively;

(b) transition from P^k to P^{k+1} consists of two independent steps. The first step is related to optimization tasks of (6.13) type, and the second one is concerned with the construction of conv$\{P^k, w^*\}$ in the form of (6.12) using convolution methods.

2.   Methods for polyhedral approximation of convex bodies

In this section, we describe iterative methods for approximation of compact convex bodies (CCB) by sequences of polytopes. Such methods can approximate a CCB with any given accuracy. These methods are used as the main approximation tool in the framework of the IDM technique.

The problem of polyhedral approximation of compact convex bodies is a classic problem of applied mathematics (Gruber and Wills, 1993). Let $\mathbb{R}^m$ be Euclidean space with the distance $\rho(z', z'')$ for $z', z'' \in \mathbb{R}^m$. By compact bodies we understand closed bounded sets with non-empty interior. Methods for polyhedral approximation of a convex compact body C from $\mathbb{R}^m$ are based on evaluation of its *support function* for various directions u, i.e., on computing the values of the function

$$g_C(u) = \text{conv}\{\langle u, z \rangle : z \in C\},$$

where it is convenient to assume that the directions u belong to the unit sphere of directions

$$S = \{u \in \mathbb{R}^m : \langle u, u \rangle = 1\}.$$

It is assumed that, for the approximated set C, we are able to compute the values of the support function for any direction $u \in S$. It is usually true in the IDM technique, in the framework of which we have to approximate the set $Z = f(X)$, that is,

$$g_Z(u) = \max\{< u, z >: z = f(x), \ x \in X\}.$$

Various optimization techniques can be used to solve this problem for particular sets X and mappings f. If the model is linear, linear programming techniques can compute the values of the support function.

Sometimes it is proposed to compute the support function for a given finite system of directions (for the so-called *a priori* grid $\{u^1, u^2, \ldots, u^L\}$) and then use this information for constructing the approximating polytope. However, it is clear that the above grid neglects the actual shape of the body being approximated. For this reason, the methods based on the *a priori* grids are not the best ones. This assertion was illustrated experimentally in (Samsonov, 1983). In (Sonnevend, 1983) it was proven that the methods that use *a priori* grids are not optimal. They require too many evaluations of the support function; polytopes constructed by them have too many vertices and faces. In this book, a different approach is used − we apply iterative adaptive methods. In the framework of such methods, directions are specified on the basis of the information obtained on the previous iterations. Usually, this information is provided by the description of the current approximating polytope. We start with introduction of the concept of iterative methods for polyhedral approximation of convex bodies.

Iterative methods of polyhedral approximation of convex bodies. By an *iterative method* for the polyhedral approximation of a body C we mean a method for constructing a sequence of convex polytopes

$$P^0, P^1, \ldots, P^k.$$

We consider the sequences for which the number of vertices of polytopes $P^0, P^1, \ldots, P^k$ is increased by one on an iteration. Every next polytope must be constructed on the basis of the previous one using procedures of computing the support function for the set C.

We require that a sequence P^k approximates the body C, i.e.,

$$\lim_{k \to \infty} h\left(P^k, C\right) = 0,$$

where $h(\cdot, \cdot)$ is Hausdorff metric, that is,

$$h(C_1, C_2) = \max\{\sup\{\rho(x, C_2) : x \in C_1\}, \ \sup\{\rho(x, C_1) : x \in C_2\}\},$$

where $\rho(x, C) = \inf\{\rho(x, x') : x' \in C\}$. The ability of sequences of polytopes to approximate convex compact bodies to any degree of accuracy is an important advantage of them in comparison with the methods based on approximation by a single body of a specific form such as a simplex, a parallelepiped, or an ellipsoid. However, a high price has to be paid for this advantage. As both practice and theory show, the complexity of an approximating polytope rapidly increases if accuracy of approximation and dimension of the body increase. Nevertheless, we need to construct polytopes that approximate the body fairly accurately, since the shape of the frontier of the approximated body is of interest in the IDM technique. Therefore, we look for methods that have an optimal complexity of approximating polytopes and are optimal in respect to the number of evaluations of the support function. Such methods are described in this section.

We consider polytopes whose vertices belong to the boundary ∂C of the approximated body C. Such class of polytopes is denoted by $\boldsymbol{P}(C)$. Another class of polytopes denoted as $\boldsymbol{Q}(C)$ is provided by convex polytopes with faces touching ∂C. Methods for iterative construction of sequences of polytopes used in this book are based on two general schemes: the augmentation scheme and the cutting scheme.

Augmentation Scheme. Let $P^k \in \boldsymbol{P}(C)$. The $(k+1)$-th iteration involves two steps:

Step 1. Select a point $\boldsymbol{p}^* \in \partial C$.

Step 2. Construct $P^{k+1} = \operatorname{conv}\{\boldsymbol{p}^*, P^k\}$ in the required form.

Methods based on the augmentation scheme have the following property. Let the initial polytope P^0 belong to $\boldsymbol{P}(C)$. Then the polytopes P^k belong to $\boldsymbol{P}(C)$ for any k. Denoting the number of vertices for a polytope $P \in \boldsymbol{P}(C)$ by $N(P)$, we obtain

$$N(P^k) \leqslant N(P^0) + k.$$

Cutting Scheme. Let $Q^k \in \boldsymbol{Q}(C)$. The $(k+1)$-th iteration involves two steps:

Step 1. Select a direction $\boldsymbol{u}^* \in S$.

Step 2. Construct $Q^{k+1} = Q^k \cap \{\boldsymbol{z} \in \mathbb{R}^m : \langle \boldsymbol{u}^*, \boldsymbol{z} \rangle \leq g_C(\boldsymbol{u}^*)\}$.

Various methods implement these schemes. These methods differ in the way points or directions are selected, what kind of polytope description is used and how this description is constructed.

Adaptive methods. The *adaptive methods* for approximation of compact convex bodies by polytopes compute the support function for the directions identified by using information on the current approximation. Say, in the framework of an adaptive method based on the augmentation scheme, the choice of the point $\boldsymbol{p}^* \in \partial C$ is based on the information on the polytope P^k. Actually, the choice of $\boldsymbol{p}^* \in \partial C$ is adapted to the shape of the body C to the same extent as P^k approximates C.

To the extent of our knowledge, the first adaptive method was proposed in (Cohon, 1978). The method called the Non-Inferior Set Estimation (NISE) method was proposed in 1970s for $m = 2$ and $m = 3$, but it has been implemented only for $m = 2$ that time. To be precise, the NISE method was used for approximation of a part of the frontier of a body, namely, the Pareto frontier of the feasible criterion set. In the

framework of the bi-criterion NISE method, the approximation is given
by a number of points of the frontier and by line segments connecting
the neighboring points. To evaluate the quality of an approximation,
several optimization tasks must be solved: for any line segment, the
point of the frontier is found that maximizes the deviation from the line
segment in the direction given by the vector that is orthogonal to the
line segment. It is clear that these tasks are equivalent to evaluation of
the support function for such directions. If the deviations are sufficiently
small, the approximation process is completed. In the opposite case, the
most distant point of the Pareto frontier is selected among the optimal
solutions of the above optimization tasks. It is entered into the list of
approximating points, the related segment is deleted and two new line
segments are introduced that intersect at the new point of the list. This
operation completes an iteration of the bi-criterion NISE method. In
the case of $m = 3$, approximation by a system of faces was proposed
instead of a system of line segments.

The method proved to be effective in real-life problems with $m = 2$.
However, its further development for $m = 3$ required substantial effort
and time (Solanki, Appino and Cohon, 1993). This fact is related to
complications related to dealing with faces of the Pareto frontier in the
case of $m > 2$.

The first effective *multi-dimensional adaptive* method for approximating of convex bodies, the Estimate Refinement (ER) method, was introduced in the beginning of the 1980s as the result of integration of the
ideas of the NISE method and of the PER method described in the previous section. The ER method can be considered as the method that
applies the augmentation scheme. Step 1 of the ER method, that is,
the way the point $p^* \in \partial C$ is selected, implements the idea of the NISE
method: one has to evaluate the support function for directions given
by faces of the approximation and to select, among the points found in
this way, the point that is the most distant from the associated face.
However, implementation of this idea is different in the ER method as
compared with the NISE method. This difference of the ER method is
related to the main feature of the GRS method in general (see Lotov,
1971; Lotov, 1975b; Lotov, 1980; Lotov, 1981a) and the FGM in particular. It consists in the fact that the whole body is approximated in the
GRS method instead of a part of its frontier in the NISE method. Due to
it, approximation by polytopes is used in the ER method instead of approximation by a system of faces. What is more, as has been proposed in
the PER method, sequences of polytopes are used, which are described
both as solution systems of linear inequalities and systems of vertices
(the so-called double description of the polytope). This feature makes

the task of constructing the convex hull of P^k and $\boldsymbol{p}^*$ at Step 2 of the augmentation scheme much easier, since the algorithms for convolution of linear inequality systems can be applied.

The ER method was introduced in this form in (Lotov, 1985a). At the first stages of the development of the ER method (in 1982–1985), the algorithm of Bushenkov (described in the previous section) was used at Step 2 of the ER method. However, a new robust algorithm for solving this task was developed by O. Chernykh (see Chernykh, 1986; Chernykh, 1987; Chernykh, 1988). This algorithm is used now in the framework of the ER method. The reasons for it are described in the second part of this section. O. Chernykh coded the software of the ER method in 1987. Finally, G. Kamenev re-formulated the ER method in the form of the augmentation scheme (Kamenev, 1986b).

Estimate Refinement (ER) method. We assume that the body C is approximated by the ER method. Let us consider the $(k+1)$-th iteration of the method. Prior to the $(k+1)$-th iteration of the ER method, the polytope $P^k \in \boldsymbol{P}(C)$ has to be constructed in two forms simultaneously: as the solution set of a linear inequality system and as a list of vertices. In the recent form of the ER method it is assumed that for any of the vertices of the polytope it is specified, to which hyperplanes it belongs. As usual in methods that apply the augmentation scheme, the $(k+1)$-th iteration of the ER method consists of two steps. At Step 1, for all directions given by the linear inequalities that describe the polytope $P^k \in \boldsymbol{P}(C)$, the support function is evaluated. To be precise, the support function is evaluated for the finite list of outer normals to hyperfaces of the polytope P^k, which is denoted by $U(P^k)$. Then, such a direction $\boldsymbol{u}^* \in U(P^k)$ is selected, for which the difference $g_C(\boldsymbol{u}) - g_{P^k}(\boldsymbol{u})$ is maximal. If this value is sufficiently small, the approximation process is completed. In the opposite case, Step 2 is started, in the framework of which the convex hull of the polytope P^k and the point $\boldsymbol{p}^* \in \partial C$ is constructed, where the point $\boldsymbol{p}^*$ that satisfies $\langle \boldsymbol{u}^*, \boldsymbol{p}^* \rangle = g_C(\boldsymbol{u}^*)$ is found in the process of evaluating the support function for the direction $\boldsymbol{u}^*$. The convex hull of the polytope P^k and the point $\boldsymbol{p}^*$ is used for the new approximation P^{k+1} of the body C. The polytope P^k is constructed in two forms: as the solution set of a linear inequality system and as a list of vertices, while for any of the vertices it is specified, to which hyperplanes it belongs.

The illustration of an iteration of the ER method for $m = 2$ is given in Figure 6.1. The approximating polytope is given by the triangle. The direction $\boldsymbol{u}^*$ was found on Step 1 of the iteration as the direction of the maximal value of $g_C(\boldsymbol{u}) - g_{P^k}(\boldsymbol{u})$ among three directions defined by the

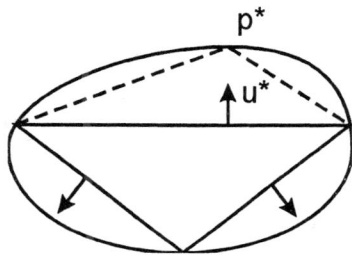

Figure 6.1. An iteration of the ER method for $m = 2$.

edges of the triangle. The new vertex p^* associated with the direction u^* is included in the description of the approximation on Step 2.

In Chapter 8 the ER method is given in the form of the augmentation scheme, which does not contain the description of the implementation of the method, but describes only those features that are important for study of convergence of the method.

Surely, the initial polytope $P^0 \in \boldsymbol{P}(C)$ must be specified to start the first iteration of the method. Its construction is discussed in (Chernykh, 1991; Chernykh, 1992), and we do not dwell on the matter here.

The ER method, as the PER method, has the following important property: evaluation of the support function for all directions given by hyperfaces of the polytope P^k provides a rough estimate of the distance $h(P^k, C)$. Indeed, the polytope

$$\hat{P}^k = \left\{ y \in \mathbb{R}^m : \langle u, z \rangle \leq g_C(u), \ u \in U(P^k) \right\}$$

contains the body C. Due to it, we have internal and external estimates for the approximated body C at any iteration

$$P^k \subset C \subset \hat{P}^k.$$

Therefore, it is possible to evaluate the distance $h(P^k, C)$ visually, by depicting two-dimensional slices of the polytopes P^k and $\hat{P}^k$. This property of the method can be used in real-life problems for human decision to complete the approximation procedure.

At the same time, the external polytope $\hat{P}^k$ plays a passive role in the ER method. It helps to evaluate the distance $h(P^k, C)$, but does not influence the iterations of the method. However, an interactive method for polyhedral approximating the multiple-dimensional compact convex bodies does exist, in the framework of which external polytopes play an active role. It is the Mutually Converging Polytopes (MCP) method proposed in (Kamenev, 1986b).

Method of Mutually Converging Polytopes (MCP). We provide the MCM method in the form of the augmentation scheme, since the features of its implementation to a large extent coincide with the ER method. Prior to the $(k+1)$-th iteration of the method, we should have two polytopes constructed: the polytope $P^k \in \boldsymbol{P}(C)$ and the polytope $Q^k \in \boldsymbol{Q}(C)$, both in the form of the solution set of linear inequality systems. Then, the iteration consists of two steps:

Step 1. Find $\boldsymbol{u}^* \in U(P^k)$ which solves
$$\max \left\{ \left(g_{Q^k}(\boldsymbol{u}) - g_{P^k}(\boldsymbol{u}) \right) : \boldsymbol{u} \in U(P^k) \right\};$$
compute a point $\boldsymbol{p}^* \in C$ such that $\langle \boldsymbol{u}^*, \boldsymbol{p}^* \rangle = g_C(\boldsymbol{u}^*)$.

Step 2. a) Find $U(P^{k+1})$ for $P^{k+1} = \operatorname{conv}\{\boldsymbol{p}^*, P^k\}$
upon constructing a linear inequality system, which solution set coincides with $P^{k+1} = \operatorname{conv}\{\boldsymbol{p}^*, P^k\}$;
b) Let
$$Q^{k+1} = Q^k \cap \{\boldsymbol{y} \in \mathbb{R}^m : \langle \boldsymbol{u}^*, \boldsymbol{z} \rangle \le g_C(\boldsymbol{u}^*)\}.$$

The vertices of the original polytope $P^0 \in \boldsymbol{P}(C)$ are assumed to be located on the faces of the original polytope $Q^0 \in \boldsymbol{Q}(C)$. For this reason, the resultant polytopes also possess this property.

The illustration of an iteration of the MCM method for $m = 2$ is given in Figure 6.2. Once again, the approximating internal polytope is given by the triangle. However, the external polytope is depicted as well (the external triangle in the figure). The direction $\boldsymbol{u}^*$ is found on Step 1 of the iteration as the direction of the maximal value of the distance between two polytopes, i.e., the value of $g_{Q^k}(\boldsymbol{u}) - g_{P^k}(\boldsymbol{u})$, among all directions defined by the edges of the internal triangle. The new vertex $\boldsymbol{p}^*$ associated with the direction $\boldsymbol{u}^*$ is included in the description of the approximation on Step 2. The external polytope is changed as well — a new plane (supporting hyperplane in the point $\boldsymbol{p}^*$) is included in its description.

As one can see, the MCP method is based on the augmentation scheme. However, due to the active role of the external polytope Q^k, the MCP method requires only one evaluation of the support function for the body C per iteration! In this sense, it differs from the ER method, which is related to a fairly large number of such evaluations. The direction $\boldsymbol{u}^*$, for which the support function of the body C is computed, is selected on the basis of the polytope Q^k. Due to this, the direction can be found faster than in the case of the ER method. However, it can result in an inefficient selecting of the new vertex. Experiments support the anxiety. To avoid the problem, the Modified MCP (MMCP) method was proposed (Bourmistrova, 2000a; Bourmistrova, 2000b). The modification looks fairly simple: one has to introduce a threshold β, $0 < \beta \le 1$,

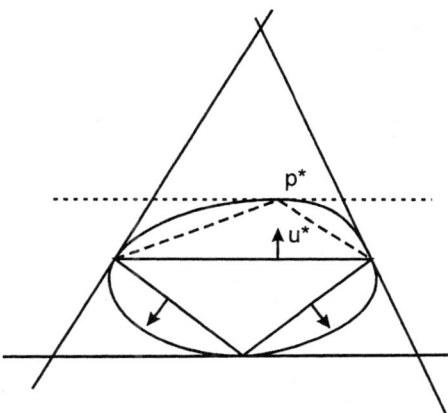

Figure 6.2. An iteration of the MCM method for $m = 2$.

and to check the inequality

$$g_C(u^*) - g_{P^k}(u^*) > \beta \left(g_{Q^k}(u^*) - g_{P^k}(u^*) \right) \qquad (6.14)$$

before Step 2 of the method is started. The condition (6.14) means that the potential vertex $p^* \in C$ is sufficiently distant from the polytope P^k. If the condition (6.14) is satisfied, the new vertex p^* is included in the polytope P^k and the related inequality $\langle u^*, z \rangle \le g_C(u^*)$ is included in the description of the polytope $Q^{k\cdot}$. In the opposite case, the vertex p^* is not included in the polytope P^k, but the related inequality $\langle u^*, z \rangle \le g_C(u^*)$ is still included in the description of the polytope $Q^{k\cdot}$. Then, the process returns to Step 1. The iteration is completed only after such a direction u^* is found that satisfies (6.14). So, in contrast to the MCP method, the support function of the body C can be evaluated several times at an iteration of the MMCP method, and several new inequalities may be included in the description of the polytope $Q^{k\cdot}$.

Note that for $\beta = 0$ the modified MCP method coincides with the MCP method; on the other hand, in the case of $\beta = 1$, the generated sequence of polytopes coincides with the polytope sequence constructed by the ER method. However, the ER method does not require constructing of the sequence of the polytopes $Q^{k\cdot}$ and searching for the best direction u^* for the polytope $Q^{k\cdot}$. By the way, the last problem is not so simple as it looks at first glance. Though the polytopes of $Q^{k\cdot}$ type belong to the space of the small dimension m, they are described by linear inequality systems that usually have only a small number of zeros among their coefficients. For this reason, solution of optimization problems at Step 1 of

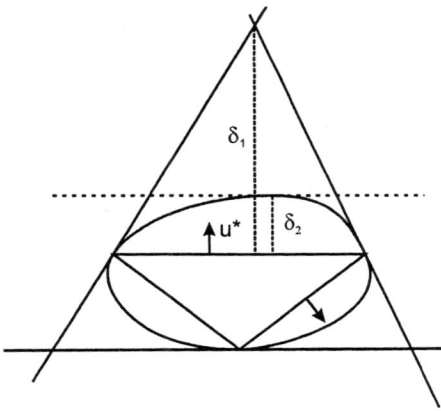

Figure 6.3. First outcome of the iteration of the MMCM method for $m = 2$ and $\beta = 0.3$.

the modified MCP method may require computational time comparable with computational time of Step 1 in the ER method.

The illustration of an iteration of the MCP method for $m = 2$ and $\beta = 0.3$ is given in Figures 6.3–6.4. Once again, the approximating internal polytope is given in Figure 6.3 by the triangle; the external polytope is depicted as well. The direction u^* was found on Step 1 of the iteration as the direction of the maximal value of the distance between two polytopes, i.e., the value of $\delta_1 = g_{Q^k}(u) - g_{P^k}(u)$, among all directions defined by the edges of the internal triangle. However, now the value of $\delta_2 = g_C(u) - g_{P^k}(u)$ plays its role, too.

Since $\delta_2/\delta_1 < 0.3$, the new vertex p^* associated with the direction u^* is *not* included in the description of the approximation. In contrast, the external polytope is changed − a new plane (supporting hyperplane in the point p^*) is included in its description. The iteration continues (see Figure 6.4). The new external approximation results in a new direction u^* on the repeated Step 1 of the iteration. For this direction we have $\delta_2/\delta_1 > 0.3$. For this reason, the new vertex p^* associated with the direction u^* is included in the description of the approximation. The external polytope is changed once again − a new plane (supporting hyperplane in the point p^*) is included in its description.

The above small modification of the MCP method results in new important features, which will be discussed later, after we introduce several basic concepts of the theory of adaptive methods for polyhedral approximation of multiple-dimensional compact convex bodies.

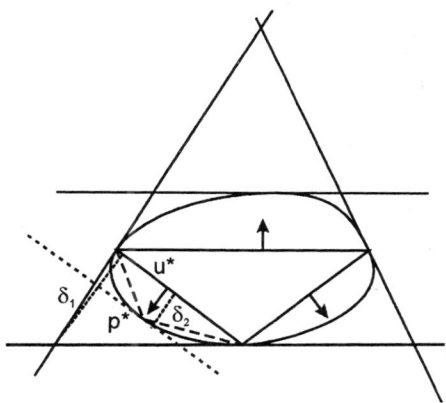

Figure 6.4. Second outcome of the iteration of the MMCM method for $m = 2$ and $\beta = 0.3$.

Theory of the augmentation schemes. To evaluate the quality of iterative methods, a "reference" sequence of polytopes must be considered that gives the best approximation of the convex compact body C. It is known (Gruber, 1983) that among the polytopes with a given number of vertices N there always exists a polytope $P_N \in \mathbf{P}(C)$ with the minimum of distance $h(C, P_N)$. The polytope P_N is denoted as the *polytope of best approximation* (PBA). It is known that $h(C, P_N) \to 0$ while $N \to \infty$. If the body C has two times continuously differentiable boundary with positive principal curvatures (or, speaking in short, C belongs to the class $\mathbf{C}^2$), there exist positive constants k_C and K_C such that

$$k_C/N^{2/(m-1)} \leq h(C, P_N) \leq K_C/N^{2/(m-1)} \tag{6.15}$$

where m is the dimension of the space (Bronshtein and Ivanov, 1975; Schneider and Wieacker, 1981; Gruber and Kendrov, 1982). So, the distance of the PBA from such a body decreases with the order $2/(m-1)$. For example, for the case of $m = 5$, which is the most interesting from the point of view of applications, $h(C, P_N)$ decreases proportionally to $1/N^{1/2}$, that is, fairly slowly.

The PBA can be found for extremely simple bodies only. Therefore, one can use their sequence for theoretical purposes, say as the "reference" sequence of approximating polytopes. Moreover, it is important to remember that the PBA-based "reference" sequence of polytopes provides an ideal that is not feasible in reality. Indeed, vertices of P_N are not related to vertices of P_{N-1}, and so one cannot even dream about an iterative procedure that constructs the sequence of PBA.

Several convergence characteristics can be used in the process of comparing a sequence of polytopes generated by an approximation method with the sequence of the PBA. Methods from the *Hausdorff classes* introduced in (Kamenev, 1992) construct polytopes that are close to the sequence of the PBA in the sense of the convergence. Hausdorff classes of methods can be defined for both augmentation and cutting schemes (see Chapter 8 for details), but here we restrict ourselves to the methods based on the augmentation scheme. An augmentation method is called a Hausdorff method with a constant $\gamma > 0$ for a body C, if it results in a sequence of polytopes $\{P^k\}$ for which for all k greater than some k_0 it holds that

$$h(P^k, P^{k+1}) \geq \gamma h(P^k, C).$$

It was shown that, for any convex compact body C, a Hausdorff method results in a sequence of polytopes that approximates the body. Moreover, for convex compact bodies C from the class $\mathbf{C}^2$ the distance $h(C, P^k)$ has the convergence rate $2/(m-1)$, i.e., for sufficiently large numbers N_k of vertices of the polytope P^k it holds that

$$h(C, P^k) \sim 1/N_k^{2/(m-1)}.$$

When compared with (6.15), this estimate shows that a sequence of polytopes generated by a Hausdorff method has the same rate of convergence $2/(m-1)$ as the sequence of the PBA! This statement means that Hausdorff methods are *asymptotically optimal with respect to the order of the number of vertices*. Since the number of vertices of the polytope P^k is related to the number of iterations, Hausdorff methods are *asymptotically optimal with respect to the order of the number of iterations*.

For methods, which are optimal with respect to the order of the number of vertices, it is important to know about another characteristic of convergence − the ratio of distances from the approximated body C for approximating polytopes $h(C, P^k)$ and for the PBA with the same number of vertices $h(C, P_{N_k})$ for sufficiently large numbers k. Main results in this field are provided in Chapter 8. We only outline them here.

Let us consider a sequence of polytopes $F = \{P^k\}_{k=0,1,\dots}$ that approximates a compact convex body C. The value

$$\eta(F) = \liminf_{k \to \infty} \frac{h(C, P_{N_k})}{h(C, P^k)}$$

is denoted as the *lower asymptotic efficiency* of the method that generated the sequence F. Evidently, the value $\eta(F) = 1$ could be achieved by the sequence of the PBA only. For a sequence of polytopes that is asymptotically not optimal, it holds that $\eta(F) = 0$. For a sequence of polytopes which is asymptotically optimal with respect to the order of the number of vertices, it holds that

$$0 < \eta(F) < 1.$$

In (Efremov and Kamenev, 2002) it is shown that for a sequence of polytopes produced by a Hausdorff method for convex compact bodies from the class C^2 it holds that

$$\eta(F) \geq \frac{\left(1 - \sqrt{1 - \gamma}\right)^2}{4}.$$

A subclass of Hausdorff methods (H_1-methods) was introduced in (Kamenev, 1996). This subclass has better convergence properties. In particular, for a sequence of polytopes produced by a H_1-method for $C \in C^2$ it holds that (Efremov and Kamenev, 2002)

$$\eta(F) \geq \gamma/4.$$

It is especially important that H_1-methods can approximate a CCB of a general type, that is, $C \in C$, with the convergence rate $2/(m-1)$. It means that for a broad class of CCBs, which PBA sequences have the convergence rate $2/(m-1)$, the H_1-methods are asymptotically efficient. Note that for any CCB which boundary contains a point where the boundary is two times continuously differentiable and has positive principal curvatures, the PBA sequences possess this property. Therefore, the H_1-methods are asymptotically efficient for such a CCB.

Asymptotic properties of the ER and MCP methods. The first example of a Hausdorff method is provided by the ER method. In (Kamenev, 1994) it is proved that, for compact convex bodies from the class C^2, the ER method is an asymptotically Hausdorff method with some constant γ, which is asymptotically close to 1. Due to this, we can assert that the ER method has the optimal convergence rate and it holds that $\eta \geq 1/4$ for such bodies.

Though the MCP method itself does not belong to the Hausdorff class, it has optimal convergence rate for the bodies from the class C^2 (Kamenev, 1996). It means that the MCP method is asymptotically optimal with respect to the order of the number of vertices. Because of this fact and the fact that only one evaluation of the support function is needed on an iteration, the MCP method has the optimal number of evaluations of the support function for bodies from the class C^2.

The MMCP method, however, belongs to the class of Hausdorff methods (Bourmistrova, 2000a). For this reason, it is asymptotically optimal with respect to the order of the number of vertices. Though the MMCP method, in contrast to the MCP method, requires more than one evaluation of the support function on an iteration, it was proved (Bourmistrova, 2000b) that it has the asymptotically optimal number of evaluations of the support function for bodies from the class C^2, too.

Computational experiments and practical use of the methods show that they provide an effective tool for approximation of compact convex bodies for $m < 8$, if the approximated body is not too flat. Experimental approximation of ellipsoids with different asphericities was carried out in (Dzholdybaeva and Kamenev, 1992; Bourmistrova, 1999; Efremov, 2002; Bourmistrova, 2003). Ellipsoids exemplify the bodies, polyhedral approximation of which is a fairly complicated task. Therefore, a study of the properties of their approximation is very educational. In particular, the class of ellipsoids includes the sphere, which is the most challenging object for approximation in Hausdorff metric. Experimental estimates for the asymptotic efficiency of the method were computed and compared with theoretical values. The results of theoretical analysis were confirmed, and the convergence constants were estimated. In particular, it turned out that the experimental lower asymptotic efficiency of the ER method is independent of the shape of the body being approximated and is greater than 0.5 for $m > 2$. Therefore, the above estimate of the asymptotic efficiency hopefully can be improved.

Experimental comparison of the MMCP method with the ER method and the original MCP method was carried out, too. It was clear from the very beginning that the ER method is able to construct polytopes that are better than the polytopes constructed by the modified MCP method. However, it turned out that the asymptotic efficiency of all the methods is just the same. As to the number of evaluations of the support function per iteration, the experiment showed that, for $\beta < 0.7$, the MMCP method requires only about 3.5 evaluations. Note that this number is much greater in the ER method. Experiments with bodies from the class C^2 result in recommendation to use the value $\beta = 0.4$.

As we have already said, the problem that is solved at Step 1 of the modified MCP method may require much time. The comparison of the MMCP method with the ER method was carried out on the basis of problems studied in the framework of the DSS described in Chapter 3. It turned that at least for some of the multiple criteria problems the MMCP method required less time than the ER method. For example, in the five-criterion problem that was used to describe implementation of the DSS, approximating of the EPH with the precision of 1% required about four times less time in the case of MMCP method than in the case of the ER method.

It is worth mentioning that a new modification of the ER method for linear problems was developed recently that applies interim information obtained in the process of solving linear optimization problems on the previous iterations. To be precise, optimal bases of linear programming problems are used on the next iterations of methods for speeding up the process of solving the optimization problems. Experiments carried out by A. Pospelov proved that the modified ER method has approximately the same time efficiency as the MMCP method.

Once again, important results were obtained for approximation of bodies with non-smooth boundary. Say, the ER, MCP and MMCP methods have optimal convergence rate and positive asymptotic efficiency for large classes of multi-dimensional bodies with non-smooth boundary. In particular, it was shown that the ER and MMCP methods are asymptotically H_1-methods. Due to it, for a broad class of CCBs, which PBA sequences converge with the rate $2/(m-1)$, the ER and MMCP methods are asymptotically efficient. This topic is considered in Chapter 8.

We can add that adaptive methods for polyhedral approximation of convex compact bodies offer a way to estimate several characteristics of convex bodies, such as volume, surface volume, and other Minkowski measures (Leichtweiss, 1980). However, this matter is far beyond the scope of the book.

Constructing the convex hull of a polytope and a point. We now consider the second step of the augmentation scheme; that is, constructing the convex hull of a polytope and a point. The first algorithm for constructing of conv $\{p^*, P^k\}$ was introduced in (Bushenkov, 1981), in the framework of the PER method (see Section 1). The algorithm is based on the ideas that are now called the beneath–beyond method (Preparata and Shamos, 1985). The beneath–beyond method is based on the McMullen–Shephard Theorem (McMullen and Shephard, 1971). We describe the idea of the McMullen–Shephard Theorem on the basis of a three-dimensional example.

Suppose that one needs to construct the faces of the convex hull of polytope *ABCDEF* and point *G* (Figure 6.5a). It is clear that the convex hull will include all faces of the original polytope invisible from point *G* (that is, *ABCD*, *BCE*, *CDE*, *DEF* and *ADF*). However, it will not include any of the faces visible from point *G* (that is, *ABF* and *BEF*, which are shown shaded in the figure). Instead, the convex hull will acquire new faces (Figure 6.5b). These new faces are the faces of a minimal cone whose apex is located at *G* and contains the polytope *ABCDEF*. Each such face lies in a plane passing through the edge of the polytope lying at the boundary between the visible and invisible

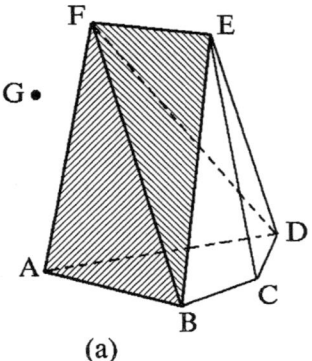 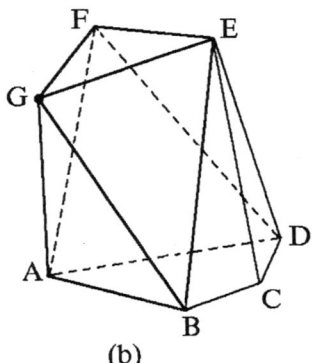

Figure 6.5. The beneath-beyond method

parts of the polytope's surface. It is clear that the boundary between the visible and invisible parts consists of intersections of pairs of adjacent faces of the polytope, one being visible and another being invisible. For example, the boundary in Figure 6.5 consists of the edges **AB**, **BE**, **EF** and **FA**. Say, the edge **BE** is the intersection of the visible face **BEF** and of the invisible face **BCE**. The new face **BEG** passes through the edge **BE**. This algorithm can be easily generalized to a space of an arbitrary dimension m if we recall that the boundary between the hyperfaces of the polytope (that is, $(m-1)$-dimensional faces) that are visible and invisible from a point in $\mathbb{R}^m$ consists of $m-2$ dimensional faces.

Particular methods that implement the beneath–beyond method differ in the way they solve the three following problems:

- How to determine whether a hyperface is visible from a point;

- How to determine whether two hyperfaces are adjacent or not; and

- How to transform the representation of the polytope into the convex hull.

The way, in which the above problems can be solved, depends on the representation of the polytope. Within the framework of the IDM technique, we need to construct a polytope that is described as the solution set of an inequality system. Such representation of polytopes was used by Bushenkov as early as in 1980 in the PER method (Bushenkov, 1981), which included constructing the multidimensional convex hulls (see Section 1). The algorithm of Bushenkov seems to be the first published algorithm based on the beneath–beyond method and implemented in the form of working software.

A bit later than Bushenkov, but independently of him, alternative algorithms based on the beneath–beyond method were proposed in unpublished papers of Kallay (Kallay, 1981) and Seidel (Seidel, 1981) who solved the above three problems in a different way. A description of the algorithms can be found in (Preparata and Shamos, 1985). To represent a polytope, Kallay stores the coordinates of its vertices and a complete combinatorial structure of the polytope as an incidence graph showing the affiliation of polytope faces of all dimensions. To say whether a face is visible or not, it is necessary to determine the orientation of the half-space whose boundary passes through the vertices of the face relative to the point being attached. The adjacency of hyperfaces can be readily determined from the incidence graph. A rather sophisticated combinatorial procedure is proposed for transforming the representation of the polytope. The algorithm of Seidel is the dual of the algorithm of Kallay; it is optimal with respect to the number of operations in spaces of even dimensions. We have no information about how successful these two algorithms have been in practical applications.

A new algorithm based on the beneath–beyond method is used now in the framework of the IDM technique as a part of the ER, MCP and MMCP methods. The new algorithm was proposed by Chernykh, 1988. Though the algorithm of Chernykh can be applied in the case of a sequence of vertices arriving one after another, we start its description with the task of constructing the convex hull for a pre-specified collection of s points $\{v^1, v^2, ..., v^s\} \subset \mathbb{R}^m$. Once again, the convex hull must be constructed in the form of the solution set of a linear inequality system. By definition, the point $y \in \mathbb{R}^m$ belongs to a convex hull of the above points if there exist values $\lambda_1, \lambda_2, ..., \lambda_s$ such that

$$y = \sum_{i=1}^{s} \lambda_i v^i, \quad \sum_{i=1}^{s} \lambda_i = 1, \quad \lambda_i \geq 0, \quad i = 1, 2, ..., s. \tag{6.16}$$

Consider the space $\mathbb{R}^{s+m}$ of variables $\lambda_1, \lambda_2, ..., \lambda_s$ and y. Then, the system (6.16) specifies a polyhedral set in this space. According to the definition of the projection, the convex hull of points $\{v^1, v^2, ..., v^s\}$ is the projection of the set into the space $\mathbb{R}^m$ of variables y. Therefore, in order to construct the desired convex hull, one can eliminate the variables λ from the system (6.16).

The variables λ are eliminated from the system (6.16) in the same order as they are numbered. The first $m + 1$ variables $\lambda_1, \lambda_2, ..., \lambda_{m+1}$ can be eliminated from the system by expressing them in terms of the other variables and using the equalities that the system (6.16) contains.

The remaining variables can be eliminated by the method of reduced fundamental convolution described in Section 1.

Let us consider the resulting system obtained after the variables λ_1, $\lambda_2, ..., \lambda_q$, where $q \geq m + 1$, have been eliminated. The system provides, in effect, the description of the convex hull of the points $v^1, v^2, ..., v^q$: it is only needed to let the variables $\lambda_{q+1}, ..., \lambda_s$ be equal to zero. Excluding the next variable λ_{q+1} from the resulting system is equivalent to attaching a point v^{q+1} to the convex hull of the points $v^1, v^2, ..., v^q$. So, when constructing a convex hull, information about the points that have yet to be attached is not needed. Therefore, it is not necessary to know the future points themselves or even their number. Thus, this method can be used for constructing the polytope sequentially.

Application of the reduced fundamental convolution method (see Section 1) for excluding the variables $\lambda_1, \lambda_2, ..., \lambda_s$ transforms the described method into a method based on the beneath–beyond method. The inequality index storage used in the reduced fundamental convolution method is equivalent to a partial storage of combinatorial structure of the intermediate polytope. Therefore, a polytope is, in effect, stored as a system of inequalities, each of which corresponds to a hyperface of the polytope. Also, each inequality is stored along with information that gives the numbers of the vertices which belong to a particular hyperface. This information helps to solve three problems listed above. For example, the question as to whether or not a given face is visible from a new point is answered by inserting the point being attached in the linear inequality corresponding to that hyperface.

The crucial point of the method is now the choice of adjacent hyperfaces in the convex hull of points $v^1, v^2, ..., v^q$ with a view to constructing a new hyperface passing through their intersection and the point v^{q+1}. If the variables λ were eliminated from system (6.16) by Fourier method, this would produce combinations of all pairs of inequalities corresponding to one visible and one invisible hyperface (irrespective of their adjacency). Naturally, the number of incidental inequalities would then increase catastrophically, and the method would no longer be of the beneath–beyond type. When the reduced fundamental convolution is used, the pairs that are combined correspond to adjacent faces; that is, they are combined in compliance with the beneath–beyond method. For this reason, the use of the reduced fundamental convolution in this method cannot produce redundant inequalities. The representation of a polytope is a fairly simple task. The transformation of indices poses no difficulty at all, and the coefficients of the inequality corresponding to a new hyperface are calculated as a linear combination with positive coefficients of the two inequalities representing two adjacent hyperfaces.

The algorithm of Chernykh has a higher theoretical order of time complexity than the purely combinatorial algorithm due to Kallay, because all possible pairs of visible and invisible faces have to be tested in order to establish the adjacency. Within the IDM technique, however, the time required for operation of the algorithm of Chernykh is quite satisfactory because the matter of adjacency is resolved by a small number of fast operations. Since m is relatively small in the IDM technique (not larger than seven or eight), the algorithm permits convex hulls for sets consisting of many hundreds of points to be constructed even on an old-fashioned computer.

With the algorithm of Bushenkov, in order to construct new hyperfaces, one has to select the vertices $V = \{v^1, v^2, ..., v^q\}$, which belong to the visible hyperfaces of the polytope. In the original version of the algorithm described in Section 1, the storing of the combinatorial structure of the polytope was not used. Therefore, the collection of vertices was found by a direct inserting of all vertices of the polytope in violated inequalities of the system. Then, the hyperfaces of the cone were computed upon excluding the auxiliary variables $\lambda_1, \lambda_2, ..., \lambda_r$ from the system of equalities and inequalities defining the cone,

$$z - z^* = \sum_{i=1}^{r} \lambda_i \left(v^i - y^* \right), \quad \lambda_i \geq 0, \quad i = 1, 2, ...r, \qquad (6.17)$$

where $v^1, v^2, ..., v^r$ are the vertices that belong to the violated inequalities. The reduced fundamental convolution method was used for excluding the variables $\lambda_1, \lambda_2, ..., \lambda_r$.

Let us compare the algorithms due to Bushenkov and Chernykh. The algorithm of Chernykh has several advantages. First, it has a simpler structure and requires less time and memory. In contrast, the attachment of one vertex to the hull in the algorithm of Bushenkov makes it necessary to eliminate the entire set of variables λ_i whose number is equal to the number of vertices belonging to the visible hyperfaces from the auxiliary system (6.17). Secondly, the algorithm of Chernykh is more reliable because a technique was developed to control the inaccuracy due to the rounded-off errors (Chernykh, 1988). In studying this matter, it turned out to be convenient to use the classical concepts of combinatorial topology and to consider the collection of hyperfaces of a polytope as a cycle of an abstract simplicial complex. Due to this device, it was possible to prove the validity of the simplest *a posteriori* estimate of the error in the convex hull construction. It can be done by inserting all of its vertices in all inequalities of the resultant system approximately describing the convex hull.

However, the algorithm of Bushenkov has several advantages, too. It does not require implementing the adjacency test for all pairs composed of visible and invisible faces. In effect, when one more vertex is attached to the hull, only the visible part and not the entire polytope needs to be considered in the algorithm of Bushenkov. Because of this, the task of constructing the hull of a system of points is decomposed into a sequence of local subtasks. In solving each of these subtasks, there is no need to consider the current polytope as a whole. This property of the algorithm of Bushenkov can be important when the number of hyperfaces is great. However, this decomposition of the task makes it difficult to maintain the correct combinatorial structure of the polytope, which is essential to controlling the accuracy when round-off errors are present.

3.   Approximating the Edgeworth–Pareto Hull in the convex case

Methods for constructing or approximating an EPH are outlined in this section for the convex case. Once again, we assume that the user is interested in decreasing the criterion values, i.e., a criterion point z' dominates a criterion point z, if and only if $z' \leq z$ and $z' \neq z$. In this case the non-dominated frontier $P(Z)$ of the set Z is of interest and the EPH of Z can be used instead of Z. Once again, the EPH is defined as

$$Z^* = Z + R_+^m,$$

where R_+^m is the non-negative cone of $\mathbb{R}^m$. Exploration of the EPH instead of Z is based on the important fact that the non-dominated frontiers of the sets Z and Z^* coincide, but the dominated frontiers disappear in Z^*. Note that the set Z^* can be represented in an equivalent form as

$$Z^* = \{y \in \mathbb{R}^m : z \leq f(x), \quad \text{where} \quad x \in X\}.$$

Constructing the EPH for linear models.   We start with constructing the EPH for the systems studied in Section 1, i.e., the set X is polyhedral

$$X = \{x \in \mathbb{R}^n : Hx \leq h\},$$

where H is a given matrix and h is a given vector, and the mapping f is linear

$$f(x) = Fx,$$

where F is a given matrix. Therefore, we need to construct the set

$$Z^* = \{z \in \mathbb{R}^m : z \leq Fx, \ Hx \leq h\}$$

in the form
$$Q^* = \{z \in \mathbb{R}^m : D^* y \leq d^*\}.$$
It means that we need to construct the matrix D^* and the vector d^*. One can see that the only difference of the above formula for Z^* from (6.3) consists in the presence of inequality instead of the equality between the vector z and the matrix F. Therefore, as in Section 1, we can consider the graph
$$Y^* = \{(x, z) \in \mathbb{R}^n \times \mathbb{R}^m : z \leq Fx, \ Hx \leq h\}$$
and construct its projection onto $\mathbb{R}^m$, which coincides with the set Z^*. Therefore, the set Z^* can be found by the convolution methods that have already been described in Section 1. All advantages and disadvantages of the convolution techniques discussed in Section 1 are valid here. Such an approach to constructing an EPH for linear models was proposed in (Lotov, 1983). However, it has not found a broad application because of the fast development of methods for approximating the EPH on the basis of polyhedral approximation of convex compact bodies. The advantages of convolution methods, which consist in the opportunity to consider a large number of criteria and non-bounded polyhedral sets, turned out to be not so important in real-life problems. This is why we proceed to methods based on polyhedral approximation of convex compact bodies without further discussing the convolution-based methods.

Polyhedral approximating of a convex EPH. The methods described in this sub-section are based on approximation of Z^* by the sum of a polytope and of the cone R_+^m. It is assumed that the set Z is bounded, and so the set Z^* can be approximated in this form. Note that the convexity of the set Z is not required, but the set Z^* is assumed to be convex. In addition, we assume that it is possible to evaluate the support function of the set Z^* for any finite number of directions u that belong to both the unit sphere of directions and the non-positive cone (it is clear in advance that for the directions that do not belong to the non-positive cone, the support function of Z^* equals infinity). As in Section 2, adaptive iterative methods can be used that are based, say, on the augmentation scheme. At an iteration, a non-dominated point on the frontier of the set Z^* is selected. Selecting the point can be based on the ideas of the ER or another adaptive method. Then, the convex hull of the selected point and of the current approximation, which is given by the sum of a polytope and the cone R_+^m, is constructed in the form of a linear inequality system. The same ideas are used as in the case of constructing the convex hull of a point and of a polytope. Details of the method for constructing the convex hull can be found in (Chernykh,

1995). In particular, it was shown that any convergent augmentation scheme can be used without a change if the initial approximation is given by the cone R_+^m with a vertex located in a point of the Pareto frontier.

It is worth noting that a fast method for on-line display of slices of the convex EPH (decision maps) has been developed in (Chernykh and Kamenev, 1993).

4.   Approximating the FCS for non-linear models

Here we describe the mathematical basis of the FGNL method. As in the previous section, the set X of feasible decisions belongs to the decision space $\mathbb{R}^n$. However, the set X is now described by non-linear constraints and is non-convex. The objective functions given by the mapping $f : \mathbb{R}^n \to \mathbb{R}^m$ that relates decisions to m criterion values is non-linear, too. Therefore, the set $f(X)$ is usually non-convex, and this evidence is the crucial difficulty in its approximation and visualization in the non-linear case. As we have said already in Chapter 5, the method described herein is based on simulation of random decisions, i.e., on computing outputs $f(x)$ of uniformly distributed random points x from the set X. These outputs provide the basis for evaluation and display of the set $Z = f(X)$.

The FGNL method applies the approximation of the set Z by a finite system of rectangular parallelepipeds with edges parallel to the axes (or, speaking in short, boxes). Boxes with equal lengths of edges are used here. To be able to approximate the set Z with a finite number of boxes, we have to assume that the set is confined. This method was introduced in (Kamenev and Kondrat'ev, 1992).

It is important to stress from the very beginning that the computational procedure for approximating the set Z in the non-linear case (except relatively simple problems) requires some training. Therefore, it must be carried out by a specialist in computational methods that is denoted as the computing expert in this book. In contrast, software for visualization of the approximation and identification of the feasible goal can be mastered by any computer-literate person. For this reason, it is assumed to be used by the decision maker directly. Such an interface of decision maker with computer has already been exemplified in the previous chapter (Section 1). The task described in this section is supposed to be performed by a computing expert.

Main concepts. It is assumed that the Tchebycheff metric $\rho(v, w)$ is used to measure distance between the points v and w in $\mathbb{R}^m$,

$$\rho(v, w) = \max_{i=1,2,...,m} \lambda_i |v_i - w_i|,$$

where λ_i for $i = 1, 2, .., m$ are positive values. Let $\varepsilon > 0$. We denote by $(Q)_\varepsilon$ the ε-neighborhood of the set Q; that is, the set of points distant from Q less than ε. In the case of the Tchebycheff metric, the ε-neighborhood of a vector z from $\mathbb{R}^m$ is a box, that is a parallelepiped with edges parallel to the axes, with center in z and length of the edges equal to 2ε.

The set $X \in \mathbb{R}^n$ is assumed to be compact. Then, the notion of the volume of the set X has a sense. We assume that the set X is bodily, i.e., its volume is not zero. The mapping $f : \mathbb{R}^n \to \mathbb{R}^m$ is assumed to be continuous on X. Then, the set $f(X)$ is compact, too. Note that one can study problems with a finite number of sets at which the mapping f is continuous. However, this topic is beyond the scope of this book.

As it was said earlier, a particular case of approximation of the set $Z = f(X)$ is considered here — the approximation by a finite variety of boxes, which centers are located in the points of Z. The collection of centers of the boxes is named the *approximation base* and is denoted by T. The set $(T)_s$, that is, the collection of s-neighborhoods of the points in T, is then the *approximation* of the set Z. In order to find criterion points that can provide a good approximation of Z, a global sampling of the set X is carried out. To be more precise, uniformly distributed random points of the set X are generated. Then, their outputs are computed, and a small number of them are selected as the centers of boxes. The length of the edges of the boxes must be specified in such a way that the resulting system of boxes approximates the set Z with a desired degree of accuracy. Collections of two-criterion slices of such an approximation can be then displayed on-line by personal computers and workstations reasonably fast.

To evaluate the quality of an approximation, two indicators are used that describe the deviation of $(T)_s$ from Z and the deviation of Z from $(T)_s$. For the first indicator the value of s is used. The base T belongs to Z, and the deviations of the points of $(T)_s$ from Z are not greater than s (remember that the Tchebycheff metric is used). As an indicator of the deviation of Z from $(T)_s$, a bit more complicated characteristic is used.

For a given value of s, the notion of the *completeness* $\eta_T(s)$ of the approximation $(T)_s$ is defined as the probability that the output of a random point from S belongs to $(T)_s$. For a given approximation base T, $\eta_T(s)$ as a function of s is understood as the completeness function of the approximation. It is clear that the completeness function $\eta_T(s)$ depends on the value of s in a monotonic way — the value of the function is not decreasing while the value of s is increasing. Moreover, since the set Z is bounded, the function $\eta_T(s)$ achieves its maximal value (one) at some value of s. Thus, the completeness function $\eta_T(s)$ describes the deviation of the set Z from the approximation $(T)_s$.

Let ε and η (with $\eta < 1$) be some positive numbers that characterize the requirements set to the deviation of the set $(T)_s$ from the set Z and the deviation of the set Z from the set $(T)_s$, respectively. Then, the task of constructing an approximation of the set Z can be reduced to finding a base T that

$$\eta_T(s) \geq \eta \quad \text{for some} \quad s \leq \varepsilon. \tag{6.18}$$

Since the value of $(T)_s$ does not decrease while the value of s is increasing, we can write the same requirement (6.18) in a shorter form

$$\eta_T(\varepsilon) \geq \eta. \tag{6.19}$$

Note that the application of the ε-neighborhood of the base T as the approximation implies that, for any point z^* from $(T)_\varepsilon$, it is possible to find a feasible solution x^* from X such that

$$\rho(z^*, \ f(x^*)) \leq \varepsilon.$$

This statement means that the approximation is not too fuzzy. The property (6.19) implies that the approximation $(T)_\varepsilon$ contains at least the η-th portion of the output of the set X. It means that the $(T)_\varepsilon$ is not too poor in comparison with Z.

Statistical method for constructing and testing an approximation base. Several conflicting goals must be taken into account in the process of constructing an approximation of the set Z. First of all, as was said earlier, the approximation is supposed to be used in the process of on-line visualization of the set Z using its bi-criterion slices. That is why one needs an approximation that consists of a relatively small number of boxes. On the other hand, it is desirable to construct a base T, for which the requirement (6.19) is satisfied for sufficiently small values of ε and $1 - \eta$. Surely, these two goals are naturally conflicting.

To construct a sufficiently good approximation base T, an iterative adaptive method is used. At any iteration of the method, N random points are generated and the associated output points are computed. Then, the output point that is the most distant from the current approximation base T is included in it. This feature means that the approximation method has an adaptive nature. Note that including of the most distant points in the approximation is the basic feature of the approach named the Deep Holes Method (Kamenev, 2001).

Before the construction of an approximation base T can be started, the computing expert has to specify the number N of random points that are generated at each iteration and the maximum number M of points in the approximation base. At the initial stage, an arbitrary point of the set X is generated, and the initial base T is formed as its output.

Note that the same strategy of including the most distant point in the approximation is used in methods for approximating the convex bodies. However, in contrast to the convex case, a theory for specifying the value of M on the basis of the desired values of ε and η and, perhaps, the dimension m of the criterion space, has not been developed yet. For this reason, validity of the requirement (6.19) is tested at any iteration of the method. The approximation process can be stopped after an approximation with the desired properties has been found (in addition to the stopping rule based on the pre-specified value of M). To test the requirement (6.19), the FGNL applies statistical tools. The test is based on using the same N random points from the set X that have already been generated.

In order to apply statistical tests that provide results with a certain reliability χ with $0 < \chi < 1$, one has to transform the requirement (6.19) into a different form. Instead of (6.19), the following condition is tested,

$$\mathbf{P}\left\{\eta_T(\varepsilon) > \eta\right\} \geq \chi. \tag{6.20}$$

It is important that using of N random points from the set X provides an opportunity not only to test the condition (6.20), but also to estimate the completeness function $\eta_T(s)$ for all positive values of s. To be precise, a function $\eta(s)$ is constructed at an iteration such that

$$\mathbf{P}\left\{\eta_T(s) > \eta(s)\right\} \geq \chi \tag{6.21}$$

for any $s \geq 0$. Let us next consider a method for constructing $\eta(s)$.

We denote a sample of N random uniformly distributed points from the set X by $H_N = \{x_1, \ldots, x_N\}$. Let us introduce the empirical completeness function of s as

$$\eta_T^{(N)}(s) = n(s)/N$$

where $n(s)$ is the number of points $x \in H_N$, for which $\rho(f(x), T) < s$.

For a given s, the value $\eta_T^{(N)}(s)$ is a non-biased estimate of the completeness $\eta_T(s)$. At the same time, it can be used for constructing the value of the estimate $\eta(s)$.

THEOREM 6.13 *For any $s > 0$ it holds that*

$$\mathbf{P}\left\{\eta_T(s) > \eta_T^{(N)}(s) - \Delta\right\} \geq \chi \qquad (6.22)$$

where Δ is given by

$$\Delta(\chi, N) = \sqrt{-\ln(1 - \chi) / 2N}. \qquad (6.23)$$

The proof of Theorem 6.13 is given in Appendix 6.A.

When comparing the expression (6.22) with the expression (6.21), one can note that the difference $\eta_T^{(N)}(s) - \Delta$ can be used for the value of the estimate $\eta(s)$.

It is important that the confidence interval Δ does not depend on the approximation base. Due to this, the expression (6.23) can be used for evaluation of the number N of random uniformly distributed points of the set X required to estimate the completeness with a given accuracy Δ and reliability χ. From (6.23) it follows that one can use the minimal integer N that satisfies

$$N \geq -\ln(1 - \chi) / (2\Delta^2). \qquad (6.24)$$

Table 6.1 contains the values of N depending on the values of Δ and χ.

Table 6.1. Values of $N(\Delta, \chi)$

	$\Delta = 0.10$	$\Delta = 0.05$	$\Delta = 0.01$
$\chi = 0.90$	116	461	11513
$\chi = 0.95$	150	600	14979
$\chi = 0.99$	231	922	23026

So, one can specify the value of N based on the desired χ and Δ of statistical testing of the approximation. In turn the parameters χ and Δ (as well as the parameters M, η and ε) depend on the importance of the problem, computing cost and time (including time needed to compute $f(x)$, etc. The choice of the parameters of the approximation procedure must be made by the computing expert and the decision maker together.

Approximation Algorithm. The statistical method for constructing an approximation base and estimating the completeness function has been implemented in the form of the following approximation algorithm. It is assumed that the values of $M, \eta, \varepsilon, \chi$ and Δ have already been specified, and so the number N of random points of the set X generated at any iteration can be found using (6.24). In the framework of this algorithm, three stopping rules are applied. Two of them have already been discussed — these rules are based on the maximal number M of points in the approximating base and on the testing (6.20). The third one can be based on exploration of the graphs of the empirical completeness function and of the interval estimate. These graphs help the computing expert to assess the convergence speed. The computing expert can stop the approximation process at some iteration if the values of the empirical completeness function and interval estimate for some value of s are already satisfactory. Such an option is very convenient since the form of the set Z is not known in advance, and so the values of M, ε and η can be specified only roughly. Due to possible human intervention into the approximation process, the a priori values of the process parameters M, ε and η play a minor role.

On the other hand, it is possible to carry out the process without an intervention. In this case, the values of the process parameters M, ε and η define the termination conditions. It was proven (Kamenev, 2001) that for any compact set S and any continuous mapping $f(\boldsymbol{x})$, with a properly specified number N and a sufficiently large number M, the method can construct an approximation base T that satisfies (6.20) in a finite number of iterations for any values of ε, χ and η.

Iteration of the algorithm. The current approximation base T is assumed to be constructed on the previous iterations of the algorithm.

1 Generate N random uniformly distributed points from the set X and compute their outputs.

2 Compute the empirical completeness function $\eta_T^{(N)}(s)$ for all values of s.

3 Display graphs of $\eta_T^{(N)}(s)$ and of $\eta(s) = \eta_T^{(N)}(s) - \Delta$ as functions of s.

4 Stop if (6.20) is satisfied. Stop if the number of points in T is equal to M. Stop if the computing expert is satisfied with the completeness achieved for some value of s. Otherwise, augment the current approximation base T by the most distant output point among the N points and go to step 1.

Other algorithms, more sparing in computing the output, may be proposed for constructing the approximation base T (Kamenev and Kondrat'ev, 1992).

Imagine that we have constructed an approximation $(T)_s$ of the set Z. The approximation has a simple explicit description as a system of boxes with centers that belong to the set Z and edges of length $2s$. Since the number of boxes is relatively small, bi-criterion slices of the approximation can rapidly be computed and displayed by computer graphics. For each point $(T)_s$, it is possible to find the nearest point of the approximation base T and to reconstruct the corresponding decision. In practice, the approach is at its best with up to five criteria. In the case of more than three criteria, scroll-bars are used that help to fix or to control the values (or ranges) of the other criteria and to study the influence of them on the three-criterion pictures. It is important to stress that though the approximation process may take a fairly long time, the visualization procedure can be carried out by the decision maker on-line without waiting after the request. Application of the display in the framework of the FGM for non-linear models has already been described in Chapter 5.

Application of the FGNL method on parallel and meta- computing platforms. Approximating the FCS with sufficiently good completeness and precision may require massive computing, and so parallel computing may be very helpful in the case of using the FGNL method. Fortunately, the FGNL method can be implemented in the form of software for parallel computing fairly easily. We describe this feature of the FGNL method for the most complicated form of parallel computing — application of meta-computing platforms. Surely, the FGNL technique can be applied in a more simple case of a given computer cluster.

A meta-computing platform is based on application of a collection of computers (and possibly other resources such as storage devices) that are geographically distributed, but networked in various ways. Meta-computing platforms have the important advantage that they are inexpensive. They utilize idle time on workstations, which is essentially free. Despite their low cost, the meta-computing platforms are potentially very powerful. However, they have features that make them much more difficult to program than traditional parallel computers. These features include (see http://www-unix.mcs.anl.gov/metaneos/):

- Dynamic availability. The number of processors available to a platform may vary over time, as may the throughput one receives from an individual processor.

- Unreliability. Processors that are used for meta-computing may disappear without notice.

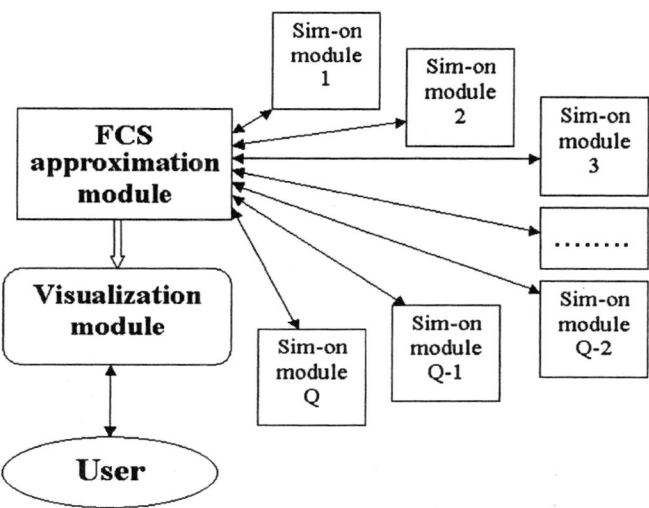

Figure 6.6. Application of the FGNL method on a meta-computing platform.

- Poor communications properties. Communication latency between any given pair of processors may be high, variable, and unpredictable.

- Heterogeneity. The processors in the meta-computer may differ in their CPU speed or amount of memory, for instance.

To implement the FGNL method at a meta-computing platform, one has to transform it into a special form oriented on meta-computing. It is important that the FGNL method has this form originally, and so it can be easily implemented in meta-computing. To do it, one loosely has to separate data generation and data analysis.

Among the computers, a coordinating computer must be selected. In contrast to other computers, it must be highly reliable. This computer is supposed to be used for interaction with other computers, for coordinating their job and, finally, for data analysis. The coordinating computer must have a fast access to a database containing millions of random numbers. The random numbers must be randomly distributed to other computers on the platform.

The scheme of the FGNL method in the framework of a meta-computing platform is given in Figure 6.6.

The data generation process consists of two parts:

- a process of computing feasible random decision alternatives; and

- a process of computing their outputs (the points of the FCS).

To carry out this job on the platform's computers, the simulation software must be located on the computers in advance. The simulation software has to transform a random number into a random point of the box that contains the variety of feasible decisions. Then, it has to test whether the generated point belongs to the variety of feasible decisions. If it holds, output of the point is computed and provided (maybe jointly with the associated decision) to the coordinating computer.

Due to the randomization of the input, particular input points (i.e., points from the box containing the variety of feasible decisions) play a minor role: the statistical analysis depends upon the number of computed random outputs collected at the coordinating computer, but not on particular output. Dynamic availability of processors, their unreliability, heterogeneity and possible poor communication play no role in this case.

The central computer collects the database of generated inputs and outputs. Since the requirement of statistical independence of inputs is satisfied, statistical analysis of the data can be performed. Note that it can be done simultaneously with database generation. The coordinating computer can evaluate the current value of the completeness function and display the approximation in a graphic form. After receiving additional outputs, the completeness function can be updated. Then, the approximation can be displayed in additional details.

So, the FGNL method can be easily implemented in parallel on multiple (say, many hundreds) heterogeneous computers of various types. Specific synchronization problems, which are met in distributed computing so often, do vanish in the case of the FGNL method.

5.   Approximating the EPH in the non-linear case

This section is devoted to a method for constructing an approximation of the EPH in the non-linear case. The method described herein was proposed in (Lotov, Kamenev and Berezkin, 2002). As in the previous section, the compact set X of feasible decisions belongs to a finite-dimensional linear decision space and is described by non-linear constraints. So, it may happen to be non-convex. The continuous mapping f that relates decisions to m criterion values may be non-linear, too. Therefore, the EPH defined as $f(X) + R_+^m$ is normally non-convex. Though the EPH may be convex even for a non-convex set $f(X)$ (this feature is often used by researchers in the field of economic models), in this section we consider a general case with a non-convex EPH.

As in the previous section, the method described here is based on simulation of random decisions, i.e., on computing output $f(x)$ of uniformly distributed random points $x \in X$. The output of a large number of random points provides the basis for approximation and display of the EPH. However, in contrast to the approximation of the set $Z = f(X)$ by boxes, the EPH is approximated by the union of cones. Such approximation can be used for interactive visualization of decision maps and even animation of them. In other words, the IDM technique can be applied for exploration of the Pareto frontier in this case.

Vertices of the cones $z + R_+^m$ used in the FGM/IDM technique for the non-linear case are located in points $z \in Z$. The variety of such points $z \in Z$ is denoted as the *approximation base* T. Note that the union of cones $z + R_+^m$ with $z \in T$ provides the EPH of the approximation base T. Therefore such an approximation of the EPH is denoted by T^*. It is clear that $T^* \subset Z^*$. For this reason, the problem is reduced to selecting such collection of a relatively small number of points $z \in Z$, the EPH of which is close to Z^*.

Note that the task of approximating the EPH by cones is fairly close to the task of approximating the Pareto frontier by points. Indeed, such an approximation of the Pareto frontier can be used as the approximation base (filtering of points may be applied, if needed). Several types of methods have been applied for approximating of the Pareto frontier by points. For example, if information on the Lipschitz constant is available, the technique can be applied that is based on covering of the set X using uniform grids (Nefedov, 1984). A more sophisticated approach applies adaptive covering of the set X by balls (Evtushenko and Potapov, 1986). Another idea consists in solving a large number of a priori formulated single-criterion optimization problems. Such methods have been proposed independently in (Krasnoshchekov, Morozov and Fedorov, 1979), and (Steuer, 1986). They can be used in the case of problems with a small number of local extrema. Recently, various genetic optimization methods were proposed for approximating of Pareto frontiers by points (see, for example, Deb, 2001).

The method for approximation of a EPH described in this section is based on an idea that is close to the idea of Section 4. So, it differs from the above mentioned methods. It is based on iterations that combine a *global random search*, i.e., generating a large number of random decision vectors, computing associated criterion vectors and filtering of them, with *local scalar optimization*. Hybridization of random search with local optimization is a standard approach in global optimization with a single criterion (Horst and Pardalos, 1995). The FGM/IDM technique applies this idea in multi-criteria problems. We show that such hybridization can be used for approximation of the EPH in the nonlinear case. Due to the global search, the method can be used even in the case of a fairly large number of local minima of optimization problems.

Application of any method for Pareto frontier approximation results in the question: is the resulting approximation sufficiently good? This question can be definitely answered in the case of the covering techniques. However, covering methods require one to evaluate the function in a large number of points of the set X, which is exponentially growing with the growth of the dimension of the set. Therefore, covering techniques may be used only if the dimension of the decision space is very small. Knowledge of the Lipschitz constants can help to estimate the precision of an approximation in the case of the methods proposed in (Krasnoshchekov, Morozov and Fedorov, 1979; Steuer, 1986), however, the Lipschitz constant-based estimates usually require (see Nefedov, 1984) to solve an enormously large number of optimization problems, too. The problem of the quality of an approximation is not studied in the case of genetic methods: instead, methods are evaluated by using simple problems with a known Pareto frontier or on the basis of mutual comparison of methods (Deb, 2001).

The method for approximation of the EPH described in this section is associated with estimation of the quality of an EPH approximation. This estimation is based on statistical tests, and so it can be carried out without any knowledge of the Lipschitz constant. It is clear that such procedures can be used for evaluating the quality of a Pareto frontier approximation developed by any other method.

Prerequisites of the method. It is assumed in the framework of the method that three codes are available. One of them must be able to generate random uniformly distributed points of the set X. This task can be solved easily in the case when one knows such a box that contains the set X while the volume of the set X is not too small in comparison with the volume of the box. However, this problem can turn out to be extremely complicated in a general case. Here, we do not discuss this topic and loosely assume that such a code has already been prepared.

The second code must be able to compute the criterion point $x = f(x)$ for any point $x \in X$. This code may be based on simulation of a decision. In a different case, it may be a module based on a finite-element method (FEM) or on a finite difference method (FDM) for solving a boundary problem in partial derivatives. It is important that we do not require any other feature of the module except its ability to find the value of $f(x)$.

The third code is assumed to be able to solve local scalar optimization problems of the form

$$\text{minimize} \quad \phi(f_1(\boldsymbol{x}), f_2(\boldsymbol{x}), \dots, f_m(\boldsymbol{x}))$$
$$\text{subject to} \quad \boldsymbol{x} \in X$$

for any scalarizing function $\phi(z_1, z_2, \dots, z_m)$ of the criterion values and any starting point $\boldsymbol{x}^0 \in X$. In other words, the third code must be able to find the local minimum $\boldsymbol{x}^*$ of the function $\phi(f_1(\boldsymbol{x}), f_2(\boldsymbol{x}), \dots, f_m(\boldsymbol{x}))$ that depends on the starting point $\boldsymbol{x}^0 \in X$ in the case of multiple minima. Application of such a local optimization algorithm to a starting point $\boldsymbol{x}^0 \in X$ will be denoted further the *improvement* of the point $\boldsymbol{x}^0$ in accordance to a scalarizing function ϕ. It may be required, because of the nature of the code for computing the criterion point, that such a local optimization code uses the values of the function $\phi(f_1(\boldsymbol{x}), f_2(\boldsymbol{x}), \dots, f_m(\boldsymbol{x}))$, but not of its gradients or second derivatives. Now, we use a local optimization code that has such a property (Evtushenko and Grachev, 1979; Evtushenko, 1982). However, if the second code is able to compute derivatives, special derivative-based local optimization codes may be used, too.

We start the description of the method with the most important feature — the statistical tests used in its framework.

Statistical tests. It is clear that the concept of completeness described in the previous section can be used for estimating the quality of an approximation of the EPH. However, the completeness test by itself proved to be not effective in the case of a precise approximation of the EPH, since the output of all generated points is usually located deep inside of the approximation. Because of it, it turned out that the experimental completeness η_T^N equals 1 and so the confidence interval depends on the number of sampled points only. The maximal deviation from the set T^* of outputs $f(\boldsymbol{x})$ for the points of a sample equals zero in this case. However, such a result of testing does not guarantee a high quality of approximation, since we are interested in the frontier, but not in the set as a whole. Therefore, the completeness test cannot be used for testing the quality of an approximation of the EPH: it is used for triggering another statistical test — the test of the optimization-based completeness, which develops the concept of completeness and can be used in the case of the Pareto frontier approximation.

Let us assume that a given T^* is an approximation of an EHP for a problem. We consider the scalarizing function $\phi(\boldsymbol{z})$ that describes the distance of a point from T^*. The optimization-based completeness $\eta_T(\phi)$ is defined as the probability of $f(\boldsymbol{x}^*) \in T^*$ for the point $\boldsymbol{x}^*$, which is the result of local improvement of a random point $x^0 \in X$. As for ordinary completeness, it holds that $0 \leq \eta_T(\phi) \leq 1$. Note that simpler scalarizing functions that describe a proxy distance from T^* can be used, too.

One can see that the concept of optimization-based completeness $\eta_T(\phi)$ has a constructive form − it is based on our ability to solve local single-criterion optimization problems for a random point $x^0 \in X$. As in the case of ordinary completeness, we look for an estimate $\eta^*(\phi)$ of the value of $\eta_T(\phi)$ with a certain confidence level χ, i.e.,

$$\mathbf{P}\{\eta^*(\phi) \leq \eta_T(\phi)\} \geq \chi. \tag{6.25}$$

To find such an estimate, the random sample H_N from X is generated, local optimization procedure is applied to the points $x^0 \in H_N$ and the statistics $\eta_T^N(\phi) = n/N$ is computed where n is the number of such outputs $f(x^*)$, for which it holds that $f(x^*) \in T^*$. Once again this statistics is an unbiased estimate of the value of $\eta_T(\phi)$. To specify the value of N and calculate the value of $\eta^*(\phi)$ in (6.25), Theorem 6.13 can be used. Therefore, the dependence of N on confidence and on interval is the same as in Section 4. If the estimates of $\eta^*(\phi)$ and $\eta_T^N(\phi)$ are sufficiently close to 1, the quality of the approximation T^* is considered to be sufficiently good.

In the process of testing the quality of an approximation, along with the values $\eta^*(\phi)$ and $\eta_T^N(\phi)$, the maximal distance of points $f(x^*)$ from the set T^* may be taken into account (the so-called deviation test).

It is worth stressing that computing of $f(x^*)$ may require substantially more computer time than computing $f(x)$. For example in the steel cooling problem described in Section 2 of Chapter 5, about 10,000 computations of $f(x)$ were needed in the case of simulation-based optimization (note that formulae for computing gradients were not used). Therefore, different statistical tests may be of interest, too. For example, in (Kamenev, 2001) it is shown how the tests that are based on the theory of ordered statistics (Zhigljavsky, 1991) can be applied for estimating the deviation of the approximated set Z^* from the approximation T^*. However, application of ordered statistics still waits for its implementation.

The algorithm. Now we can proceed to a description of the approximation algorithm that consists of the initial iteration, a sequence of repeating iterations and the final iteration.

Initial iteration. The initial iteration is devoted to the constructing of the initial approximation base T_0. It is started by generating a random sample $H_{ini} = \{x^{(l)}, ..., x^{(L)}\}$ from the set X and further computing of output of the points of H_{ini}. This information helps to understand the problem, to select proper scaling of it.

For all the particular criteria f_j, $j = 1, 2, ..., m$, the optimization code solves about a dozen local optimization tasks. In these tasks, local minima are found for starting points that belong to a random subset of H_{ini}. The best criterion values found in such a process provide an estimate of the *ideal criterion vector* $z^\star$, components of which are formed by values that minimize separately every f_i. Since local optimization is used in this process, one cannot assert that the obtained vector coincides with the actual ideal point. Therefore, it is considered as the reference point that helps to formulate single-criterion optimization problems that are solved in the main part of the procedure. By computing the outputs of all points involved and by excluding the dominated criterion points we obtain the initial approximation base T_0.

Iterations of the main part of the method. Let us consider the k-th iteration. Before it starts, an approximation base T_{k-1} must be given. Any iteration consists of two phases.

First phase. The aim of the first phase is to test the quality of a given approximation T_{k-1}^* and to carry out a global search, that is, to select a small number of points to be the starting points for local optimization procedure at the second phase.

The approximation quality is evaluated using the described statistical tests. At the first iterations, when the approximation is still rough, the completeness test is used. As has already been said, it is used to check whether the current approximation is good enough to apply the time-consuming test of optimization-based completeness. If the results of testing optimization-based completeness (or of the deviation test) satisfy the user, the approximation process is completed. The user has to proceed to the final iteration and subsequent visualization of the approximation. In the opposite case, the preparation for the second phase is carried out: several criterion points, denoted as the list X_k, are selected from the points generated at the first phase. Various heuristic procedures may be used for it. We do not discuss them here, since they depend on various aspects including approximation precision. An example of application of such procedures for selecting the list X_k is given in (Berezkin, Kamenev, Lotov and Miettinen, 2003).

Second phase. The goal of the second phase is to construct a new improved approximation base T_k.

The following local optimization problems are solved for the starting points from X_k:

$$\text{minimize} \quad \left\{ \max_{j=1,2,\ldots,m} \lambda_j \left(f_j(\boldsymbol{x}) - z_j^\star \right) + \sum \varepsilon_j f_j(\boldsymbol{x}) \right\} \qquad (6.26)$$

$$\text{subject to} \quad \boldsymbol{x} \in X.$$

One can see that the Tchebycheff metric is used in optimization problem (surely, if the estimate of the ideal vector is sufficiently good). It describes maximal deviation of outputs from the estimate of the ideal vector for some positive values of parameters $\lambda_j, j = 1, 2, \ldots, m$. The parameters $\lambda_j, j = 1, 2, \ldots, m$, in our method are specified in accordance with the output of the starting point $\boldsymbol{x}_0 \in X_k$ under improvement, i.e.,

$$\lambda_j = 1/\left(f_j(\boldsymbol{x}_0) - z_j^\star \right), \qquad (6.27)$$

for $f_j(x^0) > z_j^\star$. Due to such a formulation of the optimization problem, the output of the point is moved in the direction of the ideal point $\boldsymbol{z}^\star$. It is important that the output of the optimized point dominates the output of the initial point or, in the worse case, coincides with it. If $f_j(x^0)$ is close to $z_j^\star$ at least for some j, more sophisticated local optimization procedures can be used. They are not discussed here.

Note that the estimate $\boldsymbol{z}^\star$ of the ideal point may be improved as the result of local optimization, too.

The second phase completes with constructing of the new approximation base. First, the points from the base T_{k-1} are merged with outputs of the decision points obtained in the process of local optimization (6.26). The dominated criterion vectors are excluded from the list, and the list is filtered to avoid points that are located close to each other in decision space. The list of points obtained in this way is denoted by T_k. The k-th iteration is completed by this.

Final iteration. The final improvement of the approximation base is based on heuristic techniques that are not described here (see, for example, Berezkin, Kamenev, Lotov and Miettinen, 2003). The quality of the final approximation can be estimated by using the optimization completeness test, if needed.

Discussing the method. Selecting of the parameters in the scalarizing function (6.26) provides an adaptive specification of the scalarizing function. It is an important advantage of the method. Experiments show that the non-adaptive specification of them often results in criterion points, which are very close one to another (Deb, 2001) and are not needed for approximating the Pareto frontier. So, computing efforts are used for nothing. Adaptive specification of the parameters of the scalarizing function helps to avoid such a negative effect.

It is important to stress that the optimization-based completeness of an EPH does not depend on the method that was used for constructing the approximation base. Therefore it can be used for testing any approximation of the Pareto frontier.

As has already been said, the software in C++ that implements the method for approximation and visualization of the EPH for large non-linear models was coded (Berezkin, 2002). This software can be interfaced with any simulation/FEM/FDM module for computing the criterion values. The EPH for the steel cooling problem described in Section 2 of Chapter 5 was constructed using this software. In this fairly complicated problem characterized by application of an FEM/FDM module with 325 input (decision) variables, 1666 internal variables and four criteria, about 300 local optimization tasks were solved in the process of constructing a fairly good approximation of the EPH. The decision maps are given in Chapter 5.

It is important to stress once again that modern computers provide an opportunity to compute millions of outputs for a fairly complicated mathematical model in a relatively short time and for a small cost. Due to this feature, methods of statistical optimization (both single- and multi-criteria) are intensively applied in various forms. Genetic algorithms and simulated annealing provide examples of such methods that are very popular now. However, the more traditional statistical methods (as random search) can compete with them and even provide more information given in the form of statistical tests. The method described in this section, which is based on hybridization of random global search (statistical testing) and local optimization, proves it.

APPENDIX 6.A: Proof of the Theorem 6.13

Let us consider a random variable ξ, that is equal to 1 if $f(x) \in T^*$ and equals zero in the opposite case. Let us note that the probability p of $f(x) \in T^*$ equals η_T and the probability q that $f(x)$ does not belong to T^* equals $1 - \eta_T$. It is clear that $\mathbf{M}\xi = p$, $\mathbf{D}\xi = pq$. For the sum S of a final number of N independent realizations of the random variable ξ it holds that $\mathbf{M}(S/N) = p$. Moreover, the following exponential Tchebycheff inequality is true for $\Delta > 0$ (Shiryaev, 1989, page 80)

$$\mathbf{P}\{S/N - p \geq \Delta\} \leq \exp(-2N\Delta^2).$$

Since $\eta_T^{(N)} = S/N$ and $p = \eta_T$, the exponential Tchebycheff inequality results in

$$\mathbf{P}\{\eta_T^{(N)} - \eta_T < \Delta\} \leq -\exp(-2N\Delta^2).$$

Since the estimation of the completeness η_T with a given reliability χ is required, we set

$$\chi = 1 - \exp(-2N\Delta^2).$$

This equality results in the dependence of the precision of evaluation Δ on the reliability and the number of points

$$\Delta(\chi, N) = \sqrt{-\ln(1 - \chi) / (2N)}.$$

Therefore, the probabilistic estimation of the completeness has the form

$$\mathbf{P}\{\eta_T > \eta_T^{(N)} - \Delta(\chi, N)\} \geq \chi,$$

where the precision $\Delta(\chi, N)$ is given by the above formula. This ends the proof.

Chapter 7

COMPUTATIONAL METHODS FOR DYNAMIC SYSTEMS

In this chapter we consider methods for approximation of FCSs and their EPHs for systems described by ordinary differential equations or partial derivatives. Though the methods are based on ideas which have already been described in Chapter 6, the specific form of such systems requires special attention. Now, in contrast to Chapter 6, it is assumed that the feasible decision set X belongs to infinite-dimensional linear space $\mathbb{W}$ of a general nature. However, the criterion vectors z are still assumed to be elements of the linear finite-dimensional criterion space $\mathbb{R}^m$. The criterion vectors are related to decisions by objective functions given by the mapping

$$f : \mathbb{W} \to \mathbb{R}^m.$$

As earlier, the main computational problem consists in approximation of either the set FCS given by

$$Z = f(X),$$

or its Edgeworth–Pareto Hull Z^* that is defined (decreasing of criterion values is of interest) as

$$Z^* = Z + R^m_+,$$

where R^m_+ is, as usual, the non-negative cone of $\mathbb{R}^m$. In this chapter, we show that approximation of FCSs and their EPHs can be based on the computational methods described in Chapter 6, that is, application of Fourier convolution in the linear case, polyhedral approximation of convex bodies in the convex case and generating of random decisions in the non-linear case.

In Section 1 we show that approximation of an FCS or its EPH for a dynamic system given by ordinary differential equations can be reduced to approximation of reachable sets for the system and subsequent approximation of the FCS (or its EPH) for a reachable set. Sections 2 and 3 are devoted to convex dynamic systems of this kind. Section 4 describes a method for estimation of the reachable sets for non-convex systems. In Section 5, possible implementation of the approach in the case of boundary problems for partial derivative systems is outlined.

1.　Approximation of FCSs and EPHs for dynamic systems

In this section the problem of approximation of FCSs and EPHs for dynamic systems given by ordinary differential equations is formulated and the main approaches to its solution are described.

Basic notions. Let us introduce some basic notions of the theory of differential equations that are the subject of an external impact (control in our case). We consider systems of differential equations of the type

$$\dot{x} = g(x, u, t), \tag{7.1}$$

where t is time, x is the n-dimensional phase vector, u is the r-dimensional control vector, and $g(x, u, t)$ is the n-dimensional function defined for $t \in [0, T]$ and all $u \in \mathbb{R}^r$ and $x \in \mathbb{R}^n$. It is assumed that there are constraints imposed on the pair of control and phase vectors in any moment $t \in [0, T]$,

$$(u(t), x(t)) \in Y(t), \tag{7.2}$$

where $Y(t)$ is a given family of subsets of $\mathbb{R}^r \times \mathbb{R}^n$ that depends on time. Note that in the framework of (7.2), the purely phase constraints

$$x(t) \in Q(t)$$

or the constraints imposed in control only

$$u(t) \in U(t)$$

can be considered as well. In addition, the set of initial feasible phase vectors $\Gamma(0)$ must be specified, that is,

$$x(0) \in \Gamma(0), \tag{7.3}$$

where $\Gamma(0) \subset \mathbb{R}^n$.

Let us assume that the functions of time considered in the system (7.1)–(7.3) belong to some classes, say

$$u(\cdot) \in \mathbb{L}^r_\infty[0, T] \quad \text{and} \quad x(\cdot) \in \mathbb{C}^n[0, T],$$

where $\mathbb{L}_\infty[0, T]$ is the linear space of confined functions measurable almost everywhere on $[0, T]$, and $\mathbb{C}[0, T]$ is the linear space of absolutely continuous on $[0, T]$ functions. In both cases, the maximal norm is used.

Let $\mathbb{W} = \mathbb{L}_\infty^r[0, T] \times \mathbb{C}^n[0, T]$. Let us assume that the mapping

$$\boldsymbol{f} : \mathbb{W} \to \mathbb{R}^m \tag{7.4}$$

is given that defines a criterion vector $\boldsymbol{z}$ for any $(\boldsymbol{u}(\cdot), \boldsymbol{x}(\cdot))$ from $\mathbb{W}$. Let us denote by X the feasible pairs $(\boldsymbol{u}(\cdot), \boldsymbol{x}(\cdot))$ from $\mathbb{W}$, that is, the pairs that satisfy (7.1)–(7.3). Then, the FCS for the system (7.1)–(7.4) can be given in its usual form $Z = \boldsymbol{f}(X)$, and the EPH is given as $\boldsymbol{f}(X) + R_+^m$.

There exists an alternative, sometimes more convenient description of the dynamic systems with an external impact − the description that applies the notion of *differential inclusions*. Let

$$\dot{\boldsymbol{x}} \in G(\boldsymbol{x}, t), \tag{7.5}$$

where $G(\boldsymbol{x}, t)$ is a subset of $\mathbb{R}^n$ that depends on time and $x \in \mathbb{R}^n$. It is assumed here that $G(\boldsymbol{x}, t)$ is given for $t \in [0, T]$ and all $\boldsymbol{x} \in \mathbb{R}^n$. By this a *set-valued mapping* from $\mathbb{R}^n$ into $\mathbb{R}^n$, which relates the set $G(\boldsymbol{x}, t)$ to any point $\boldsymbol{x} \in \mathbb{R}^n$, is defined for any $t \in [0, T]$.

It is clear that the system (7.1)–(7.2) is close to (7.5). Denoting $\boldsymbol{v} \equiv \dot{\boldsymbol{x}}$, we obtain the system

$$\dot{\boldsymbol{x}} = \boldsymbol{v}, \quad \boldsymbol{v} \in G(\boldsymbol{x}, t),$$

which is a partial case of (7.1)–(7.2). Therefore, one can introduce the notion of the FCS and its EPH for the system (7.3)–(7.5) in the same form as for the system (7.1)–(7.4). Note that the purely phase constraints can also be described by (7.5). It is clear that it makes a sense to consider the differential inclusion (7.5) if the trajectories $\boldsymbol{x}(\cdot)$ are of interest, but not the controls. To be precise, the objective functions (7.4) must have the form of the mapping $\boldsymbol{f} : \mathbb{C}^n[0, T] \to \mathbb{R}^m$.

Let us consider two important particular cases of the mapping $\boldsymbol{f} : \mathbb{C}^n[0, T] \to \mathbb{R}^m$. In the framework of the first case, the criteria coincide with the phase variables at some time moment $t^* \in [0, T]$, i.e.,

$$\boldsymbol{f}(\boldsymbol{x}(\cdot)) = \boldsymbol{x}(t^*).$$

The FCS is denoted in this case by $\Gamma(t^*)$ and is known as the *reachable set*. So,

$$\Gamma(t^*) = \{\boldsymbol{x} \in \mathbb{R}^n : \boldsymbol{x} = \boldsymbol{x}(t^*), \text{ where } \boldsymbol{x}(\cdot) \text{ satisfies } (7.5), (7.3)\}. \tag{7.6}$$

Another important particular case is provided by the criteria, which depend on the phase variables at some time moment $t^* \in [0, T]$, i.e. $\boldsymbol{x}(\cdot)$, i.e.,

$$f(x(\cdot)) = \tilde{f}(x(t^*)). \qquad (7.7)$$

In this case, it holds that $Z = \tilde{f}(\Gamma(t^*))$ and $Z^* = \tilde{f}(\Gamma(t^*)) + R_+^m$. If $t^* = T$, the criteria of this kind are known as the *terminal criteria*. To reduce a problem with the criteria of (7.7) type to a problem with the terminal criteria, it is sufficient to consider the shortened time period $[0, t^*]$.

It is possible to consider more complicated criteria that depend on trajectories. However, usually they can be reduced to the terminal criteria. Therefore, the terminal criteria play an important role. It is clear that approximation of an FCS or EPH for the terminal criteria can be based on the preliminary approximation of the reachable set $\Gamma(T)$ and application of one of the methods described in Chapter 6 (for both linear and non-linear objective functions). Say, convolution methods can be used in the case of linear objective functions and polyhedral sets $\Gamma(T)$. Methods for polyhedral approximation can be used in the case of convex bounded sets $\tilde{f}(\Gamma(T))$. In the case of non-linear objective functions, approximation of an FCS by a system of boxes or approximation of its EPH by a system of cones can be used in the case of bounded sets $\Gamma(T)$ and $\tilde{f}(\Gamma(T))$. It is important that the method of approximating the FCS by boxes, which is effective in the case of relatively small dimension of the space $\mathbb{R}^n$, can be applied for approximation of the set $\tilde{f}(\Gamma(T))$ for non-linear objective functions.

Now let us assume that we have already approximated the reachable sets $\Gamma(t)$ for a large, but finite number of time-moments $t_i = i\tau$, where $\tau = T/N$, $i = 0, 1, \ldots, N$. Then, we can approximate, say, the set $Z^* = \tilde{f}(\Gamma(t_i)) + R_+^m$ for all $i = 0, 1, \ldots, N$. It means that we can display the Pareto frontier for any time moment $t_i = i\tau$, $i = 0, 1, \ldots, N$, by its decision map. If the value of τ is sufficiently small, we can provide animation of the picture by changing the decision map in accordance with the monotonic automatic increment of the value of i, that is, of the time moment t_i, for which the decision map is displayed. By this, the user can obtain information on the *dynamic Pareto frontier*, that is, dynamics of the Pareto frontier for the problem.

Now let us consider the differential inclusion (7.5), which is a subject of a controlled momentary impact on the system at some given time moments $0 < t^{(1)} < t^{(2)} < \ldots < t^{(K)} = T$. It is assumed that the momentary impact is described by a mapping that differs from the differential inclusion (7.5). In this case one can approximate the reachable set $\Gamma(t^{(1)})$ and apply then one of the methods described in Chapter 6 for transformation of $\Gamma(t^{(1)})$ into the new set $\Gamma_+(t^{(1)})$, defined in accordance with the description of the momentary impact. By this, the new set $\Gamma_+(t^{(1)})$ can be approximated. Then, it can be used as the initial set on the interval $[t^{(1)}, t^{(2)}]$, at which the reachable set $\Gamma(t^{(2)})$ is approximated, etc. So, one can see that methods for approximation of the reachable sets can be used in this case as well.

These examples prove that approximation of the reachable sets can play an important role in the case of dynamic systems described by controlled differential equations or differential inclusions. Various methods for approximation or estimation of the reachable sets have been proposed (see, for example, Chernous'ko, 1988; Kurzhanski and Valyi, 1996). We concentrate on this problem in the next sections where we show how the methods described in Chapter 6 can be adapted for approximation of the reachable sets.

2. Approximation of reachable sets for convex systems

Let us consider the system (7.1)–(7.3) with linear differential equations, that is,

$$\dot{x} = A(t)x + B(t)u + a(t),$$

where $A(t)$, $B(t)$ and $a(t)$ are matrices and vectors given for $t \in [0, T]$. Let the sets $Y(t)$, $t \in [0, T]$, and $\Gamma(0)$ be convex. Then, the reachable sets $\Gamma(t)$ are convex for $t \in [0, T]$.

However, the reachable sets can be convex in a more general case, too. Let us consider the differential inclusion (7.5). It is known that the set $G(x, t)$ in (7.5) can be substituted for its convex hull without changing the reachable set $\Gamma(T)$ if some simple assumptions are satisfied (Blagodatskikh and Filippov, 1985). So, we can consider conv $G(x, t)$ instead of $G(x, t)$ in (7.5). For any $t \in [0, T]$, the graph $V(t) \subset \mathbb{R}^n \times \mathbb{R}^n$ of the set-valued mapping from $\mathbb{R}^n$ to subsets of $\mathbb{R}^n$ given by conv $G(x, t)$ can be considered, i.e.,

$$V(t) = \{(v, x) \in \mathbb{R}^n \times \mathbb{R}^n : v \in \text{conv } G(x, t), \ x \in \mathbb{R}^n\}.$$

Then, the system

$$\dot{x} = v, \ (v, x) \in V(t) \tag{7.8}$$

can be considered instead of the differential inclusion (7.5). The controlled system of differential equations (7.1)–(7.2) or differential inclusion (7.5) is called convex if it can be reduced to the form (7.8) with convex sets $V(t)$, $t \in [0, T]$. A sufficient condition for the convexity of the sets $V(t)$, $t \in [0, T]$ is given, for example, in (Blagodatskikh and Filippov, 1985). In (Kondrat'ev and Lotov, 1990) several examples of non-linear systems are given that can be reduced to the system (7.8) with a convex set $V(t)$. Say, the systems described by the equation

$$\dot{x} = \varphi(x)\psi(u) \tag{7.9}$$

with a convex $\varphi(x)$ belong to such systems under simple conditions.

It is important that for the convex systems (7.1), (7.2) or (7.5) the set $\Gamma(T)$ is convex in the case of a convex initial set $\Gamma(0)$. This feature is used in several methods for approximation of reachable sets. Such methods are given in this and the next sections. Some of them require a reduction of the system to a finite-dimensional one; other techniques do not require it. We start with a method that does not require such a reduction. It can be applied in the case of systems which are linear in respect to the phase vector and for which the constraints imposed on the control vector do not depend on the phase vector. The methods that require a reduction are described in the next section.

Polyhedral approximation of the reachable sets without finite-dimensional reduction. We consider the differential inclusion (7.5) that is linear in respect to the phase vector, i.e., $G(x, t) = A(t)x + U(t)$. Therefore,

$$\dot{x} = A(t)x + U(t). \tag{7.10}$$

As we have already mentioned, the set $U(t)$, $t \in [0, T]$, can be assumed to be convex. So, the systems described by the linear differential equation, for which the constraint (7.2) is imposed on the controls only, can be reduced to such a class of differential inclusions. Various methods have been proposed for approximation and estimation of the reachable set in this case (see, for example, Pecsvaradi and Narendra, 1971; Chernous'ko, 1988; Kurzhanski and Valyi, 1996).

Let $U(t)$, $t \in [0, T]$, be compact. Moreover, we assume that the initial set $\Gamma(0)$ is convex and compact. Then, reachable set $\Gamma(T)$ is convex and bounded. If the reachable set is bodily, the methods for polyhedral approximation of convex sets based on evaluation of the support function described in Section 2 of Chapter 6 can be applied. The support function for the reachable set $\Gamma(T)$ for the system (7.10) and (7.3) is defined for a direction $w \in \mathbb{S}^n$ as

$$g(w) = \max\left\{\langle w, x \rangle : x \in \Gamma(T)\right\}.$$

One can see that a support function of this kind can be found by solution of the terminal optimization problem

$$\max\left\{\langle w, x(T) \rangle, \text{where } x(\cdot) \text{ satisfies (7.10), (7.3)}\right\}.$$

Such a problem can be solved using the Pontryagin maximum principle introduced in (Pontryagin, Boltyanskii, Gamkrelidze and Mishchenko, 1961). One has first to solve the conjugated problem

$$\dot{\psi} = -A(t)\psi(t), \ t \in [0, T], \ \psi(T) = w.$$

Then, one has to select $x^*(0) \in \Gamma(0)$ and $u^*(t) \in U(t)$ such that

$$\langle \psi(0), x^*(0) \rangle = \max_{x \in \Gamma(0)} \langle \psi(0), x \rangle,$$

$$\langle \psi(t), u^*(t) \rangle = \max_{u(t) \in U(t)} \langle \psi(t), u(t) \rangle.$$

Then the differential equation

$$\dot{x} = A(t)x + u^*(t)$$

is solved for the initial condition $x(0) = x^*$. The end $x^*(T)$ of the computed trajectory provides the value of the support function

$$g(w) = \langle w, x^*(T) \rangle.$$

The described procedure of evaluating the support function is used in the framework of a method for polyhedral approximation of convex bodies to find a sequence of polytopes $P^0, P^1, \ldots, P^k$ that approximates the reachable set $\Gamma(T)$ with any desired precision. It is worth noting that the sequence has the optimal rate of convergence and its efficiency is not less than $1/4$ if the ER method is used. This approach to approximation of reachable sets was introduced in (Lotov, 1985a) and developed further in (Kondrat'ev and Lotov, 1990).

The method described here approximates the reachable set for the final moment T. However, the intermediate reachable sets $\Gamma(t^{(k)})$, $k = 1, 2, \ldots, K$, for time moments $0 < t^{(1)} < t^{(2)} < \ldots < t^{(K)} = T$ can be constructed for any K. It may make a sense to construct the next reachable set using the approximation of the previous set. In this case the problem of specification of the required approximation precision for intermediate reachable sets must be solved. This topic is, however, outside of the scope of our book.

Note that the method described here can be applied in the case of a differential inclusion, which is linear in respect to the phase vector, while the initial set is convex. In the case of more complicated differential inclusions some sophisticated techniques are required. They are described in the next section.

3.   Reduction-based methods for constructing reachable sets

In this section, reduction-based approaches are described. They implement two basic techniques described in Chapter 6 — convolution of linear inequality systems and polyhedral approximation of compact convex bodies.

Reduction to finite-dimensional systems.   The reduction of the systems (7.8) given by

$$\dot{x} = v, \ (v, x) \in V(t),$$

to a finite-dimensional problem consists in substitution of the system by a multi-step system. The sets $\Gamma(0)$ and $V(t), [0, T]$, are assumed to be convex, the set must depend on time continuously. Let us split the interval $[0, T]$ into N equal parts by points $t_i = i\tau$, where $\tau = T/N$, $i = 0, 1, \ldots, N$. Instead of the system (7.8), we consider the linear system

$$\frac{x^{i+1} - x^i}{\tau} = v^i, \quad (v^i, x^i) \in V^\Delta(t_i), \tag{7.11}$$

for $i = 0, 1, \ldots, N-1$, where $V^\Delta(t_i)$ is a polyhedral set that approximates the set $V(t_i)$ with a positive precision Δ. Let $\Gamma^\Delta(0)$ be a polyhedral set that approximates $\Gamma(0)$ with the same precision Δ. For the system (7.11), one can introduce the notion of the reachable set for a particular step k as

$$\Gamma_k = \Big\{ x \in \mathbb{R}^n : x = x^k,$$

$$\text{where } \left(x^0, x^1, \ldots, x^N\right) \text{ satisfies (7.11) and } x^0 \in \Gamma^\Delta(0) \Big\}.$$

Note that if the sets $V(t_i), \ i = 0, 1, \ldots, N - 1$, and $\Gamma(0)$ are not bounded, the opportunity to approximate them by polyhedral sets with any desired positive precision Δ is not guaranteed and must be assumed. As an approximation of the reachable set $\Gamma(T)$, the set Γ_N of all reachable vectors x^N for the system (7.11) and $x^0 \in \Gamma^\Delta(0)$ is used. The sufficient conditions of approximation of the set $\Gamma(T)$ by the sets Γ_N while $N \to \infty$ and $\Delta \to 0$ are given in the paper (Lotov, 1979) and repeated in the book (Lotov, Bushenkov and Kamenev, 1999).

So, we have reduced the problem of approximation of the reachable set for (7.8) to the problem of approximation of the reachable set for the multi-step system (7.11). It is clear in advance that such an approach has a negative feature — an additional error is brought by the reduction to a multi-step system. However, the reduction-based methods can be applied for a large scope of dynamic systems. Here, we consider several types of methods for constructing an approximation of the reachable set Γ_N for the multi-step system (7.11) — methods based on convolution of linear inequality systems and methods based on polyhedral approximation of convex bodies.

Methods based on convolution techniques (direct method).
The first ideas concerning the approximation of the reachable sets for convex systems (7.1)–(7.3) on the basis of convolution of linear inequality systems were introduced and studied as early as in the 1970s (Lotov, 1972a; Lotov, 1972b; Lotov, 1973a; Lotov, 1975b; Lotov, 1978; Lotov, 1979). New convolution-based methods have been proposed later (Lotov, 1985a). All these methods are based on the reduction to multi-step systems. Here we consider the direct method, while an alternative approach (method of tied phase vectors) is described in the next subsection.

The direct method is based on the idea to use convolution for step-by-step constructing of the reachable sets for the system (7.11) for which $x^0 \in \Gamma^\Delta(0)$. Let us consider the first step, that is, constructing the set Γ_1. This set is given by the variety of vectors x^1 that satisfy (7.11) for $i = 0$ and $x^0 \in \Gamma^\Delta(0)$. In other words, for a vector $x^1 \in \Gamma_1$, there exist such vectors x^0 and v^0 that

$$\frac{x^1 - x^0}{\tau} = v^0, \quad (v^0, x^0) \in V^\Delta(0), \quad x^0 \in \Gamma^\Delta(0). \tag{7.12}$$

In turn, Γ_2 is the variety of vectors x^2, for which such vectors x^0, x^1, v^0, v^1 can be found that

$$\frac{x^2 - x^1}{\tau} = v^1, \quad \frac{x^1 - x^0}{\tau} = v^0,$$

$$(v^1, x^1) \in V^\Delta(0), \quad (v^0, x^0) \in V^\Delta(\tau), \quad x^0 \in \Gamma^\Delta(0).$$

Note that we can modify the definition of the set Γ_2, if we use the set Γ_1. We can re-write the above system in the following manner: vector x^2 belongs to the set Γ_2, if there exist such x^1 and v^1 that

$$\frac{x^2 - x^1}{\tau} = v^1, \quad (v^1, x^1) \in V^\Delta(t_1), \quad x^1 \in \Gamma_1. \tag{7.13}$$

In a totally analogous way, we can find that the reachable set Γ_{i+1} contains such vectors x^{i+1}, for which x^i and v^i can be found that

$$\frac{x^{i+1} - x^i}{\tau} = v^i, \quad (v^i, x^i) \in V^\Delta(t_i), \quad x^i \in \Gamma_i. \tag{7.14}$$

Therefore, for the set Γ_i, which is given in the form of the solution set of a linear inequality system, the problem of constructing the set Γ_{i+1} can be solved by application of Fourier convolution. Let us assume that the approximation $\Gamma^\Delta(0)$ of the initial set $\Gamma(0)$ as well as the sets $V^\Delta(t_i)$ are given in the form of solution sets of linear inequality systems. Then, we can apply the convolution techniques to exclude the vectors x^0 and v^0 from the linear inequality system (7.12). The set Γ_1 is obtained in the desired form. Continuing this procedure, we can use (7.13) to obtain the set Γ_2. Then, step by step, we obtain the sets $\Gamma_3, \Gamma_4, \ldots, \Gamma_N$. At the $(i+1)$-th step, devoted to constructing Γ_{i+1}, we have to exclude the vectors x^i and v^i from the linear inequality system (7.14). It is important that linear equalities can be used for excluding the vector x^i. Therefore, only the vector v^i must be excluded by means of the convolution techniques.

The above step-by-step procedure for constructing the reachable set Γ_N has the following advantages:

1 in contrast to the method of polyhedral approximation described in the previous section, now the systems can be studied, which include phase constraints and the constraints that combine control and phase vectors;

2 the intermediate sets $\Gamma_1, \Gamma_2, \ldots, \Gamma_{N-1}$ are constructed as an adjoint product;

3 the sets $V(t_i)$ and $\Gamma^\Delta(0)$ may be not confined.

The negative features of the method are related to the disadvantages of the convolution techniques that have already been discussed in Chapter 6. To improve the situation, one can use approximation of intermediate reachable sets by simplified polyhedral sets on some steps. By this, a bulky description of an intermediate reachable set can be simplified.

Several convolution-based methods for approximation of the reachable sets for multi-step systems were coded in the framework of the package of programs POTENTIAL outlined in Chapter 6.

The method described in this subsection has recently been applied for constructing the reachable sets in the framework of a study of the stability in dynamic control problems (Bushenkov and Smirnov, 1992; Bushenkov and Smirnov, 1997).

The method of tied phase vectors. The convolution techniques can be applied for constructing the reachable sets for the system (7.11) in a different way in the case of autonomous systems. The system (7.8) is autonomous if the sets $V(t)$ do not depend on time, that is $V(t) \equiv V$

for all $t \in [0, T]$. As a result, the sets V are the same on all steps in the multi-step system (7.11). Let us consider the set

$$\Gamma_{k,k+1} = \left\{ (x^k, x^{k+1}) \in \mathbb{R}^{2n} : \frac{x^{k+1} - x^k}{\tau} = v^k, \ (v^k, x^k) \in V^\Delta \right\}.$$
$$(7.15)$$

The $2n$-dimensional set $\Gamma_{k,k+1}$ contains such phase vectors (x^k, x^{k+1}) that one can bring the system (7.11) from x^k to x^{k+1} using feasible control v^k. To obtain the set $\Gamma_{k,k+1}$ in an explicit form that does not contain the control v^k, one has to exclude the vector v^k by using the equality in (7.15). As the result, we obtain

$$\Gamma_{k,k+1} = \left\{ (x^k, x^{k+1}) \in \mathbb{R}^{2n} : \left(\frac{x^{k+1} - x^k}{\tau}, x^k \right) \in V^\Delta \right\}. \qquad (7.16)$$

Since the set V^Δ does not depend on time, it holds that $\Gamma_{k,k+1} = \Gamma_{0,1}$ for any $k = 0, 1, \ldots, N - 1$, that is, all the sets of this kind are equal.

Now let us consider the set $\Gamma_{k,k+2}$ that is defined as

$$\Gamma_{k,k+2} = \left\{ (x^k, x^{k+2}) \in \mathbb{R}^{2n} : \begin{array}{c} (x^k, x^{k+1}) \in \Gamma_{k,k+1}, \\ (x^{k+1}, x^{k+2}) \in \Gamma_{k+1,k+2} \end{array} \right\}$$

for $k = 0, 1, \ldots, N - 2$. Since $\Gamma_{k,k+1} = \Gamma_{0,1}$ for any $k = 0, 1, \ldots, N - 1$, it holds that

$$\Gamma_{k,k+2} = \left\{ (x^k, x^{k+2}) \in \mathbb{R}^{2n} : \begin{array}{c} (x^k, x^{k+1}) \in \Gamma_{0,1}, \\ (x^{k+1}, x^{k+2}) \in \Gamma_{0,1} \end{array} \right\}. \qquad (7.17)$$

Once again, to obtain the set $\Gamma_{k,k+2}$ in the explicit form of the solution set of a linear inequality system, one has to exclude the vector x^{k+1} from the description (7.17) by using convolution techniques. Note that $\Gamma_{k,k+2} = \Gamma_{0,2}$ for $k = 0, 1, \ldots, N - 2$.

In the same way we can define $\Gamma_{k,k+2l}$ for any differences of step numbers $l = 2^s$, where $s = 0, 1, 2, \ldots$:

$$\Gamma_{k,k+2l} = \left\{ (x^k, x^{k+2l}) \in \mathbb{R}^{2n} : \begin{array}{c} (x^k, x^{k+l}) \in \Gamma_{0,l}, \\ (x^{k+l}, x^{k+2l}) \in \Gamma_{l,2l} \end{array} \right\}$$

and can obtain the explicit form of $\Gamma_{k,k+2l}$ by using the convolution techniques.

If the difference of step numbers l can not be represented in the form 2^s, the related sets $\Gamma_{k,k+l}$ can also be found by means of convolution techniques. For example, for $l = 3$ we have to consider its implicit form

$$\Gamma_{k,k+3} = \left\{ (x^k, x^{k+3}) \in \mathbb{R}^{2n} : \begin{array}{c} (x^k, x^{k+2}) \in \Gamma_{0,2}, \\ (x^{k+2}, x^{k+3}) \in \Gamma_{0,1} \end{array} \right\}$$

and use convolution techniques to exclude the vector x^{k+2} to obtain its explicit description.

After a finite number of exclusions, we find, say, the set $\Gamma_{0,N}$. Now, to obtain the reachable set Γ_N, one has to note that the set Γ_N can be represented in the implicit form

$$\Gamma_N = \{x^N \in \mathbb{R}^n : (x^0, x^N) \in \Gamma_{0,N}, \ x^0 \in \Gamma_0\}.$$

If the set Γ_0 is given by an inequality system, it is sufficient to exclude the vector x^0 from the above description to obtain the reachable set Γ_N in the same form.

Note that the sets $\Gamma_{k,k+l}$ are usually not bounded. However, it is not a problem in this case, since the convolution techniques can be applied even in the case of non-bounded sets.

Method based on polyhedral approximation. In this method it is required that the approximated set Γ_N is bounded. Note that the set Γ_N is defined implicitly by the system of linear equalities and inequalities

$$\frac{x^{i+1} - x^i}{\tau} = v^i, \quad (v^i, x^i) \in V^\Delta(t_i), \quad x^0 \in \Gamma^\Delta(0),$$

where $i = 0, 1, \ldots, N - 1$. The support function for it can be evaluated using linear programming software. Due to it, the set Γ_N can be approximated by a polytope with any precision by using the polyhedral approximation techniques described in Section 2 of Chapter 6. The advantage of this method consists in the opportunity to approximate the reachable set for convex differential inclusions with relatively large dimension of the phase vector (up to eight) and with practically any number of constraints in the description of a multi-step system.

4.   Approximating and estimating reachable sets for non-convex systems

Several approaches can be used in the case of non-convex dynamic systems. We consider two of them. The first one is aimed at the approximation of the reachable sets for non-convex systems by collections of boxes. The second one applies estimating the reachable set for a non-convex system by approximating the reachable set for an auxiliary convex system, which reachable sets contain the reachable sets for the original non-convex system.

Approximating the reachable sets by collections of boxes. Since a reasonable approximation of an FCS by a collection of boxes can be obtained only in the case of a relatively small dimension of the decision space, it is impossible to apply such an idea directly to differential equation (7.1) with the infinite-dimensional decision space. However, it is often possible to decrease the dimension of the decision space by parameterization of the variety of feasible control functions. Assume that the set of feasible controls can be parameterized by S parameters, that is

$$u(t) = \sum_{s=1}^{S} \alpha_s u^{(s)}(t),$$

where α_s, $s = 1, 2, \ldots, S$, are undefined parameters and $u^{(s)}(t)$, $s = 1, 2, \ldots, S$, are given vector-functions of time. If such parameterization is possible, the method for approximation of an FCS based on covering by boxes (Section 1 of Chapter 5) can be applied directly. This approach has been used in several studies of non-linear dynamic economic models (Lotov, Bushenkov and Kamenev, 1999).

Estimation of the reachable set by using an auxiliary convex system. Let us consider the system (7.8) with the initial set (7.3). Both the system and the initial set are not assumed to be convex. Let us consider an auxiliary system

$$\dot{x} = v, \ (v, x) \in \operatorname{conv} V(t) \qquad (7.18)$$

with the initial condition

$$x(t_0) \in \operatorname{conv} \Gamma(0). \qquad (7.19)$$

Denoting by $D(t)$ the reachable set for the system (7.18)–(7.19), we can assert that $\Gamma(t)$ belongs to $D(t)$ for $t \in [0, T]$. Methods for constructing the set $D(t)$ have already been described in this chapter. Therefore, we can use these methods for estimating the set $\Gamma(t)$ for $t \in [0, T]$. An example of estimation of the reachable set for a particular non-convex system is given in (Kondrat'ev and Lotov, 1990).

5. Approximating the FCS for systems in partial derivatives

Abstract statement of the problem. Let us consider a confined set D from the metric space $\mathbb{R}^r$ with the boundary ∂D. Let us assume that linear normed spaces Φ, U and V are given that contain functions

$$\varphi \in \Phi, \quad u \in U, \quad v \in V$$

defined at $D+\partial D$, D and ∂D correspondingly. Let us consider the linear equation in partial derivatives defined on D,

$$A\varphi = u, \qquad (7.20)$$

where A is a linear operator from Φ into U, and let us assume that the boundary condition is given on ∂D,

$$a\varphi = v, \qquad (7.21)$$

where a is a linear operator that maps Φ into V. The relations (7.20)–(7.21) form the boundary problem, which is assumed to be stated correctly, i.e., there exists the only function $\varphi \in \Phi$ that satisfies (7.20)–(7.21) for any pair $u \in U$ and $v \in V$, which depends continuously on $u \in U$ and $v \in V$.

For the boundary problem (7.20)–(7.21), the problem of the choice of functions $u \in U$ and $v \in V$ can be stated. In such problems, the right-hand functions u and v in (7.20) and (7.21) are not given in advance, they represent the control functions that must satisfy the constraint

$$\{\varphi, u, v\} \in \Psi, \qquad (7.22)$$

where Ψ is a set of the space $\Phi \times U \times V$. Let us assume here that the set Ψ is convex.

Let us introduce the notion of FCS for this problem(Lotov, 1981c). Let linear continuous operators F_1, F_2, F_3, be given that map Φ, U and V into the linear normed criteria space B. Let us denote

$$f(\varphi, u, v) = F_1\varphi + F_2u + F_3v. \qquad (7.23)$$

Note that the space B may be infinite-dimensional, that is, functions may be its elements.

DEFINITION 7.1 *FCS for the system (7.20)–(7.23) is given by*

$$Z = \{z \in B : z = f(\varphi, u, v), \ A\varphi = u, \ a\varphi = v, \ \{\varphi, u, v\} \in \Psi\}. \qquad (7.24)$$

Note that the set Z is convex.

Approximation of the FCS. Approximation of the above set Z can be based on substitution of the system (7.20)–(7.22) by a finite-dimensional system and constructing of an FCS for it. Say, finite differences can be applied (though finite element methods can be used, too). Let us consider uniform meshes with the step h in the vicinity of the set D. Let us denote by ∂D_h the set of mesh points that are used for approximation of the boundary condition (7.21). The internal mesh points are denoted by D_h. Analogously to the definition of the spaces Φ, U and V,, we can define the spaces of mesh functions Φ_h, U_h and V_h on $D_h + \partial D_h$, D_h and ∂D_h correspondingly. The following equations are considered in the space of such functions:

$$A^h \varphi_h = u_h \quad \text{on} \quad D_h, \tag{7.25}$$

$$a^h \varphi_h = v_h \quad \text{on} \quad \partial D_h, \tag{7.26}$$

where A^h and a^h are the linear operators, $\varphi_h \in \Phi_h, u_h \in U_h, v_h \in V_h$. Let us require that φ_h, u_h and v_h satisfy

$$\{\alpha_h, u_h, v_h\} \in \Psi_h, \tag{7.27}$$

where Ψ_h is a polyhedral set of $\Phi_h \times U_h \times V_h$. If the space B is infinite-dimensional, we consider a finite-dimensional space B_h, in the opposite case we let $B_h = B$. Let F_1^h, F_2^h, F_3^h be the linear operators from Φ_h, U_h and V_h into B_h. Let

$$y = F_1^h \varphi_h + F_2^h u_h + F_3^h v_h. \tag{7.28}$$

DEFINITION 7.2 *The FCS for the system (7.25)–(7.28) is given by*

$$Z^h = \left\{ z \in B_h : z = F_1^h \varphi_h + F_2^h u_h + F_3^h v_h, \tag{7.29} \right.$$

$$\left. A^h \varphi_h = u_h, \quad a^h \varphi_h = v_h, \quad \{\varphi_h, u_k, v_k\} \in \Psi_h \right\}.$$

Note that the set Z^h is polyhedral. Since (7.25)–(7.27) is a finite system of linear equalities and inequalities imposed on the functions that are given in the finite number of mesh points, the set Z^h can be approximated by the finite-dimensional methods described in Chapter 6. The theoretical problems of the approximation of the set Z by the sets Z^h were studied in (Lotov, 1987); see (Lotov, Bushenkov and Kamenev, 1999), too. Visualization of the finite-dimensional sets Y^k can be based on the methods developed for the finite-dimensional systems. Let us consider one example provided by the evolutionary equation.

Example: FCS for the evolutionary equation. The evolutionary equation, which is a partial case of equations (7.20)–(7.21), is given by

$$\frac{\partial \varphi}{\partial t} + A\varphi = u \quad \text{on} \quad D \equiv H \times (0, T) \tag{7.30}$$

with the boundary condition

$$a\varphi = v \quad \text{on} \quad \partial H \times (0, T) \tag{7.31}$$

and the initial condition

$$\varphi = w \quad \text{at} \quad H, \tag{7.32}$$

where H is a confined set of a linear metric space. In the system (7.30)–(7.32), the controls are provided by u, v, and w.

The constraint (7.22) is given in a particular form

$$\{\varphi|_{t=\theta}, u|_{t=\theta}, v|_{t=\theta}\} \in \Psi_\theta, \quad \theta \in (0,T), \tag{7.33}$$

where Ψ_θ is a convex set of $\Phi \times U \times V$ for all $\theta \in [0,T]$, and

$$w \in \Gamma(0) \subset \Psi_0, \tag{7.34}$$

where Ψ_0 is the space of initial conditions. The set $\Gamma(0)$ is assumed to be convex.

In addition to the FCS of the general type given by (7.23), one can consider its partial case − usual reachable set $\Gamma(T)$, that is, the set of such functions ψ defined on H, for which such a solution $\varphi \in \Phi$ of the system (7.30)–(7.34) exists that satisfies $\psi = \varphi|_{t=T}$. The set $\Gamma(T)$ provides a generalization of the reachable set for controlled differential equations or differential inclusions. Such a set $\Gamma(T)$ can be approximated using methods based on the reduction to finite-dimensional problems.

An application of the FGM to a multi-criteria problem described by a linear boundary problem in partial derivatives (control of forest plantation) is provided in (Lotov and Stolyarova, 1984). An application of the FGM/IDM technique in the case of a non-linear dynamic multicriteria decision problem described by equations in partial derivatives has already been outlined in Section 2 of Chapter 5.

III

SEVERAL TOPICS RELATED TO MATHEMATICAL BASIS OF IDM

Chapter 8

THEORY OF ITERATIVE METHODS FOR POLYHEDRAL APPROXIMATION OF CONVEX BODIES

As was shown in Chapter 6, iterative methods for polyhedral approximation of multi-dimensional convex compact bodies (CCB) play now an important role in real-life applications of the Interactive Decision Maps technique. We have demonstrated in Chapter 7 that such methods can be used for approximating the reachable sets of dynamic systems as well. In addition to the approximation of the feasible criterion sets in multi-criteria decision problems and reachable sets in dynamic problems, methods for polyhedral approximation of CCBs can be applied to mathematical programming (see, for example, Sonnevend, 1977; Vasil'yev, 1983, etc.).

Several theoretical results concerning iterative methods for polyhedral approximation of CCBs were already given in Section 2 of the Chapter 6. Here, we provide a detailed description of the theory of the methods. We restrict ourselves to asymptotic theory that describes the behavior of the methods for a sufficiently large number of iterations. Most of the results are given without proofs that have already been published. Only the most important results are proved here.

1. Polyhedral approximation of compact convex bodies

The most important distinctive feature of polyhedral approximation methods compared to methods that use a single convex body for approximation (such as a single simplex, a single cube or a single ellipsoid) is that they are able to approximate a CCB with any degree of accuracy. This advantage, however, is quite costly: practical experience and theoretical estimates (Gruber, 1983; Gruber, 1994) show that the complexity of the description of the approximating polytope increases rapidly as the required accuracy and the dimension of the CCB increase. Moreover, in many applications, each measurement of a characteristic of a CCB that is needed in the process of implementation of the method (for instance, computing its support or distance function) may be quite complicated and time consuming. All this requires estimating the rate of convergence of the existing methods and using of methods that are optimal in the sense of the complexity of the description of the approximating polytopes, and of convergence rate and of the number of computations required for the support or distance functions.

We assume that the approximated CCB belongs to linear space $\mathbb{R}^d$ (instead of letter m that denoted the dimension of the space in Chapter 6, we use letter d, which is more common in approximation theory). Note that methods for polyhedral approximation of a CCB depend on the way the approximated body is described. Except Minkowski's classical method (Minkowski, 1903) as well as the quite specific method of (Gordon, Meyer and Reisner, 1994), where the descriptions of the volumes of all l-dimensional, $1 \le l < d$, slices of the approximated body must be pre-specified, the methods for polyhedral approximation of CCBs can be divided into two large groups:

(a) non-iterative methods;

(b) iterative methods, which improve the accuracy of approximation on an iteration by including a certain number of new vertices (or hyperfaces) in the description of the approximating polytope.

Methods of the first group are mainly based on constructing an asymptotically optimal covering of the surface of the approximated body (see Schneider, 1983; Schneider, 1987; Gruber, 1988; Gruber, 1993a; Gruber, 1993b; McClure and Vitale, 1975). Such methods are non-constructive in the case of more than two dimensions, though some asymptotic approaches for $d > 2$ are suggested in (Gruber, 1993a) for an analytically defined CCB.

A different non-iterative approach is based on the idea that was proposed in the famous book (Pontryagin, Boltyanskii, Gamkrelidze and Mishchenko, 1961). The idea consists in considering the convex reachable sets of linear dynamic systems as a CCB defined implicitly by its support function. In this approach, the reachable set can be approximated by a circumscribed polytope whose hyperfaces belong to the supporting hyperplanes of the reachable set, while the normals of these hyperplanes have been pre-specified. The guaranteed approximation accuracy is not always known.

This approach was developed further in the method described in (Vasil'yev, 1983; Vasil'yev, 1988). It starts with a required accuracy and a priori information concerning the inclusion of the approximated body in a sphere of a known radius. An ε-grid is chosen on the surface of the sphere for collection of the normal directions. The grid is constructed as a uniform grid in spherical coordinates and provides directions for evaluating the support function. A similar idea is used in the method of (Samsonov, 1983), which suggests a heuristics to eliminate hyperfaces that are irrelevant for the achievement of the required accuracy.

The iterative methods can be divided into two classes: non-adaptive and adaptive. Adaptive iterative methods include the methods of (Cohon, 1978; Sonnevend, 1980; Sonnevend, 1983; Appino, 1984; Solanki, Appino and Cohon, 1993) as well as the methods described in this book. In adaptive methods, the optimization tasks solved on an iteration depend on the approximating polytope that was constructed on previous iterations.

The features of adaptive and non-adaptive methods for approximation of CCBs were theoretically studied in (Sonnevend, 1980; Sonnevend, 1983). It was shown that, in the absence of smoothness information about the surface of the approximated body, the rate of convergence of non-adaptive methods is essentially lower than the rate of convergence of adaptive methods. Experimental data (Samsonov, 1983) confirm this theoretical property.

Let us consider the adaptive methods. In Chapter 6 we have already discussed the problem of asymptotic optimality of the convergence rate $n^{2/(1-d)}$ for the number of iterations, support function evaluations and vertices (or hyperfaces) of the approximating polytope (here n is, for example, the number of iterations). In (Sonnevend, 1983) it was shown that his approximation method proposed in (Sonnevend, 1980; Sonnevend, 1983) is asymptotically sub-optimal in respect to the order of the number of iterations, hyperfaces of the approximating polytope and support function evaluations, where sub-optimality implies optimality apart from an irremovable logarithmic multiplicative factor, that is, convergence of the type

$$\mathcal{O}\left(n^{2/(1-d)}\log n\right).$$

In contrast, the methods used in this book have optimal convergence rate $n^{2/(1-d)}$ at least for the number of iterations and vertices (hyperfaces). As we have already said in Chapter 6, this property is related to an important feature of the methods — they are based on the Hausdorff adaptive schemes. This topic is considered in this chapter along with other important topics related to the methods of polyhedral approximation of CCBs. We start with a review of the asymptotic theory of polyhedral approximation of CCBs.

## 2.	Asymptotic theory of polyhedral approximation of compact convex bodies

Let $\mathbb{R}^d$ be the Euclidian space with scalar product $< \cdot, \cdot >$, distance $\rho(\cdot, \cdot)$, norm $|| \cdot ||$ and volume measure μ. Let $B_r(z)$ be the ball centered in z with radius r, B^d be the unit ball cantered in the origin, S^{d-1} be the boundary of B^d. Let $\pi_d = \mu(B^d)$. Let

$$C_1 + C_2 = \{z \in \mathbb{R}^d : z = x + y, \quad x \in C_1, \ x \in C_2\}$$

for sets C_1 and C_2. Let

$$C_\lambda = C + B_\lambda(0) = \bigcup_{x \in C} B_\lambda(x).$$

Let ∂C be the boundary of C, $\omega(C)$ be its asphericity (the minimum ratio of the radii of the concentric balls inscribed in C ($r(C)$) and circumscribed around C ($R(C)$)), $\omega_0(C)$ be the minimum ratio of the radii of the balls inscribed in C and circumscribed around C and centered in the origin.

Let C be the class of CCB. For $C \in \mathsf{C}$, let $\mu(C)$ be its volume and $\sigma(C)$ its surface volume (see Leichtweiss, 1980). Let C^s be the class of CCB with s times continuously differentiable boundaries and positive principal curvatures, $s \geq 2$. For $s \geq 2$ we denote by $r_{min}(C)$ and $r_{max}(C)$, respectively, the minimum and maximum radii of curvature of the surface of ∂C, $C \in \mathsf{C}^s$.

Let P be the class of convex bodily polytopes $\mathsf{P} \subset \mathsf{C}$. For $P \in \mathsf{P}$ we denote
the set of its vertices by $M^t(P)$;
the number of its vertices by $m^t(P)$;
the set of the unit normals to its hyperfaces (faces of the maximum dimension) by $M^f(P)$;
the number of its hyperfaces by $m^f(P)$.

For $C \in \mathsf{C}$ we define the class $\mathsf{P}^i(C)$ of inscribed polytopes whose vertices belong to ∂C, and the class $\mathsf{P}^c(C)$ of circumscribed polytopes, whose hyperfaces touch ∂C. We also define the classes

$$\mathsf{P}^i_m(C) = \{P \in \mathsf{P}^i(C) : m^t(P) \leq m\},$$

$$\mathsf{P}^c_m(C) = \{P \in \mathsf{P}^c(C) : m^f(P) \leq m\}.$$

We consider the traditional metrics on C:

the Hausdorff metric

$$\delta^H(C_1, C_2) = \max\{\sup\{\rho(x, C_2) : x \in C_1\}, \sup\{\rho(x, C_1) : x \in C_2\}\}$$

and the symmetric difference volume metric

$$\delta^S(C_1, C_2) = \mu(C_1 \triangle C_2),$$

where $C_1 \triangle C_2 = (C_1 \backslash C_2) \cup (C_2 \backslash C_1)$.

Wherever possible, indices will be omitted. Thus, we use δ to denote both δ^H and δ^S, $\mathsf{P}_m(C)$ to denote $\mathsf{P}_m^i(C)$ and $\mathsf{P}_m^c(C)$, $\mathsf{P}(C)$ to denote $\mathsf{P}_m(C)$, $M(P)$, $m(P)$ to denote $M^t(P)$, $m^t(P)$ for $P \in \mathsf{P}^i(C)$, and to denote $M^f(P)$, $m^f(P)$ for $P \in \mathsf{P}^c(C)$. We also denote

$$\delta(C, \mathsf{P}_m) = \inf\{\delta(C, P) : P \in \mathsf{P}_m(C)\}.$$

Let $\mathbb{R}_0^d = \mathbb{R}^d \backslash \{0\}$, $\mathsf{C}_0 = \{C \subset \mathsf{C} : \{0\} \in \operatorname{int} C\}$, $\mathsf{P}_0 = \mathsf{C}_0 \cap \mathsf{P}$.

For $u \in \mathbb{R}_0^d$ we define the support functions

$$g(u, C) = \max\{\langle u, x \rangle : x \in C\},$$

the supporting half-space

$$L(u, C) = \{x \in \mathbb{R}^d : \langle u, x \rangle \leq g(u, C)\},$$

the supporting hyperplane

$$l(u, C) = \left\{x \in \mathbb{R}^d : \langle u, x \rangle = g(u, C)\right\},$$

the set of points of tangency

$$T(u, C) = \{p \in \partial C : \langle u, p \rangle = g(u, C)\},$$

and the cone of the unit external normals in $p \in \partial C$,

$$S(p, C) = \left\{u \in S^{d-1} : \langle u, p \rangle = g(u, C)\right\}.$$

For $p \in \mathbb{R}^d$ we will use the notation

$$g(u, p) = \langle u, p \rangle,$$

$$l(u, p) = \{x \in \mathbb{R}^d : \langle u, x \rangle = \langle u, p \rangle\},$$

$$L(u, p) = \{x \in \mathbb{R}^d : \langle u, x \rangle \leq \langle u, p \rangle\}.$$

For $C \in \mathsf{C}_0$ and $x \in \mathbb{R}^d$ we define the distance function of C (Leichtweiss, 1980; Rockafellar, 1970)

$$g^*(x, C) = \min\{\lambda \geq 0 : x \in \lambda C\}.$$

From the definition of this function it immediately follows that

$$g^*(x, C) = ||x||/||x_0||,$$

where $x_0 = [0, x) \cap \partial C$ — the point of intersection of the ray from the origin to the point x and the boundary of C (Leichtweiss, 1980). We denote this point x_0 as

$$t(x, C) = x/g^*(x, C) \in \partial C, \ x \in \mathbb{R}_0^d.$$

Let $C \in \mathsf{C}$. According to (Gruber, 1994) for δ (i.e., δ^H and δ^S) in $\mathsf{P}_m(C)$ (i.e., in $\mathsf{P}_m^i(C)$ and $\mathsf{P}_m^c(C)$) there exists a polytope (best approximation polytope) P_m with the property

$$\delta(C, P_m) = \delta(C, \mathsf{P}_m).$$

Unfortunately, very little is known about these polytopes. A classical result (Minkowski, 1903) of the theory of convex sets states that for any $C \in \mathsf{C}$,

$$\lim_{m \to \infty} \delta(C, \mathsf{P}_m) = 0. \tag{8.1}$$

According to (Dudley, 1974; Bronshtein and Ivanov, 1975):

$$\delta(C, \mathsf{P}_m) \leq \frac{\mathrm{const}_{C,d,\delta}}{m^{2/(d-1)}}. \tag{8.2}$$

For $C \in \mathsf{C}^2$ there exists the result (Schneider and Wieacker, 1981; Gruber and Kendrov, 1982):

$$\delta(C, \mathsf{P}_m) \geq \frac{\mathrm{const}_{C,d,\delta}}{m^{2/(d-1)}}. \tag{8.3}$$

Moreover according to (Schneider, 1983; Schneider, 1987, for metric δ^H and $C \in \mathsf{C}^3$) and (Gruber, 1988; Gruber, 1993a; Gruber, 1993b, for metric δ^H and $C \in \mathsf{C}^2$, and metric δ^S) for $C \in \mathsf{C}^2$ it follows:

$$\delta(C, \mathsf{P}_m) \sim \frac{\mathrm{const}_{C,d,\delta}}{m^{2/(d-1)}}. \tag{8.4}$$

In it important that the rate of convergence $2/(d-1)$ in (8.1)–(8.4) for $C \in \mathsf{C}^2$ can not be improved (Schneider and Wieacker, 1981).

Let us consider the asymptotic values for constants in (8.4). Let $C \in \mathsf{C}^2$. Then according to (Schneider, 1983; Schneider, 1987; Gruber, 1993a), it follows that

$$\lim_{m \to \infty} \delta^H(C, \mathsf{P}_m)\, m^{\frac{2}{d-1}} = \frac{1}{2} \left(\frac{\vartheta_{d-1}}{\pi_{d-1}} \int_{\partial C} k_C(x)^{1/2} d\sigma(x) \right)^{\frac{2}{d-1}}, \tag{8.5}$$

where ϑ_l denotes the minimum covering density of $\mathbb{R}^l$ by unit balls (Rogers, 1964), π_d denotes the volume of unit ball,

$$\pi_d = \pi^{d/2}/\Gamma((d/2)+1),$$

$k_C(x)$ is the Gauss–Kronecker curvature at $x \in \partial C$ and $\sigma(x)$ is the surface volume element at x. Note that $\vartheta_1 = 1$ and $\vartheta_2 = 2\pi/\sqrt{27}$, but the values of ϑ_l are not known precisely for $l > 2$.

Let $C \in C^2$. Then according to (Gruber, 1988; Gruber, 1993a; Gruber, 1993b), there exist constants del_{d-1} and div_{d-1}, which depend only on d, such as

$$\lim_{m\to\infty} \delta^S\left(C, \mathsf{P}^i_m\right) m^{\frac{2}{d-1}} = \frac{1}{2}\,\mathrm{del}_{d-1}\left(\int_{\partial C} k_C\left(x\right)^{\frac{1}{d+1}} d\sigma(x)\right)^{\frac{d+1}{d-1}}, \quad (8.6)$$

and

$$\lim_{m\to\infty} \delta^S\left(C, \mathsf{P}^c_m\right) m^{\frac{2}{d-1}} = \frac{1}{2}\,\mathrm{div}_{d-1}\left(\int_{\partial C} k_C\left(x\right)^{\frac{1}{d+1}} d\sigma(x)\right)^{\frac{d+1}{d-1}}. \quad (8.7)$$

Only $\mathrm{del}_1 = 1/6$, $\mathrm{del}_2 = 1/(2\sqrt{3})$, $\mathrm{div}_1 = 12$, $\mathrm{div}_2 = 5/(18\sqrt{3})$ are known.

Let $C \in C$ and $P \in \mathsf{P}(C)$. We introduce the efficiency of polyhedral approximation of C by P as

$$\eta\left(P\right) = \frac{\delta\left(C, \mathsf{P}_{m(P)}\left(C\right)\right)}{\delta\left(C, P\right)}.$$

Let $\mathrm{F} = \{P^n\}_{n=1,2,\dots}$ be a sequence of polytopes from $\mathsf{P}(C)$ converging to $C \in C$. Then we denote

$$\underline{\eta}\left(\mathrm{F}\right) = \liminf_{n\to\infty}\,\eta(P^n)$$

and

$$\overline{\eta}\left(\mathrm{F}\right) = \limsup_{n\to\infty}\,\eta(P^n)$$

as, respectively, the lower and the upper asymptotic efficiency of the sequence F. The asymptotic efficiency of a method of polyhedral approximation is defined as the asymptotic efficiency of the sequence of polytopes generated by it.

Obviously

$$0 \le \underline{\eta}\left(\mathrm{F}\right) \le \overline{\eta}\left(\mathrm{F}\right) \le 1.$$

For any P_m, $m = d+1, d+2, \dots$, the efficiency of polyhedral approximation is equal to 1. For this reason the hypothetical method generating the sequence of the best approximation polytopes we will consider as the optimal one.

The method for polyhedral approximation that generates the sequence of polytopes with an asymptotic efficiency greater than zero is denoted as asymptotically efficient.

If the method for polyhedral approximation generates the sequence of polytopes with the same rate of convergence as the sequence of best approximation polytopes, then we define it as optimal in respect to the order of the number of vertices (hyperfaces).

For the sequence $\{P^n\}_{n=0,1,\dots}$ generated by an asymptotically efficient method from (8.2)–(8.3) it follows that

$$\delta(C, P^n) \leq \frac{\text{const}}{m\left(P^n\right)^{2/(d-1)}}.$$

Moreover if $C \in \mathsf{C}^2$, then it follows that

$$\delta(C, P^n) \geq \frac{\text{const}'}{m\left(P^n\right)^{2/(d-1)}}.$$

Therefore, the definitions of asymptotically efficient method and the method optimal in respect to the order of the number of vertices (hyperfaces) coincide for any $C \in \mathsf{C}^2$, as it follows from (Schneider and Wieacker, 1981). They coincide for any CCB with a boundary point, in a neighborhood of which the boundary is two-times differentiable and has positive Gaussian curvature, too.

3.　Augmentation and cutting schemes

Let us now define general schemes of polyhedral approximation.

Augmentation scheme

Let $P^n \in \mathsf{P}^i(C)$. Then, the $(n+1)$-th iteration consists of two steps:
Step 1. Select $p_n \in \partial C$.
Step 2. Construct $P^{n+1} = \text{conv}\,\{p_n, P^n\}$.
Particular methods based on this scheme are defined by the way of solving two problems:

1　how to select $p_n \in \partial C$;

2　how to construct $P^{n+1} = \text{conv}\,\{p_n, P^n\}$ in the form needed for step 1 of the next iteration.

Cutting scheme

Let $P^n \in \mathsf{P}^c(C)$. Then, the $(n+1)$-th iteration consists of two steps:

Step 1. Select $u_n \in S^{d-1}$;
Step 2. Construct $P^{n+1} = P^n \cap L(u_n, C)$.
Particular methods based on this scheme are defined by the way of solving these two problems:

1 how to select $u_n \in S^{d-1}$;

2 how to construct $P^{n+1} = P^n \cap L(u_n, C)$ in the form needed for step 1 of the next iteration.

The concept of the cutting scheme was introduced in (Button and Wilker, 1978), and the concept of the augmentation scheme was introduced in (Kamenev, 1986a; Kamenev, 1986b). Its detailed description was given in (Kamenev, 1988; Kamenev, 1992).

If a method based on the augmentation (or cutting) scheme satisfies $P^n \in \mathsf{P}^i(C)$ (or $P^n \in \mathsf{P}^c(C)$), then $P^{n+1} \in \mathsf{P}^i(C)$ (or $P^{n+1} \in \mathsf{P}^c(C)$). If the initial polytope P^0 belongs to $\mathsf{P}^i(C)$ ($\mathsf{P}^c(C)$), then it holds that

$$m(P^n) \le m(P^{n+1}) \quad \text{and}$$

$$m^t(P^n) \le m^t(P^0) + n,$$

$$m^f(P^n) \le m^f(P^0) + n.$$

Thus, a method that implements the augmentation scheme is an iterative method with non-decreasing number of vertices. It is defined by the way of choice of the point p_n in Step 1 and by the way Step 2 is realized. Similarly, a method that implements the cutting scheme is an iterative method with non-decreasing number of hyperfaces. It is defined by the choice of the direction u_n in Step 1 and by the way Step 2 is realized. We say that an iterative method that implements the augmentation (cutting) scheme is adaptive if the choice of the point (of the direction) on Step 1 is based on information about the approximated body. If for any $C \in \mathsf{C}$, an augmentation (cutting) method generates a sequence of polytopes that converges to C in a metric δ, then the method (and its generated sequence) are called the approximating method and sequence.

Let $C \in \mathsf{C}$. Denote by $\mathbf{A}^i(C)$ the class of methods that approximate C using the augmentation scheme. Similarly define the class $\mathbf{A}^c(C)$ for the cutting scheme.

4.   *H*-classes of adaptive methods

Now we consider classes of methods that have some important features.

DEFINITION 8.1 *(Kamenev, 1988; Kamenev, 1992). The sequence of polytopes* $\{P^n\}_{n=1,2,...}$ *generated for* $C \in \mathsf{C}$ *and* $P^0 \in \mathsf{P}^i(C)$ *by an augmentation method is called an* $H(\gamma, C)$-*sequence of augmentation (or a Hausdorff sequence of augmentation) if there exists a constant* $\gamma > 0$ *such that for any* $n = 0, 1, ...$ *it holds that*

$$\delta\left(P^n, P^{n+1}\right) \geq \gamma\delta^H\left(P^n, C\right). \tag{8.8}$$

Since it holds that $\delta^H(P^n, P^{n+1}) = \rho(p_n, P^n)$ for an augmentation method, where p_n is the point used at the $(n + 1)$-th iteration of the augmentation scheme, the equation (8.8) may be re-formulated as

$$\rho(p_n, P^n) \geq \gamma\delta^H\left(P^n, C\right).$$

DEFINITION 8.2 *(Kamenev, 1999).   The sequence of polytopes* $\{P^n\}_{n=1,2,...}$ *generated for* $C \in \mathsf{C}$ *and* $P^0 \in \mathsf{P}^i(C)$ *by an augmentation method is called an* $H_1(\gamma, C)$-*sequence of augmentation if there exists a constant* $\gamma > 0$ *such that for any* $n = 0, 1, ...$ *for some* $u_n \in S(p_n, C)$ *it holds that*

$$g(u_n, C) - g(u_n, P^n) \geq \gamma\delta^H(P^n, C). \tag{8.9}$$

Since $u_n \in S(p_n, C)$, then $p_n \in T(u_n, C)$ and therefore

$$g(u_n, p_n) - g(u_n, P^n) = g(u_n, C) - g(u_n, P^n).$$

Therefore, for an augmentation method, the expression (8.9) follows from

$$g(u_n, p_n) - g(u_n, P^n) \geq \gamma\delta^H(P^n, C).$$

Since

$$\rho(p_n, P^n) \geq g(u_n, p_n) - g(u_n, P^n),$$

$H_1(\gamma, C)$-sequence of augmentation is also an $H(\gamma, C)$-sequence.

DEFINITION 8.3 *(Kamenev, 1992).   The sequence of polytopes* $\{P^n\}_{n=1,2,...}$ *generated for* $C \in \mathsf{C}$ *and* $P^0 \in \mathsf{P}^i(C)$ *by a cutting method is called an* $H(\gamma, C)$-*sequence of cutting (or a Hausdorff sequence of cutting) if there exists a constant* $\gamma > 0$ *such that for any* $n = 0, 1, ...$ *it holds that (8.8).*

Since

$$\delta^H(P^{n+1}, P^n) \geq g(u_n, P^n) - g(u_n, P^{n+1}) = g(u_n, P^n) - g(u_n, C),$$

then from

$$g(u_n, P^n) - g(u_n, C) \geq \gamma \delta^H(P^n, C)$$

follows (8.8).

DEFINITION 8.4 *(Kamenev, 2002).* The sequence of polytopes $\{P^n\}_{n=1,2,\dots}$ *generated for $C \in \mathsf{C}$ and $P^0 \in \mathsf{P}^i(C)$ by a cutting method is called an $H_1(\gamma, C)$-sequence of cutting if there exists a constant $\gamma > 0$ such that for any $n = 0, 1, \dots$ for some $p_n \in T(u_n, C)$ it holds that*

$$g(u_n, t(p_n, P^n)) - g(u_n, C) \geq \gamma \delta(P^n, C).$$

Since $g(u_n, P^n) \geq g(u_n, t(p_n, P^n))$ and

$$\delta(P^{n+1}, P^n) \geq g(u_n, P^n) - g(u_n, C),$$

then an $H_1(\gamma, C)$-sequence of cutting is at the same time $H(\gamma, C)$-sequence of cutting.

Methods that generate H-sequences (H_1-sequences) of polytopes with a constant γ are called H-methods (H_1-methods) with the constant γ or $H(\gamma, C)$-methods ($H_1(\gamma, C)$-methods). Let $\{P^n\}_{n=0,1,\dots}$ be the sequence generated for $C \in \mathsf{C}$ by a method that implements the augmentation (or cutting) scheme. This method is called an asymptotically $H(\gamma, C)$-method ($H_1(\gamma, C)$-method) if there exists $N \geq 0$, such that $\{P^n\}_{n=N,N+1,\dots}$ is an $H(\gamma, C)$-sequence ($H_1(\gamma, C)$-sequence). Note that (asymptotically) H-methods (H_1-methods) for $C \in \mathsf{C}$ with the constant γ_1 are at the same time (asymptotically) H-methods (H_1-methods) for $C \in \mathsf{C}$ with the constant γ_2, where $\gamma_2 < \gamma_1$.

Let $C \in \mathsf{C}$. Denote by $\mathbf{A}_H^i(\gamma, C)$ the class of asymptotically H-augmentation methods for C with the constant γ. Similarly define $\mathbf{A}_H^c(\gamma, C)$ for the cutting method.

Analogously denote by $\mathbf{A}_{H_1}^i(\gamma, C)$ the class of asymptotically H_1-augmentation methods for C with the constant γ. Similarly define $\mathbf{A}_{H_1}^c(\gamma, C)$ for the cutting method. Examples of asymptotically H-methods (H_1-methods) are given later.

Let us consider the convergence of the sequences generated by H-methods.

THEOREM 8.5 *Let $\{P^n\}_{n=0,1,\dots}$ be an H-sequence for $C \in \mathsf{C}$. Then*

$$\lim_{n \to \infty} \delta(P^n, C) = 0.$$

Proof. First of all we prove the theorem for the Hausdorff metric. The statement of the theorem follows for augmentation methods from the compactness of C, and for cutting methods from compactness of P^0.

Indeed, for the augmentation method we can consider the sequence of points $\{p_n\}_{n=0,1,\ldots}$ from the definition of the scheme, and for the cutting method we can consider the sequence of points p_n, $n = 0, 1, \ldots$, where $p_n \in P^n$:

$$\rho(p_n, P^{n+1}) = \delta^H(P^{n+1}, P^n).$$

By compactness of the sets C (augmentation case) or P^0 (cutting case) from (8.8) it follows that

$$\lim_{n \to \infty} \rho\left(p_n, \{p_i\}_{i=1,\ldots,n-1}\right) = 0.$$

But

$$\rho(p_n, \{p_i\}_{i=1,2,\ldots n-1}) \geq \delta^H(P^{n+1}, P^n),$$

hence we get

$$\lim_{n \to \infty} \delta^H(P^n, C) = 0.$$

From convergence in the Hausdorff metric follows convergence in the volume metric. Indeed, according to (Leichtweiss, 1980), for any $C \in \mathsf{C}$ we have

$$\sigma(C) = \lim_{\lambda \to +0} \frac{\mu(C_\lambda) - \mu(C)}{\lambda}, \qquad (8.10)$$

For augmentation schemes

$$P^n \subset C \subset P^n_\lambda,$$

where $\lambda = \delta^H(C, P^n)$. So

$$\delta^S(P^n, C) = \mu(C) - \mu(P^n) \leq \mu(P^n_\lambda) - \mu(P^n)$$

and the convergence in the symmetric difference volume metric follows from the evidence that the limit (8.10) is a finite value. Quite analogously for cutting schemes

$$C \subset P^n \subset C_\lambda$$

and

$$\delta^S(P^n, C) = \mu(P^n) - \mu(C) \leq \mu(C_\lambda) - \mu(C),$$

from which follows the convergence in the symmetric difference volume metric. ∎

Thus, H-methods (and H-sequences) are approximating methods (sequences).

5.   Optimality and efficiency of H-methods

Let us start with the analysis of the asymptotic convergence of H-sequences.

THEOREM 8.6 *Let* $\{P^n\}_{n=0,1,...}$ *be an* $H(\gamma, C)$-*sequence of augmentation (of cutting) for* $C \in \mathbf{C}$. *Then for any* ε, $0 < \varepsilon < 1$, *there exists* N *such that for* $n \geq N$ *we have*

$$\delta^S(P^n, C) \leq (1 + \varepsilon)\lambda_1(\gamma)k(n)^{1/(1-d)},$$

$$\delta^H(P^n, C) \leq (1 + \varepsilon)\lambda_2(\gamma)k(n)^{1/(1-d)},$$

where $k(n)$ *denotes* n *or* $m^t(P^n)$ $(m^f(P^n))$ *and*

$$\lambda_1(\gamma) = \left\{ \frac{d}{(d-1)\,\pi_{d-1}} \left(\frac{\sigma(C)}{\gamma} \right)^d \right\}^{\frac{1}{d-1}} \omega(C),$$

$$\lambda_2(\gamma) = \left\{ \frac{d}{(d-1)\,\pi_{d-1}} \frac{\sigma(C)}{\gamma^d} \right\}^{\frac{1}{d-1}} \omega(C).$$

THEOREM 8.7 *Let* $\{P^n\}_{n=0,1,...}$ *be an* $H(\gamma, C)$-*sequence of augmentation (of cutting) for* $C \in \mathbf{C}^2$. *Then for any* ε, $0 < \varepsilon < 1$, *there exists* N *such that for* $n \geq N$ *we have*

$$\delta^S(P^n, C) \leq (1 + \varepsilon)\lambda_3(\gamma)n^{2/(1-d)},$$

$$\delta^H(P^n, C) \leq (1 + \varepsilon)\lambda_4(\gamma)n^{2/(1-d)},$$

where $k(n)$ *denotes* n *or* $m^t(P^n)$ $(m^f(P^n))$ *and*

$$\lambda_3(\gamma) = \left\{ \frac{2d}{(d-1)\,\pi_{d-1}} \frac{\sigma(C)^{\frac{d+1}{2}}}{\gamma^d} \right\}^{\frac{2}{d-1}} \frac{2}{r_{min}(C)},$$

$$\lambda_4(\gamma) = \left\{ \frac{(d+1)\,d}{(d-1)\,\pi_{d-1}} \frac{\sigma(C)}{\gamma^d} \right\}^{\frac{2}{d-1}} \frac{2}{r_{min}(C)}.$$

These theorems are proved in (Kamenev, 1992).
From Theorem 8.7 follows the next corollary.

COROLLARY 8.8 *Let* $C \in \mathbf{C}^2$ *and* $A \in \mathbf{A}_H^i(C)$ $(A \in \mathbf{A}_H^c(C))$. *Then* A *is an asymptotically efficient method in the Hausdorff and symmetric difference volume metric.*

The next estimate on the asymptotic efficiency of H-sequences in the Hausdorff metric follows from Theorem 8.7 and the asymptotic formula (8.5).

COROLLARY 8.9 *Let* $F = \{P^n\}_{n=0,1,..}$ *bean* $H(\gamma, C)$-*sequence for* $C \in$ $\mathbf{C}^2$. *Then*

$$\eta^H(F) \geq \eta_d^H \; \gamma^{2d/(d-1)} \left(r_{min}(C)/r_{max}(C)\right),$$

where

$$\eta_d^H = \frac{1}{4} \left(\frac{d-1}{d+1} \frac{\vartheta_{d-1}}{d}\right)^{\frac{2}{d-1}}$$

and $\lim\limits_{d \to \infty} \eta_d^H = \frac{1}{4}$.

Note that the estimate depends on the asphericity $r_{min}(C)/r_{max}(C)$ of the approximated body. More precise estimations are given in the next section for H_1-sequences.

Finally, note that an asymptotically efficient method is optimal in respect to the order of the number of vertices (hyperfaces) for $C \in \mathbf{C}^2$. Hence, the next corollary follows from Corollary 8.8.

COROLLARY 8.10 *Let* $C \in \mathbf{C}^2$ *and* $A \in \mathbf{A}_H^i(C)$ $(A \in \mathbf{A}_H^c(C))$. *Then* A *is optimal in respect to the order of the number of vertices (hyperfaces) in Hausdorff and symmetric difference volume metric for* C.

6.   Optimality and efficiency of H_1-methods

Let us study the asymptotic convergence of H_1-methods (sequences). We start with approximation of the bodies $C \in \mathbf{C}^2$, and only then do we turn to the main results of the section related to CCBs of general type.

Since any $H_1(\gamma, C)$-sequence is an $H(\gamma, C)$-sequence, corollaries 8.8–8.9 are valid for these sequences. In particular, the next corollary follows from Theorem 8.7.

COROLLARY 8.11 *Let* $C \in \mathbf{C}^2$ *and* $A \in \mathbf{A}_{H_1}^i(C)$ $(A \in \mathbf{A}_{H_1}^c(C))$. *Then* A *is an asymptotically efficient method in the Hausdorff and symmetric difference volume metric.*

An estimate of the asymptotic efficiency has been developed in (Efremov and Kamenev, 2002).

THEOREM 8.12 *Let* $F = \{P^n\}_{n=0,1,..}$ *be an* $H_1(\gamma, C)$-*sequence of augmentation for* $C \in \mathbf{C}^2$. *Then*

$$\underline{\eta}^H(F) \geq \gamma/4.$$

It is important that this estimate is precise (attainable): a related example was constructed in (Kamenev, 1993). The analogous result for cutting methods (sequences) was proved in (Efremov, 2003).

Note that, in contrast to the estimate given in Corollary 8.9, this estimate of the asymptotic efficiency of H_1-sequences does not depend on the asphericity or any other properties of the approximated body. This evidence is confirmed by computer experiments on numerical polyhedral approximation of multi-dimensional ellipsoids (Kamenev, 1986a; Dzholdybaeva and Kamenev, 1991; Dzholdybaeva and Kamenev, 1992; Bourmistrova, 1999; Efremov, 2002).

More precise estimates on the rate of convergence of H_1-methods follow from Theorem 8.12, asymptotic result (8.5) and the property of surface volume of a CCB (8.10).

COROLLARY 8.13 *Let $\{P^n\}_{n=0,1,...}$ be an $H_1(\gamma, C)$-sequence of augmentation for $C \in \mathsf{C}^2$. Then*

$$\limsup_{n\to\infty} \delta^H(C, P^n)\, k(n)^{\frac{2}{d-1}} \le \frac{2}{\gamma} \left(\frac{\vartheta_{d-1}}{\pi_{d-1}} \int_{\partial C} k_C(x)^{\frac{1}{2}} d\sigma(x) \right)^{\frac{2}{d-1}},$$

$$\limsup_{n\to\infty} \delta^S(C, P^n)\, k(n)^{\frac{2}{d-1}} \le \frac{2\sigma(C)}{\gamma} \left(\frac{\vartheta_{d-1}}{\pi_{d-1}} \int_{\partial C} k_C(x)^{\frac{1}{2}} d\sigma(x) \right)^{\frac{2}{d-1}},$$

where $k(n)$ denotes n or $m^t(P^n)$.

Now let us turn to the general case of $C \in \mathsf{C}$. According to (8.2), we have $\delta(C, \mathsf{P}_m) \le \mathrm{const}\, m^{2/(1-d)}$. In Theorem 8.6, a weaker estimate was obtained for the convergence rate of H-sequences for non-smooth bodies: $\delta(P^n, C) \le \mathrm{const}\, m^t(P^n)^{1/(1-d)}$. In this section, a theorem is proved, which estimates the optimal convergence rate of an H_1-sequence of polytopes as $\delta(P^n, C) \le \mathrm{const}\, m^t(P^n)^{2/(1-d)}$ for the bodies $C \in \mathsf{C}$. These statements were proved in (Kamenev, 1996) for augmentation and in (Kamenev, 2002) for cutting sequences. Here we combine results for both types of schemes in one statement.

THEOREM 8.14 *Let $\{P^n\}_{n=0,1,2,...}$ be an $H_1(\gamma, C)$-sequence of augmentation (cutting) for $C \in \mathsf{C}$. Then for any ε, $0 < \varepsilon < 1$, there exists N such that for $n \ge N$ it holds that*

$$\delta^H(P^n, C) \le (1+\varepsilon)\lambda_5(\gamma, C)\, k(n)^{2/(1-d)},$$

where $k(n)$ denotes n or $m^t(P^n)$ $(m^f(P^n))$ and

$$\lambda_5(\gamma, C) = \frac{16R(C)}{\gamma} \left[\frac{d\pi_d}{\pi_{d-1}} \right]^{\frac{2}{d-1}}.$$

LEMMA 8.15 *Let $x, y \in \mathbb{R}^d$ and $\beta > 0$. Then for any $u, v \in S^{d-1}$, where $\langle u, v \rangle > 0$, $y \in L(u, x)$, $x \in L(v, y)$, we have*

$$\rho(x, l(v, y)) \leq ||(x + \beta u) - (y + \beta v)||^2/\beta.$$

Proof. Let us denote $\Delta = ||(x + \beta u) - (y + \beta v)||$. Since $y \in L(u, x)$ and $x \in L(v, y)$, we have $\langle u, x - y \rangle \geq 0$, $\langle v, y - x \rangle \geq 0$ and

$$\Delta^2 \geq ||x - y||^2 + \beta^2 ||u - v||^2.$$

Therefore,

$$\Delta \geq \max\{||x - y||, \beta ||u - v||\}.$$

Let α be the angle between u and v. Since $\sin \alpha \leq ||u - v||$, we have $\sin \alpha \leq \Delta/\beta$.

Let z be the projection of x onto $l = l(u, x) \cap l(v, y)$ and m be the projection of x onto $l(v, y)$. The projection of m onto l is z (otherwise z will not be the projection of x), and so the value of the angle $\angle xzm$ is equal to α. Hence,

$$\Delta \geq \rho(x, y) \geq \rho(x, z) = \rho(x, m)/\sin \alpha \geq \beta \rho(x, l(v, y))/\Delta,$$

which entails the assertion of the lemma.   ∎

LEMMA 8.16 *Let $C \in \mathsf{C}$, $p, q \in \partial C$ and $\beta > 0$. Then for any $u \in S(p, C)$ and $v \in S(p, C)$, where $\langle u, v \rangle > 0$, it holds that*

$$\rho(p, l(v, C)) \leq ||(p + \beta u) - (q + \beta v)||^2/\beta.$$

Proof. Note that by the convexity of C it follows that $q \in L(u, p)$, $p \in L(v, q)$. Then the assertion of the lemma follows from the Lemma 8.15.   ∎

LEMMA 8.17 *Let $C \in \mathsf{C}_0$, $p \notin C$, $p' = t(p, C)$ and $\beta > 0$. Then for any $u \in S(p', C)$ and $v \in S^{d-1}$, $p \in L(v, C)$, $q \in T(v, C)$, it holds that*

$$||(p' + \beta u) - (q + \beta v)|| \geq \min\left\{\beta\sqrt{2}, \sqrt{\beta h} - \omega_0(C)h\right\},$$

where $h = g(u, p) - g(u, C)$.

Proof. If $\langle u, v \rangle \leq 0$, then $||(p' + \beta u) - (q + \beta v)|| \geq \beta\sqrt{2}$.

Let $\langle u, v \rangle > 0$. Since $u \in S(p', C)$, then $q \in L(u, C)$, from which $q \in L(u, p)$. Then, according to Lemma 8.15 we have

$$||(p + \beta u) - (q + \beta v)||^2/\beta \geq \rho(q, l(u, p)) \geq h.$$

Then,

$$\|(p + \beta u) - (p' + \beta u)\| = \rho(p, p') \le \omega_0(C) \left[g(u, p) - g(u, p') \right] = \omega_0(C)h.$$

Finally,

$$\|(p' + \beta u) - (q + \beta v)\| \ge$$

$$\ge \|(p + \beta u) - (q + \beta v)\| - \|(p + \beta u) - (p' + \beta u)\| \ge \sqrt{\beta h} - \omega_0(C)h.$$

∎

Let $\{P^n\}_{n=0,1,\dots}$ be an $H_1(\gamma, C)$-sequence of augmentation and $\beta > 0$. Define the sequence $\{Z^n\}_{n=0,1,\dots}$ by the following rule:

1 $Z^0 = \{p + \beta u(p) : p \in M^t(P^0)\}$ for a set of directions $\{u(p) \in S(p, C) : p \in M^t(P^0)\}$;

2 $Z^{n+1} = \{p_n + \beta u_n\} \cup Z^n$, $n = 0, 1, 2, \dots$, where the point p_n and the direction u_n are the same as those in the definition of an $H_1(\gamma, C)$-sequence of augmentation.

It is evident that $Z^n \subset \partial(C + \beta B)$. In addition in case of $C \in \mathsf{C} \backslash \mathsf{P}$ we have card $Z^n = m^t(P^0) + n$. The sequence $\{Z^n\}_{n=0,1,\dots}$ is called the associated sequence for $\{P^n\}_{n=0,1,\dots}$.

The set Z we call the base of an ε-packing if for any $x, y \in Z$ we have $\rho(x, y) \ge 2\varepsilon$ (in this case the set of open balls with radius ε centered in points of Z will be packing in the space $\mathbb{R}^d$).

LEMMA 8.18 *Let $\beta > 0$, $C \in \mathsf{C}$; let $\{P^n\}_{n=0,1,\dots}$ be an $H_1(\gamma, C)$-sequence of augmentation and $\{Z^n\}_{n=0,1,\dots}$ be the associated sequence for it. Then for any $n = 0, 1, \dots$ the set Z^n is the base of an $\varepsilon_n(\gamma)$-packing, where*

$$\varepsilon_n(\gamma) = \frac{1}{2} \min \left\{ \beta\sqrt{2},\ 2\varepsilon_0,\ \sqrt{\beta\gamma\delta^H (P^{n-1}, C)} \right\}$$

and $\varepsilon_0 = \min \{\rho(x, y) : x, y \in Z^0,\ x \ne y\}/2$.

Proof. The assertion of the lemma is obviously true for $n = 0$. Assume that it holds for $n - 1$. By the induction hypothesis, Z^{n-1} is the base of an $\varepsilon_{n-1}(\gamma)$-packing. However $\delta^H(P^{n-1}, C) \le \delta^H(P^{n-2}, C)$; hence Z^{n-1} is the base of an $\varepsilon_n(\gamma)$-packing. Let $z_{n-1} = p_{n-1} + \beta u_{n-1}$, where p_{n-1} and u_{n-1} are the same as those in the definition of H_1-sequence of augmentation. Since $Z^n = \{z_{n-1}\} \cup Z^{n-1}$, it remains to prove that $\rho(z_{n-1}, z) \ge 2\varepsilon_n(\gamma)$ for any $z \in Z^{n-1}$. Let $z = p + \beta v$, where $p \in M^t(P^{n-1})$, $v \in S(p, C)$.

If $\langle u_{n-1}, v \rangle \leq 0$, then $\rho(z_{n-1}, z) \geq \beta\sqrt{2}$. Let $\langle u_{n-1}, v \rangle > 0$. Then by Lemma 8.16 we have

$$\|z_{n-1} - z\|^2 \geq \beta\rho(z, l(u_{n-1}, C)) \geq \beta(g(u_{n-1}, C) - g(u_{n-1}, P^{n-1})) \geq$$

$$\geq \beta\gamma\, \delta^H(P^{n-1}, C) \geq \beta\gamma\, \delta^H(P^n, C),$$

which proves that $\rho(z_{n-1}, z) \geq 2\varepsilon_n(\gamma)$.   ∎

Let $\{P^n\}_{n=0,1,\ldots}$ be an $H_1(\gamma, C)$-sequence of cutting and $\beta > 0$. Define the sequence $\{Z^n\}_{n=0,1,\ldots}$ by the following rule:

1 $Z^0 = \{p(u) + \beta u : u \in M^f(P^0)\}$ for a set of points $\{p(u) \in T(u, C) : u \in M^f(P^0)\}$;

2 $Z^{n+1} = \{p_n + \beta u_n\} \cup Z^n, n = 0, 1, 2, \ldots$, where the point p_n and the direction u_n are the same as those in the definition of an $H_1(\gamma, C)$-sequence of cutting.

It is evident that $Z^n \subset \partial(C + \beta B)$. In addition in case of $C \in \mathsf{C}\backslash\mathsf{P}$ we have card $Z^n = m^f(P^0) + n$. The sequence $\{Z^n\}_{n=0,1,\ldots}$ is called the associated sequence for $\{P^n\}_{n=0,1,\ldots}$.

LEMMA 8.19 *Let $\beta > 0$, $C \in \mathsf{C}$; let $\{P^n\}_{n=0,1,\ldots}$ be an $H_1(\gamma, C)$-sequence of cutting and $\{Z^n\}_{n=0,1,\ldots}$ be the associated sequence for it. Then for any ε, $0 < \varepsilon < 1$, there exists N such that for $n \geq N$ the set Z^n is the base of an $\varepsilon_n^N(\gamma, \varepsilon)$-packing, where*

$$\varepsilon_n^N(\gamma, \varepsilon) = \frac{1}{2} \min\left\{\beta\sqrt{2},\ 2\varepsilon_N,\ (1 - \varepsilon)\sqrt{\beta\gamma\delta^H(P^{n-1}, C)}\right\}$$

and $\varepsilon_N = \min\{\rho(x, y) : x, y \in Z^N, x \neq y\}/2$.

Proof. The assertion of the lemma is obviously true for $n = N$. Assume that it holds for $n - 1$. By the induction hypothesis, Z^{n-1} is the base of an $\varepsilon_{n-1}^N(\gamma, \varepsilon)$-packing. However $\delta^H(P^{n-1}, C) \leq \delta^H(P^{n-2}, C)$; hence Z^{n-1} is the base of an $\varepsilon_n^N(\gamma, \varepsilon)$-packing. Let $z_{n-1} = p_{n-1} + \beta u_{n-1}$, where p_{n-1} and u_{n-1} are the same as those in the definition of an H_1-sequence of cutting. Since $Z^n = \{z_{n-1}\} \cup Z^{n-1}$, it remains to prove that $\rho(z_{n-1}, z) \geq 2\varepsilon_n^N(\gamma, \varepsilon)$ for any $z \in Z^{n-1}$.

Let $z = p + \beta v$, where $p \in \partial P^{n-1} \cap \partial C$, $v \in S(p, C)$ and $p \in T(v, C)$. If $\langle u_{n-1}, v \rangle \leq 0$, then $\rho(z_{n-1}, z) \geq \beta\sqrt{2}$. Let $\langle u_{n-1}, v \rangle > 0$ and

$$p' = t(p_{n-1}, P^{n-1}).$$ Then by Lemma 8.17 we have

$$\|z_{n-1} - z\| \geq \min\left\{\beta\sqrt{2},\ \sqrt{h\beta} - \omega_0(C)h\right\},$$

where $h = g(u_{n-1}, p') - g(u_{n-1}, C)$,

$$\delta^H \left(P^{n-1}, C\right) \geq h \geq \gamma \delta^H \left(P^{n-1}, C\right)$$

on the definition of H_1-sequence of cutting. Finally we choose N so that

$$\delta^H \left(P^{n-1}, C\right)^2 \leq \varepsilon/\omega_0(C).$$

■

LEMMA 8.20   *Let $C \in \mathbf{C}$. Then any base of an ε-packing on ∂C, $0 < \varepsilon < R(C)$, contains no more than $[N_1(\varepsilon, R)]$ elements, where*

$$N_1(\varepsilon, C) = \frac{d\pi_d}{\pi_{d-1}} \left[1 + \left(\frac{R(C)}{\varepsilon}\right)^2\right]^{\frac{d-1}{2}}.$$

Proof. Let $C \subset B_{R(C)}(z)$ and Z be the base of an ε-packing on ∂C. Let $i(q)$ be a ray issuing from $q \in \partial C$ and out of $S(q, C)$. Define

$$Z^* = \{i(q) \cap \partial B_{R(C)}(z) : q \in Z\}.$$

Then Z is the projection of Z^* onto ∂C. A projection onto the boundary of a convex compact body does not increase distances (see Aleksandrov, 1948); hence, Z^* is the base of an ε-packing on $\partial B_{R(C)}(z)$. For $q^* \in Z^*$, the $(d-1)$-ball

$$B_x(q^*), x = \varepsilon R(C)/(\varepsilon^2 + R(C)^2)^{1/2}$$

is the projection of $\partial B_{R(C)}(z) \cap B_\varepsilon(q^*)$ onto the plane that is tangent to $B_{R(C)}(z)$ at q^*. Hence,

$$\sigma(\partial B_{R(C)}(z) \cap B_\varepsilon(q^*)) \geq \pi_{d-1} x^{d-1}$$

and

$$\mathrm{card}\, Z = \mathrm{card}\, Z^* \leq \sigma(\partial B_{R(C)}(z))/(\pi_{d-1} x^{d-1}).$$

■

Proof of Theorem 8.14.   Since any $H_1(\gamma, C)$-sequence is an $H(\gamma, C)$-sequence, it holds (by Theorem 8.5) that

$$\lim_{n \to \infty} \delta^H \left(P^n, C\right) = 0.$$

Let $\beta > 0$ and $\{Z^n\}_{n=0,1,2,\dots}$ be the associated sequence of sets for $\{P^n\}_{n=0,1,2,\dots}$.

At first let us consider the augmentation method. Let N:

$$\delta^H\left(P^{N-1}, C\right) \leq \frac{4}{\gamma\beta} \min\left\{\frac{\beta^2}{2}, \varepsilon_0^2\right\},$$

where ε_0 is the maximal radius for the packing base Z^0. By Lemma 8.18 the set Z^n is the base of an $\varepsilon_n(\gamma)$-packing, and for $n \geq N$ is the base of an τ_n-packing on $\partial(C + B)$, where $\tau_n = \left(\beta\gamma\delta^H(P^n, C)\right)^{1/2}/2$. Since $Z^n \in \partial(C + \beta B)$, by Lemma 8.20 we have

$$Z^n \leq N_1(\tau_n, C + B_\beta).$$

In case $C \in \mathsf{P}$ the assertion of the theorem is obviously true. Let $C \notin \mathsf{P}$. Then $\delta^H(P^n, C) > 0$ for any n, and from the definition of the H-sequence of augmentation entails card $Z^n = m^t(P^0) + n$. Hence,

$$n < \operatorname{card} Z^n,$$

$$m^t(P^n) < \operatorname{card} Z^n.$$

Thus, in the theorem notation,

$$k(n) < \frac{d\pi_d}{\pi_{d-1}}\left[1 + \frac{4\left(R(C) + \beta\right)^2}{\beta\gamma\delta(P^n, C)}\right]^{\frac{d-1}{2}}.$$

the minimum of the right part of this inequality corresponds to the value $\beta = R(C)$, which entails the assertion of the theorem for the augmentation method.

Now let us consider the cutting method. By Lemma 8.19 for any ε^*, $0 < \varepsilon^* < 1$, there exists N^* such that for $n \geq N^*$ the set Z^n is the base of an $\varepsilon_n^{N^*}(\gamma, \varepsilon^*)$-packing. Let $N^{**} \geq N^*$ be such that

$$(1 - \varepsilon^*)\left(\beta\gamma\delta^H(P^{N^{**}-1}, C)\right)^{1/2}/2 \leq \varepsilon_{N^*}^{N^*}(\gamma, \varepsilon^*).$$

Then for $n \geq N^{**}$ the set Z^n is the base of a τ_n-packing, where

$$\tau_n = (1 - \varepsilon^*)\left(\beta\gamma\delta^H(P^n, C)\right)^{1/2}/2.$$

Since $Z^n \in \partial(C + \beta B)$, by Lemma 8.20 we have

$$\operatorname{card} Z^n \leq N_1(\tau_n, R(C) + \beta).$$

In case $C \in \mathsf{P}$ the assertion of the theorem is obviously true. Let $C \notin \mathsf{P}$. Then $\delta(P^n, C) > 0$ for any n, and from the definition of the H-sequence of cutting entails card $Z^n = m^{\,f}(P^0) + n$. Hence,

$$n < \operatorname{card} Z^n,$$

$$m^f(P^n) < \operatorname{card} Z^n.$$

Thus, in the theorem notation,

$$k(n) < \frac{d\pi_d}{\pi_{d-1}} \left[1 + \frac{4\left(R(C)+\beta\right)^2}{(1-\varepsilon^*)^2 \beta \gamma \delta(P^n, C)} \right]^{\frac{d-1}{2}}.$$

The minimum of the right part of this inequality corresponds to the value $\beta = R(C)$, which, under suitable choice of N^{**} and ε^*, entails the assertion of the theorem for the cutting method. ∎

Theorem 8.14 proves the optimal convergence rate of H_1-sequences for $C \in \mathsf{C}$. Therefore, from Theorem 8.14 follows

COROLLARY 8.21 *For $C \in \mathsf{C}$ let it holds that $\delta(C, \mathsf{P}_m) \geq \operatorname{const} m^{2/(1-d)}$, and $A \in \mathbf{A}^i_{H_1}(C)$ $(A \in \mathbf{A}^c_{H_1}(C))$. Then A is optimal in respect to the order of the number of vertices (hyperfaces) in the Hausdorff and symmetric difference volume metric.*

For example, from (Schneider and Wieacker, 1981) it follows that this corollary holds for any $C \in \mathsf{C}^2$ and for any CCB with a boundary point, having a neighborhood with two times continuously differentiable hypersurface with positive principal curvatures.

7. Several optimal methods for polyhedral approximation of CCB

In this section we describe several H_1-methods for polyhedral approximation of a CCB. Two following methods proposed in (Kamenev, 1986b; Kamenev, 1992) have a simple description, but it is impossible to implement them practically.

Basic Augmentation method
Let for $C \in \mathsf{C}$ and $P^0 \in \mathsf{P}^i(C)$ the polytope $P^n \in \mathsf{P}^i(C)$ be constructed. The following procedure is used to construct P^{n+1}:
Step 1. Find $p_n \in \partial C : \rho(p_n, P^n) = \delta^H(P^n, C)$.
Step 2. Construct $P^{n+1} = \operatorname{conv}\{p_n, P^n\}$.
This method is denoted as $\mathbf{A}_{BA}$.

Basic Cutting method
Let for $C \in \mathsf{C}$ and $P^0 \in \mathsf{P}^c(C)$ the polytope $P^n \in \mathsf{P}^c(C)$ be constructed. The following procedure is used to construct P^{n+1}:
Step 1. Find $u_n = \arg \max \left\{ g(u, P^n) - g(u, C) : u \in S^{d-1} \right\}$.
Step 2. Construct $P^{n+1} = P^n \cap L(u_n, C)$.
This method is denoted as $\mathbf{A}_{BC}$.

According to (Leichtweiss, 1980), for any $C_1, C_2 \in \mathsf{C}$ it holds that

$$\delta^H(C_1, C_2) = \max\left\{ |g(u, C_1) - g(u, C_2)| : u \in S^{d-1} \right\}. \tag{8.11}$$

Hence, it follows from the definition of these two methods that for $C \in \mathsf{C}$,

$$A_{BA} \in \mathbf{A}_H^i(1, C), \quad A_{BC} \in \mathbf{A}_H^c(1, C).$$

Note that we can reformulate the Step 1 of A_{BA} as
Step 1. a) Find $u_n = \arg \max\{g(u, C) - g(u, P^n) : u \in S^{d-1}\}$.
b) Find $p_n \in T(u_n, C)$.
From this formulation and the expression (8.11) we can easily deduce that for $C \in \mathsf{C}$,

$$A_{BA} \in \mathbf{A}_{H_1}^i(1, C).$$

Note that A_{BC} is an H-cutting method (but not an H_1-cutting method) in the general case. Examples of H_1-cutting methods are described in this section, too.

Let us consider and discuss several H_1 methods. One of them has already been described in Section 3 of Chapter 6 and applied throughout this book.

Estimate Refinement method
Let for $C \in \mathsf{C}$ and $P^0 \in \mathsf{P}^i(C)$ the polytopes $P^n \in \mathsf{P}^i(C)$ be constructed in the form of the solution set describing $M^f(P^n)$. The following procedure is used to construct P^{n+1}:
Step 1. a) Find $u_n = \arg \max \left\{ g(u, C) - g(u, P^n) : u \in M^f(P^n) \right\}$.
b) Find $p_n \in T(u_n, C)$.
Step 2. Construct $P^{n+1} = \mathrm{conv}\{p_n, P^n\}$ in the form of the set $M^f(P^{n+1})$.
The Estimate Refinement method is denoted here as A_{ER}.
Let $C_1, C_2 \in \mathsf{C}$ and $P \in \mathsf{P}$. The study of A_{ER} is based on the properties of the function

$$\delta_P(C_1, C_2) = \max\left\{ |g(u, C_1) - g(u, C_2)| : u \in M^f(P) \right\}.$$

First of all note that for $\{P^n\}_{n=1,2,\dots}$ generated for $C \in \mathsf{C}$ by the Estimate Refinement method, the equation

$$g(u_n, C) - g(u_n, P^n) = \delta_{P^n}(P^n, P^{n+1}) = \delta_{P^n}(P^n, C)$$

holds for all $n = 0, 1, \dots$.

LEMMA 8.22 *Let $C \in \mathsf{C}$ and $P \in \mathsf{P}^i(C)$. Then*

$$\delta^H(P, C)/\omega(P) \leq \delta_P(P, C) \leq \delta^H(P, C).$$

Furthermore, let $C \in \mathsf{C}^2$ and $\delta^H(P,C) < r_{min}(C)$, then

$$\delta^H(P,C) - \delta^H(P,C)^2/r_{min}(C) \leq \delta_P(P,C) \leq \delta^H(P,C).$$

The proof of this Lemma is given in (Kamenev, 1986b; Kamenev, 1994). Lemma 8.22 was used to prove the following theorems (Kamenev, 1994; Kamenev, 1999).

THEOREM 8.23 *For any $C \in \mathsf{C}$ and ε, $0 < \varepsilon < 1$, it holds that*

$$A_{ER} \in \mathbf{A}^i_{H_1}\left((1 - \varepsilon)/\omega(C), C\right). \tag{8.12}$$

THEOREM 8.24 *For any $C \in \mathsf{C}^2$ and ε, $0 < \varepsilon < 1$, it holds that*

$$A_{ER} \in \mathbf{A}^i_{H_1}\left(1 - \varepsilon, C\right). \tag{8.13}$$

The following cutting method was proposed and studied in (Kamenev, 2002). It can be considered as a dual analog of the Estimate Refinement method.

External Estimate Refinement method

Let for $C \in \mathsf{C}_0$ and $P^0 \in \mathsf{P}^c_0(C)$ the polytope $P^n \in \mathsf{P}^c_0(C)$ be constructed in the form of the set $M^t(P^n)$. The following procedure is used to construct P^{n+1}:

Step 1. a) Find $p_n = \arg\max\{g(u(p), p) - g(u(p), C) : p \in M^t(P^n)$,

where $u(p) \in S(p/g^*(p, C), C)\}$.

b) Find $u_n = u(p_n) \in S(p_n/g^*(p_n, C), C)$.

Step 2. Construct $P^{n+1} = P^n \cap L(u_n, C)$ in the form of the set $M^t(P^{n+1})$.

The External Estimate Refinement method we denote as A_{EER}. The study of this method is based on following two lemmas.

LEMMA 8.25 *Let $C \in \mathsf{C}_0$, $p \notin C$, $u \in S(p/g^*(p, C), C)$. Then*

$$g(u, p) - g(u, C) \geq \rho(p, t(p, C))/\omega_0(C).$$

LEMMA 8.26 *Let $C \in \mathsf{C}_0 \cap \mathsf{C}^2$, $p \notin C$, $u \in S(p/g^*(p, C), C)$ and $h = g(u, p) - g(u, C)$. Then*

$$\rho(p, t(p, C)) \leq \left[(r + h)^2 + h^2\left(\omega_0(C)^2 - 1\right)\right]^{\frac{1}{2}} - r.$$

These lemmas are used in the following theorems (Kamenev, 2002).

THEOREM 8.27 *For any $C \in \mathsf{C}_0$ we have*

$$A_{EER} \in \mathbf{A}^c_{H_1}\left(1/\omega_0(C), C\right). \tag{8.14}$$

THEOREM 8.28 *For any $C \in \mathsf{C}_0^2$ and ε, $0 < \varepsilon < 1$, we have*

$$A_{EER} \in \mathbf{A}_{H_1}^c (1 - \varepsilon, C). \tag{8.15}$$

Using the estimates of the convergence rate and of asymptotic efficiency of H_1-sequences of polytopes (Theorems 8.14 and 8.12, Corollaries 8.21, 8.11 and 8.13) as well as the statements (8.12)–(8.13) and (8.14)–(8.15), one can prove the following results for methods described in this section (see for details Kamenev, 1994; Kamenev, 1999; Kamenev, 2002; Efremov and Kamenev, 2002; Efremov, 2003).

THEOREM 8.29 *Let $\{P^n\}_{n=0,1,2,...}$ be the sequence of polytopes generated by A_{ER} (A_{EER}) for $C \in \mathsf{C}$ ($C \in \mathsf{C}_0$). Then for any ε, $0 < \varepsilon < 1$, there exists N such that for $n \geq N$ we have*

$$\delta^H(P^n, C) \leq (1 + \varepsilon)\lambda_6(C)\ k(n)^{2/(1-d)},$$

where $k(n)$ denotes n or $m^t(P^n)$ ($m^f(P^n)$) and $\lambda_6(C) = \lambda_5(1, C)$, that is

$$\lambda_6 = 16R(C) \left[\frac{d\pi_d}{\pi_{d-1}} \right]^{\frac{2}{d-1}}.$$

COROLLARY 8.30 *Let for $C \in \mathsf{C}$ it holds that $\delta(C, \mathsf{P}_m) \geq \mathrm{const}\, m^{2/(1-d)}$. Then A_{ER} (A_{EER}) is optimal in respect to the order of the number of vertices (hyperfaces) in Hausdorff and symmetric difference volume metric.*

COROLLARY 8.31 *Let $C \in \mathsf{C}^2$. Then A_{ER} (A_{EER}) is the asymptotically efficient method in Hausdorff and symmetric difference volume metric.*

THEOREM 8.32 *Let $\{P^n\}_{n=0,1,..}$ be the sequence of polytopes generated by A_{ER} for $C \in \mathsf{C}^2$ ($C \in \mathsf{C}_0^2$). Then*

$$\underline{\eta}_t^H(F) \geq 1/4, \quad (\underline{\eta}_f^H(F) \geq 1/4).$$

COROLLARY 8.33 *Let $\{P^n\}_{n=0,1,..}$ be the sequence of polytopes generated by A_{ER} for $C \in \mathsf{C}^2$ ($C \in \mathsf{C}_0^2$). Then*

$$\limsup_{n\to\infty} \delta^H(C, P^n)\, k(n)^{\frac{2}{d-1}} \leq 2 \left(\frac{\vartheta_{d-1}}{\pi_{d-1}} \int_{\partial C} k_C(x)^{\frac{1}{2}} d\sigma(x) \right)^{\frac{2}{d-1}},$$

$$\limsup_{n\to\infty} \delta^S(C, P^n)\, k(n)^{\frac{2}{d-1}} \leq 2\sigma(C) \left(\frac{\vartheta_{d-1}}{\pi_{d-1}} \int_{\partial C} k_C(x)^{\frac{1}{2}} d\sigma(x) \right)^{\frac{2}{d-1}},$$

where $k(n)$ denotes n or $m^t(P^n)$ ($m^f(P^n)$).

8. Related topics

In this chapter we have considered the asymptotic theory of Hausdorff methods for polyhedral approximation of convex bodies. Let us mention several other topics of the theory of optimal methods of polyhedral approximation.

To construct an approximation polytope, one needs to carry out measurements using the approximated body. As a rule, values of support or distance functions are computed. For this reason, it is desirable to use methods that, for a given accuracy, use the minimal number of computing the support or distance function. In (Kamenev, 1986b; Kamenev, 1996; Kamenev, 2003; Bourmistrova, 2000a) some methods of this kind are proposed. New ideas in this field are provided by the duality theory of H-sequences and methods of augmentation and cutting in (Kamenev, 2002). In particular, a method is proposed in (Kamenev, 2003), the so-called self-dual method, that is optimal in respect to the order of the numbers of vertices and hyperfaces of the approximating polytopes as well as of computing the value of support and distance functions even for the general class C of CCB.

As can be seen in the expression (8.2), the estimate of the minimal number of vertices (hyperfaces) is independent of the smoothness of the approximated body. For example, in case of two-dimensional convex disks it is given by the square root of the required accuracy. In (Kamenev, 2000), an H-method is constructed and a more precise estimate is obtained for non-smooth convex disks.

In conclusion we note that the class of H-methods that was proposed as a generalization of the Estimate Refinement method, turned out to be an excellent tool for theoretical study. Moreover, it resulted in new practical methods for polyhedral approximation of convex bodies.

Chapter 9

PERTURBATIONS OF SOLUTION SETS OF LINEAR SYSTEMS

This chapter is devoted to the analysis of stability and perturbations of the feasible criterion set (FCS). This problem is extremely important since the data are usually disturbed to a certain extent in real-life applications. It is required that the resulting perturbations of an FCS are not too substantial. In other words, an FCS must depend on the data disturbances in a continuous way. This chapter deals with the estimation of such perturbations in the linear case. The problem is reduced to estimating perturbations of the feasible decision set given by a system of linear inequalities and equalities. To be precise, the distance between two sets, the original set and the perturbed set is estimated.

We start the exploration of stability and perturbations of the feasible decision set with the case of an abstract Banach space, and only then do we concentrate on finite-dimensional linear systems. The study of the problem in an abstract Banach space provides an opportunity to apply the results in the case of functional decision spaces where decisions are modeled by functions that satisfy ordinary or partial differential equations (examples are given in Chapter 7). The finite-dimensional case, however, is important as well since more precise results can be obtained for it and applied in the case of systems considered in Chapter 6.

The problem of estimating perturbations in the solution sets of linear inequality systems is well known: it has been discussed, for example, in (Hoffman, 1952; Daniel, 1973a; Mangasarian, 1981; Robinson, 1973; Mangasarian and Shiau, 1987). In these papers, relations were established between the deviation of a point from the solution set of an inequality system and the discrepancy of the inequalities for this point. Such results make it possible to estimate the deviation of any point of the perturbed solution set from the original solution set as a function of parameter perturbation. In turn, this result provides an opportunity to estimate the influence of perturbations on solutions of linear programming problems. In (Mangasarian and Shiau, 1987; Daniel, 1973b), the Lipschitz continuity problem was studied in the same sense.

It is necessary to stress in advance that the perturbation measure used in this chapter differs from the measure considered in the above papers: instead of the deviation of the perturbed set we use the Hausdorff distance between the sets. This is related to different real-life problems that gave rise to the study — we construct the whole FCS instead of solving a single optimization problem. In addition, we look for estimates that can be effectively calculated in real-life problems.

Robinson (Robinson, 1973) suggested using the convex processes technique for estimating the deviation of a point from the original solution set. Here we show that the same technique of convex processes can be applied for evaluating the Hausdorff distance between the original and perturbed sets. To do it, alternative convex processes must be introduced and studied. Along with Robinson, we obtain our main results by considering the problem in partially ordered Banach space, and only then do we obtain results for the finite dimensional case, both for inequality systems and combined equality and inequality systems.

Theoretical results are provided here without proofs, which can be found in the paper (Lotov, 1995). These results were partially published earlier in (Lotov, 1984b; Lotov, 1985b) but with minor inaccuracies that were improved in (Lotov, 1995). Note that the English translations of (Lotov, 1984b) and (Lotov, 1985b) are not adequate in several important parts of the text.

1.   Perturbations of convex bodies given by linear mappings

Let $\mathbb{W}$ be a real Banach decision space, let the set X be the feasible decision set. Let the criterion vectors, $z \in \mathbb{R}^m$, be given by

$$z = f(x),$$

where f is a linear mapping (operator) from $\mathbb{W}$ to $\mathbb{R}^m$. Let $\|z\|$ be a norm in $\mathbb{R}^m$. The FCS, denoted by Z, is defined as usual:

$$Z = \{z \in \mathbb{R}^m : \ z = f(x), x \in X\},$$

or, simply speaking,

$$Z = f(X).$$

Let $\tilde{Z}$ be a perturbed FCS, that is

$$\tilde{Z} = \{z \in \mathbb{R}^m : \ z = \tilde{f}(x), x \in \tilde{X}\},$$

where $\tilde{f}$ is a linear operator from $\mathbb{W}$ to $\mathbb{R}^m$, too, and $\tilde{X}$ that belongs to $\mathbb{W}$ is a disturbed feasible decision set.

We use the Hausdorff distance as a measure of mutual deviation of two sets. The *Hausdorff distance* between any non-empty sets G and G' in a normed linear space is defined as follows. Let us define first the distance between a point x and a set G in the same normed linear space as

$$\rho(x, G) = \inf\{\|x - x'\| : \ x' \in G\},$$

where $\|x\|$ is a norm. Then, the deviation of the set G from the set G' is defined as

$$d(G, G') = \sup\{\rho(x, G') : \ x \in G\}.$$

The Hausdorff distance between any non-empty sets G and G', denoted by $h(G, G')$, is defined then as

$$h(G, G') = \max\{d(G, G'), d(G', G)\}.$$

Using the Hausdorff distance as a measure, one can easily prove that

$$h(Z, \tilde{Z}) \leq \|f - \tilde{f}\| \sup\{\|x\| : \ x \in X\} + \|\tilde{f}\| h(X, \tilde{X}),$$

where $\|f\|$ is the norm of the linear operator f. A disturbance of the operator $\|f - \tilde{f}\|$ is often given in an explicit way. In contrast, a perturbation of the feasible decision set, that is $h(X, \tilde{X})$, is not given explicitly. Instead, disturbances of some parameters that define X are given. Therefore, the problem of estimating a perturbation of the feasible set in criterion space is reduced to estimating of $h(X, \tilde{X})$.

In this chapter we consider a particular, but very important kind of sets X, namely, the sets given by linear relations. To define a linear inequality system in some Banach space $\mathbb{W}$, we have to consider another real Banach space $\mathbb{W}_1$. Let K be a non-empty closed cone in $\mathbb{W}_1$. The cone K partially orders $\mathbb{W}_1$, i.e., $y \leq y'$ for y, $y' \in \mathbb{W}_1$, is equivalent to $y' - y \in K$. Let $B(\mathbb{W}, \mathbb{W}_1)$ be a space of linear continuous operators from $\mathbb{W}$ into $\mathbb{W}_1$ with the norm $\|A\| = \sup\{\|Ax\| : \ \|x\| \leq 1\}$.

Let C be a non-empty closed set in $\mathbb{W}$. Let the space $\mathbb{W}_1$ and the above cone $K \in \mathbb{W}_1$ be given. We consider an original linear system

$$Ax \leq b, \ x \in C \tag{9.1}$$

and perturbed linear system

$$\tilde{A}x \leq \tilde{b}, \ x \in C \tag{9.2}$$

where A, $\tilde{A} \in B(\mathbb{W}, \mathbb{W}_1)$, $b, \tilde{b} \in \mathbb{W}_1$.

If we consider X as an image of a set-valued mapping F from (A, b)-space to $\mathbb{W}$, the problem can be reformulated as a problem of estimation of

$$h(F(A, b), \ F(\tilde{A}, \tilde{b}))$$

as a function of $\|A - \tilde{A}\|$ and $\|b - \tilde{b}\|$. Special interest is related to Lipschitz continuity, i.e., to the proof of existence and to evaluating the values of positive constants γ_A, γ_b, δ_A and δ_b such as

$$h(F(A, b), F(\tilde{A}, \tilde{b})) \leq \gamma_A \|A - \tilde{A}\| + \gamma_b \|b - \tilde{b}\|$$

whenever $\|A - \tilde{A}\| \leq \delta_A$ and $\|b - \tilde{b}\| \leq \delta_b$. In this chapter, the existence of Lipschitz constants is proven and their values are obtained as a corollary of the main result that provides a nonlinear estimate of $h(F(A, b), F(\tilde{A}, \tilde{b}))$.

Note that in our case the inequality (9.1) is defined by a closed cone $K \in \mathbb{W}_1$. By considering the cone $K = \{0\}$, we can study systems of linear equalities. This means that the linear inequalities under consideration can be extended by linear equalities, which can be subjects of perturbations, too. The Hausdorff distance between solution sets of original and perturbed linear systems of this kind is studied here as well.

2. Perturbations of solution sets of linear systems in Banach space

In (Robinson, 1973) the concept of a regular system (9.1) is introduced. This is understood as a system with

$$b \in \text{int}\{A(C) + K\}. \tag{9.3}$$

It is shown that the system's regularity is the necessary and sufficient condition of stability in the sense of the deviation of the perturbed set used in (Robinson, 1973). It is clear in advance that the condition (9.3) should play an important role in the case of application of the Hausdorff distance as well.

Several concepts of the theory of convex processes will be used hereafter. The convex process S that acts from $\mathbb{W}$ into $\mathbb{W}_1$ is understood as a mapping $S : \mathbb{W} \to 2^{\mathbb{W}_1}$ whose graph

$$\text{gr}\, S = \{\{x, y\}: \quad y \in Sx\}$$

is a convex cone in $\mathbb{W} \times \mathbb{W}_1$, containing zero. Let us define

$$\text{dom}\, S = \{x \in \mathbb{W}: \quad Sx \neq \emptyset\}$$

and

$$\text{range } S = \bigcup \{Sx : \quad x \in \text{dom } S\}.$$

For any convex process S we can determine the inverse mapping S^{-1} acting from $\mathbb{W}_1$ into $2^{\mathbb{W}}$:

$$S^{-1} = \{x \in \mathbb{W} : \quad y \in Sx\}.$$

Clearly, S^{-1} is a convex process and at the same time $\text{gr } S^{-1} = \text{gr } S$, $\text{dom } S^{-1} = \text{range } S$ and $\text{dom } S^{-1} = \text{range } S$.

For a convex process S we can determine the norm:

$$\|S\| = \sup\{\inf\{\|y\| : \quad y \in Sx\} : \quad \|x\| \leq 1, \ x \in \text{dom } S\}.$$

If the value of $\|S\|$ is finite, the convex process S is referred as a normed process.

Let the set

$$X = \{x \in \mathbb{W} : \quad Ax \leq b, \quad x \in C\} \tag{9.4}$$

be non-empty and bounded. Let us consider the space $\mathbb{W} \times \mathbb{R}$ with the norm

$$\|\{x, \xi\}\| = \max\{\|x\|, |\xi|\}.$$

Let us consider a cone

$$P_0 = \{\{x, \xi\} : \quad \xi > 0, \quad Lx/\xi \in C\}$$

where L is a finite positive number such as

$$L \geq \{\|x\| : \quad x \in X\}. \tag{9.5}$$

Since $C \neq \emptyset$, we have $P_0 \neq \emptyset$. Let P be the closure of P_0. Since C is a closed set, it follows that $P \setminus P_0$ consists of points of $\{x, 0\}$ type, and that $P + P_0 = P$.

Let us introduce two mappings $T\{x, \xi\}$ and $\delta T\{x, \xi\}$ from $\mathbb{W} \times \mathbb{R}$ into $2^{\mathbb{W}_1}$ which are defined as follows:

if $\{x, \xi\} \in P$, then

$$T\{x, \xi\} = Ax - b\xi/L + K,$$

$$\delta T\{x, \xi\} = (A - \tilde{A})x - (b - \tilde{b})\xi/L,$$

elsewhere

$$T\{x, \xi\} = \delta T\{x, \xi\} = \emptyset.$$

Clearly, T and δT are convex processes. It is clear that for any given systems (9.1) and (9.2) the convex process δT is normed, and that its norm reflects the perturbation value. It is easy to obtain that

$$\|\delta T\| \leq \|A - \tilde{A}\| + \|b - \tilde{b}\|/L. \tag{9.6}$$

Though the convex processes introduced here differ only slightly from those proposed in (Robinson, 1973), they provide an opportunity to obtain a new result formulated in the following main statement.

THEOREM 9.1 *Let the system (9.1) be regular and its solution set X be non-empty and bounded. Let L satisfy (9.5). Then, the convex process T^{-1} is normed, and it holds that*

$$h(X, \tilde{X}) \le L \frac{2\|T^{-1}\|\|\delta T\|}{1 - 2\|T^{-1}\|\|\delta T\|}$$

whenever

$$2\|T^{-1}\|\|\delta T\| < 1.$$

The main Theorem 9.1 can be proven on the basis of the theory of convex processes. However, since the proof is extremely sophisticated, we skip it in this book (see Lotov, 1995).

Let us discuss the boundness assumption of the theorem. It is well known that in the finite dimensional case an unbounded solution set of a linear inequality system can be represented in the form of a sum of a polytope and of a polyhedral convex cone. Since small perturbations of inequalities describing the cone result in infinite Hausdorff distance between original and perturbed cones, boundness assumption in Theorem 9.1 seems to be inevitable. Application of this result in the study of stability and approximation of dynamic systems is described in (Lotov, 1986).

Lipschitz continuity can be easily deduced from Theorem 9.1. First of all, using (9.6) one immediately receives the following corollary of Theorem 9.1.

COROLLARY 9.2 *Under assumptions of Theorem 9.1 it holds that if*

$$2\|T^{-1}\|(\|A - \tilde{A}\| + \|b - \tilde{b}\|/L) < 1$$

then

$$h(X, \tilde{X}) \le L \frac{2\|T^{-1}\|(\|A - \tilde{A}\| + \|b - \tilde{b}\|/L)}{1 - 2\|T^{-1}\|(\|A - \tilde{A}\| + \|b - \tilde{b}\|/L)}.$$

The corollary of Theorem 9.1 results in a simple proof of a statement related to Lipschitz constants.

THEOREM 9.3 *Under assumptions of Theorem 9.1 it holds that,*
  for any

$$\delta \in (0, 1/2\|T^{-1}\|),$$

if

$$\|A - \tilde{A}\| + \|b - \tilde{b}\|/L \le \delta,$$

then

$$h(X, \tilde{X}) \le \gamma \left(\|A - \tilde{A}\| + \|b - \tilde{b}\|/L \right)$$

where

$$\gamma = \frac{2L\|T^{-1}\|}{1 - 2\|T^{-1}\|\delta}.$$

Proof. Let $z = \|A - \tilde{A}\| + \|b - \tilde{b}\|/L$. In accordance with the corollary of Theorem 9.1 we have

$$h(\tilde{X}, X) \le \frac{2L\|T^{-1}\|z}{1 - 2\|T^{-1}\|z}$$

whenever $z \in (0, 1/2\|T^{-1}\|)$. The denominator of $h(\tilde{X}, X)$ is a monotonically decreasing function of z. Therefore, if $0 \le z \le \delta$ where $\delta \in (0, 1/2\|T^{-1}\|)$, it holds that

$$h(X, \tilde{X}) \le \frac{2L\|T^{-1}\|z}{1 - 2\|T^{-1}\|\delta}.$$

Assertion of the theorem follows. ∎

Theorem 9.3 provides explicit expressions for Lipschitz constants for different perturbation ranges. For example, if $\delta = \frac{1}{8}\|T^{-1}\|$, then $\gamma = \frac{8}{3}L\|T^{-1}\|$.

Another example: for small values of δ, it approximately holds that

$$\gamma = 2L\|T^{-1}\|.$$

3. Perturbations of solution sets of finite-dimensional linear inequality systems

Let us consider a finite linear inequality system in the finite dimensional space $\mathbb{R}^n$, i.e.,

$$Ax \le b, \quad x \in \mathbb{R}^n, \tag{9.7}$$

where A is a matrix, $A: \mathbb{R}^n \to \mathbb{R}^{n_1}$, $b \in \mathbb{R}^{n_1}$. Let the cone K coincide with the non-negative orthant $R_+^{n_1}$. For any x, $x' \in \mathbb{R}^n$ we introduce the scalar product

$$\langle x, x' \rangle = \sum_{j=1}^{n} x_j x_j',$$

and the associated norm

$$\|x\| = \langle x, x \rangle^{1/2}.$$

Then, the finite system (9.7) can be re-written in the form

$$\langle a_i, x \rangle \le b_i, \quad i = 1, 2, \ldots, n_1$$

where a_i is the i-th row of A, regarded as an element of $\mathbb{R}^n$. In what follows, we shall assume that $\|a_i\| \ne 0$ for any $i = 1, 2, \ldots, n_1$.

In $\mathbb{R}^{n_1}$, we introduce the norm

$$\|y\| = \max_{i=1,\ldots,n_1} |y_i|.$$

This induces the norm

$$\|A\| = \max_{i=1,\ldots,n_1} \|a_i\|$$

in the space of linear operators from $\mathbb{R}^n$ into $\mathbb{R}^{n_1}$.

As earlier, apart from the linear constraints (9.7), we shall consider possibly non-linear constraints which are not the subject of perturbations

$$x \in C \qquad\qquad (9.8)$$

where C is a non-empty closed convex set from $\mathbb{R}^n$. Since the system (9.7), (9.8) is the special case of (9.1), Theorem 9.1 holds for it.

In (Robinson, 1973) it was shown that the regularity condition (9.3) for the system (9.7), (9.8) is fulfilled if there exists such $x^* \in C$ that

$$x^* \in \text{int } X_A$$

where $X_A = \{x \in \mathbb{R}^n: \; Ax \le b, \}$. Then, there exists such a positive value r_{min} that

$$x^* + r_{min} B \in X_A \qquad\qquad (9.9)$$

where B is the closed ball of unit radius with its center at zero in $\mathbb{R}^n$.

If the solution set $X = X_A \bigcap C$ of the system (9.7), (9.8) is bounded, then there exists such a positive value r_{max} that

$$X \in x^* + r_{max} B. \qquad\qquad (9.10)$$

This means that one can use $\|x^*\| + r_{max}$ for L in convex processes T and δT.

The estimate of the norm of the convex process T^{-1} for the system (9.7), (9.8) is constructed in the following lemma.

LEMMA 9.4 *For the system (9.7), (9.8) that satisfies (9.9) and (9.10) it holds that*

$$\|T^{-1}\| \le \frac{\|x^*\| + r_{max}}{r_{min} \min\limits_{i=1,\ldots,n_1} \|a_i\|}.$$

Proof. By definition,

$$\|T^{-1}\| = \sup_{\|y\|\le 1} \inf \big\{\max\{\|x\|, |\xi|\}: \{x, \xi\} \in T^{-1}(y)\big\} =$$

$$\sup_{\|y\|\le 1} \inf \Big\{\max\{\|x\|, |\xi|\}: y \ge Ax - \frac{b\xi}{\|x^*\|+r_{max}}, \ \{x, \xi\} \in P\Big\}.$$

It is easy to show that $b_k \ge r_{min}\|a_k\| + \langle a_k, x^*\rangle$ for $k = 1, \ldots, n_1$. Therefore,

$$\frac{b_k}{r_{min} \min_{i=1,\ldots,n_1} \|a_i\|} \ge 1 + \frac{\langle a_k, x^*\rangle}{r_{min} \min_{i=1,\ldots,n_1} \|a_i\|}.$$

Hence, given an $y \in \mathbb{R}^{n_1}$ such that $\|y\| \le 1$ (i.e. $|y_k| \le 1$, $k = 1, \ldots, n_1$), we have

$$y_k + \frac{b_k}{r_{min} \min_{i=1,\ldots,n_1} \|a_i\|} \ge \frac{\langle a_k, x^*\rangle}{r_{min} \min_{i=1,\ldots,n_1} \|a_i\|}$$

where $k = 1, \ldots, n_1$. It follows that the pair

$$\{\hat{x}, \hat{\xi}\} = \left\{\frac{x^*}{r_{min} \min_{i=1,\ldots,n_1} \|a_i\|}, \frac{\|x^*\| + r_{max}}{r_{min} \min_{i=1,\ldots,n_1} \|a_i\|}\right\}$$

satisfies the relation

$$y \ge Ax - \frac{b\xi}{\|x^*\| + r_{max}}$$

for any $y \in \mathbb{R}^{n_1}$ such that $\|y\| \le 1$. Moreover, because $x^* \in C$, we have

$$\{\hat{x}, \hat{\xi}\} \in P.$$

Therefore,

$$\inf \Big\{\max\{\|x\|, |\xi|\}: y \ge Ax - \frac{b\xi}{\|x^*\|+r_{max}}, \ \{x, \xi\} \in P\Big\} \le \max\{\|\hat{x}\|, |\hat{\xi}|\}$$

for all $y \in \mathbb{R}^{n_1}$ such that $\|y\| \le 1$. Hence, the lemma follows. ∎

The theorem below follows from Theorem 9.1, Lemma 9.4, and the plain evidence that in the case under consideration $\|\delta T\| \le \varepsilon_1$ where

$$\varepsilon_1 = \|A - \tilde{A}\| + \frac{\|b - \tilde{b}\|}{\|x^*\| + r_{max}}.$$

Note that the value of ε_1 characterizes the disturbance.

THEOREM 9.5 *For the system (9.7), (9.8) which satisfies (9.9), (9.10), for $\Delta_1 < 1$ where*

$$\Delta_1 = \frac{2(\|x^*\| + r_{max})}{r_{min} \min\limits_{i=1,\dots,n_1} \|a_i\|} \varepsilon_1, \tag{9.11}$$

it holds that

$$\frac{h(X, \tilde{X})}{\|x^*\| + r_{max}} \leq \frac{\Delta_1}{1 - \Delta_1}.$$

Note that Δ_1 given by (9.11) can be presented in an alternative, perhaps more simple form

$$\Delta_1 = 2\frac{(\|x^*\| + r_{max})}{r_{min}} \times \frac{\max\limits_{i=1,\dots,n_1} \|a_i\|}{\min\limits_{i=1,\dots,n_1} \|a_i\|} \times \frac{\varepsilon_1}{\max\limits_{i=1,\dots,n_1} \|a_i\|}. \tag{9.12}$$

Here the first multiplier is the geometric characteristic of the set, the second multiplier is the characteristic of the inequalities system, and the third one is the characteristic of relative perturbation values for the coefficients. So, one can identify three sources of the value of Δ_1. This topic is discussed in detail in Section 6. Here, we would like to attract attention to the fact that the value of Δ_1 can receive a simpler expression, if rows of the matrix A are normed, that is,

$$\Delta_1 = \frac{2(\|x^*\| + r_{max})}{r_{min}} \varepsilon_1. \tag{9.13}$$

The following statement that can be proven analogous to Theorem 9.3 gives Lipschitz constants for the particular case under consideration.

THEOREM 9.6 *For the system (9.7), (9.8) which satisfies (9.9) and (9.10), it holds that,*
   for any

$$\delta \in \left(0, \; \frac{r_{min} \min\limits_{i=1,\dots,n_1} \|a_i\|}{2(\|x^*\| + r_{max})}\right),$$

if

$$\varepsilon_1 \leq \delta,$$

then

$$h(X, \tilde{X}) \leq \gamma \varepsilon_1$$

where

$$\gamma = \frac{2(\|x^*\| + r_{max})^2}{r_{min} \min\limits_{i=1,\dots,n_1} \|a_i\| - 2\delta(\|x^*\| + r_{max})}.$$

4. Perturbations of solution sets of linear systems of equalities and inequalities

In this section, an estimate for $h(X, \tilde{X})$ is provided in the case if linear equalities are considered along with linear inequalities in finite dimensional space $\mathbb{R}^n$. We show that in this case Theorem 9.1 can be used, too.

So, in finite dimensional space $\mathbb{R}^n$, we consider the finite system of linear inequalities

$$A^{(1)}x \leq b^{(1)} \tag{9.14}$$

and the finite system of linear equalities

$$A^{(2)}x = b^{(2)} \tag{9.15}$$

where $A^{(1)}, A^{(2)}$ are matrices, $A^{(1)}\colon \mathbb{R}^n \to \mathbb{R}^{n_1}$, $b^{(1)} \in \mathbb{R}^{n_1}$, $A^{(2)}\colon \mathbb{R}^n \to \mathbb{R}^{n_2}$, $b^{(2)} \in \mathbb{R}^{n_2}$. Possibly non-linear constraints that are not a subject of perturbation are given by

$$x \in C \tag{9.16}$$

where C is a non-empty closed convex set in $\mathbb{R}^n$.

As in the previous section, we can use an alternative form of the system (9.14)

$$\left\langle a_i^{(1)}, x \right\rangle \leq b_i^{(1)}, \quad i = 1, 2, \ldots, n_1,$$

where $a_i^{(1)}$ is the i-th row of $A^{(1)}$, regarded as an element of $\mathbb{R}^n$. In what follows, we assume that $\|a_i^{(1)}\| \neq 0$ for any $i = 1, 2, \ldots, n_1$.

In $\mathbb{R}^{n_1}$, we use the same norm as in the previous section. The same norm is used in $\mathbb{R}^{n_2}$. This induces norms in the spaces of linear operators from $\mathbb{R}^n$ into $\mathbb{R}^{n_1}$ and from $\mathbb{R}^n$ into $\mathbb{R}^{n_2}$.

Since the system (9.14), (9.15) (9.16) is a special case of the system (9.1), Theorem 9.1 holds for it.

In (Robinson, 1973) it was shown that the regularity condition (9.5) for the system (9.14), (9.15), (9.16) is fulfilled if the rank of matrix $A^{(2)}$ equals the number of its rows, and if simultaneously there exists such $x^* \in X$ that

$$x^* \in \operatorname{int} X_1 \bigcap \operatorname{int} C, \tag{9.17}$$

where $X_1 = \{x \in \mathbb{R}^n \colon A^{(1)}x \leq b^{(1)}\}$. The condition that the rank of matrix $A^{(2)}$ equals n_2 is equivalent to the fact that by rearranging the columns of $A^{(2)}$, the relation (9.15) can be written in the form

$$A^{(21)}x^{(1)} + A^{(22)}x^{(2)} = b^{(2)}, \tag{9.18}$$

where $A^{(2)} = \{A^{(21)}, A^{(22)}\}$, $x = \{x^{(1)}, x^{(2)}\}$, $A^{(21)}$ is a square matrix for which $(A^{(21)})^{-1}$ does exist.

The condition (9.17) means that there exists such a positive value r_{min} that

$$x^* + r_{min}B \in X_1 \bigcap C. \tag{9.19}$$

If the solution set X of the system (9.14), (9.15), (9.16) is bounded, then there exists such a positive value r_{max} that

$$X \in x^* + r_{max}B. \tag{9.20}$$

It means in particular that one can use $\|x^*\| + r_{max}$ for L.

Note that the cone $K \subset \mathbb{R}^{n_1} \times \mathbb{R}^{n_2}$ has a bit more complicated structure in case of the system (9.14), (9.15), (9.16) than for the system (9.7), (9.8), that is

$$K = R_+^{n_1} \times \{0\}.$$

The estimate of the norm of the convex process T^{-1} for the system under study is given by the following lemma.

LEMMA 9.7 For the system (9.14), (9.15), (9.16) which satisfies (9.18), (9.19), (9.20), it holds that

$$\|T^{-1}\| \leq \frac{(\|x^*\| + r_{max})(1 + \|A^{(1)}\|\|(A^{(21)})^{-1}\|)}{r_{min} \min_{i=1,\ldots,n_1} \|a_i^{(1)}\|}.$$

We do not prove the lemma herein (see Lotov, 1985b). The idea of the proof resembles the idea of the proof of Lemma 9.4.

The theorem below follows from Theorem 9.1, from the above Lemma 9.7, and the plain evidence that in the case under consideration it holds that $\|\delta T\| \leq \varepsilon_2$ where

$$\varepsilon_2 = \max\{\|A^{(1)} - \tilde{A}^{(1)}\|, \|A^{(2)} - \tilde{A}^{(2)}\|\} + \frac{\max\{\|b^{(1)} - \tilde{b}^{(1)}\|, \|b^{(2)} - \tilde{b}^{(2)}\|\}}{\|x^*\| + r_{max}}.$$

The value of ε_2 characterizes the disturbance of the system (9.14), (9.15), (9.16).

THEOREM 9.8 For the system (9.14), (9.15), (9.16) that satisfies (9.18), (9.19), (9.20), for $\Delta_{12} < 1$ where

$$\Delta_{12} = \frac{2(\|x^*\| + r_{max})(1 + \|A^{(1)}\|\|(A^{(21)})^{-1}\|)}{r_{min} \min_{i=1,\ldots,n_1} \|a_i^{(1)}\|} \times \varepsilon_2, \tag{9.21}$$

it holds that

$$\frac{h(X, \tilde{X})}{(\|x^*\| + r_{max})} \leq \frac{\Delta_{12}}{1 - \Delta_{12}}.$$

The value of Δ_{12} has an especially simple form in the case if $\|A^{(1)}\| = 1$.

The Lipschitz constant can be easily obtained on the basis of Theorem 9.8.

THEOREM 9.9 *For the system (9.14), (9.15), (9.16) that satisfies (9.18), (9.19), (9.20) it holds that,*

for any

$$\delta \in \left(0, \ \frac{r_{min} \min\limits_{i=1,...,n_1} \|a_i^{(1)}\|}{2(\|x^*\| + r_{max})(1 + \|A^{(1)}\|\|(A^{(21)})^{-1}\|)}\right),$$

if

$$\varepsilon_2 \leq \delta,$$

then

$$h(X, \tilde{X}) \leq \gamma\varepsilon_2$$

where

$$\gamma = \frac{2(\|x^*\| + r_{max})^2(1 + \|A_1\|\|A_{21}^{-1}\|)}{r_{min} \min\limits_{i=1,...,n_1} \|a_i^1\| - 2\delta(\|x^*\| + r_{max})(1 + \|A_1\|\|A_{21}^{-1}\|)}.$$

5. Perturbations of solution sets of linear equality systems

Now let us consider the system (9.15), (9.16) that does not contain inequalities (9.14). In this case the condition (9.18) can be reformulated as follows: there exists such a positive value r_{min} that

$$x^* + r_{min}B \in C. \tag{9.22}$$

LEMMA 9.10 *For the system (9.15), (9.16) that satisfies (9.19), (9.20), (9.22), it holds that*

$$\|T^{-1}\| \leq \frac{(\|x^*\| + r_{max})\|(A^{(21)})^{-1}\|}{r_{min}}.$$

The proof of Lemma 9.10 is not provided here because of the same reasons as in the case of Lemma 9.7 (see Lotov, 1985b). Let

$$\varepsilon_3 = \|A^{(2)} - \tilde{A}^{(2)}\| + \frac{\|b^{(2)} - \tilde{b}^{(2)}\|}{\|x^*\| + r_{max}}.$$

THEOREM 9.11 *For the system (9.15), (9.16) that satisfies (9.19), (9.20) and (9.22), for $\Delta_2 < 1$ where*

$$\Delta_2 = \frac{2(\|x^*\| + r_{max})\|(A^{(21)})^{-1}\|}{r_{min}}\varepsilon_3, \qquad (9.23)$$

it holds that

$$\frac{h(X, \tilde{X})}{\|x^*\| + r_{max}} \leq \frac{\Delta_2}{1 - \Delta_2}.$$

Comparing the results of Theorems 9.5, 9.8 and 9.11, we find that their assertions look just the same. It is interesting that the values of Δ_1 given by (9.13) and of Δ_{12} given by (9.21) (both for $\|A^{(1)}\| = 1$) as well as of Δ_2 given by (9.23), differ only by the factors near coefficient perturbations (these factors are 1, $1 + \|(A^{(21)})^{-1}\|$, $\|(A^{(21)})^{-1}\|$, respectively).

The Lipschitz constant follows from Theorem 9.11.

THEOREM 9.12 *For the system (9.15), (9.16) that satisfies (9.19), (9.20) and (9.22) it holds that,*

for any

$$\delta \in \left(0, \; r_{min}/[2(\|x^*\| + r_{max})\|(A^{(21)})^{-1}\|]\right)$$

if

$$\varepsilon_3 \leq \delta,$$

then

$$h(X, \tilde{X}) \leq \gamma\varepsilon_3$$

where

$$\gamma = \frac{2(\|x^*\| + r_{max})\|(A^{(21)})^{-1}\|/r_{min}}{1 - 2\delta(\|x^*\| + r_{max})\|(A^{(21)})^{-1}\|/r_{min}}.$$

One can compare the values of γ and δ in the three Theorems 9.6, 9.9 and 9.12 related to three different linear systems.

6.   Condition number for finite-dimensional systems and its application

Let us consider the factorization of the value of Δ_1 given by (9.12). First of all, recall that $\max_{i=1,\ldots,n_1} \|a_i\| = \|A\|$. Note that all inequalities of (9.7) which do not satisfy

$$b_i \leq r_{max}\|a_i\| + \langle a_i, x^* \rangle$$

can be excluded from the system (9.7) without altering the set X. Therefore, one can assume that

$$\|b\| = \max_{i=1,\dots,n_1} |b_i| \leq (r_{max} + |x^*|)\|A\|. \qquad (9.24)$$

Let us introduce the notation:

$$\delta A = \|A - \tilde{A}\|/\|A\|,$$

$$\delta b = \|b - \tilde{b}\|/\|b\|,$$

$$\delta X = h(X, \tilde{X})/(\|x^*\| + r_{max}),$$

$$\nu_X = (\|x^*\| + r_{max})/r_{min},$$

$$\nu_A = \max_{i=1,\dots,n_1} \|a_i\|/\min_{i=1,\dots,n_1} \|a_i\|,$$

$$\nu = \nu_X \nu_A.$$

Then, taking (9.24) into account, we can write (9.12) in the form

$$\Delta_1 = 2\nu(\delta A + \delta b).$$

We can reformulate Theorem 9.5 in the following simple form.

THEOREM 9.13 *For the system (9.7), (9.8) that satisfies (9.9) and (9.10), for*

$$2\nu(\delta A + \delta b) < 1,$$

it holds that

$$\delta X \leq \frac{2\nu(\delta A + \delta b)}{1 - 2\nu(\delta A + \delta b)}. \qquad (9.25)$$

In our opinion, the value ν introduced here can be referred to as the *condition number* for the system (9.7), (9.8), since the estimate (9.25) resembles the following famous result for linear equations systems (see, for example, Voevodin, 1977): the solution x of a linear equations system $Ax = b$, where $x \in \mathbb{R}^n$, $A: \mathbb{R}^n \to \mathbb{R}^n$, $b \in \mathbb{R}^n$, and the solution $\tilde{x}$ of a perturbed system $\tilde{A}x = \tilde{b}$, where $\tilde{A}: \mathbb{R}^n \to \mathbb{R}^n$, $\tilde{b} \in \mathbb{R}^n$, are connected by the relation

$$\delta x \leq \frac{\nu(\delta A + \delta b)}{(1 - \nu \delta A)},$$

where

$$\delta x = \frac{\|x - \tilde{x}\|}{\|x\|}, \quad \delta A = \frac{\|A - \tilde{A}\|}{\|A\|}, \quad \delta b = \frac{\|b - \tilde{b}\|}{\|b\|}, \quad \nu = \|A\|\|A^{-1}\|.$$

Note that the value of ν for the system (9.7), (9.8) is the product of ν_X and ν_A that can be referred to as the condition numbers for the set and for the matrix. The matrix can easily be normed in such a way that its condition number ν_A will turn out to be 1. In this case,

$$\nu = \nu_X = \frac{\|x^*\| + r_{max}}{r_{min}}.$$

Using the concept of condition number, we obtain the following Lipschitz constants.

THEOREM 9.14 *For the system (9.7), (9.8) that satisfies (9.9) and (9.10), it holds that, for any $\delta \in (0, \ 0.5\nu)$, if $\delta A + \delta b \leq \delta$, then $\delta X \leq \gamma(\delta A + \delta b)$ where $\gamma = 2\nu/(1 - 2\nu\delta)$.*

Note that, for using the estimate (9.25), one has first to evaluate the values of r_{max} and r_{min}. The point x^* and the value r_{min} can be found by solving one special linear programming problem (see, for example, Ashchepkov, 1980). From the theoretical point of view, evaluating of r_{max} is a more complicated task. Fortunately, one usually can estimate it in real-life studies roughly on the basis of content analysis of the problem.

Applications. In the introduction we have already discussed how the results of this chapter could be used for estimating perturbations of FCSs. Other applications do exist as well. One of them is related to the process of studying of perturbations of the reachable sets for linear multi-step systems (see Chapter 7 of the book (Lotov, Bushenkov and Kamenev, 1999)). The estimate (9.25) can be used for a priori estimating of perturbation of the optimal criterion values of linear programming problems. Moreover, the theory developed herein has been applied for a priori estimating of rounding errors in Fourier convolution processes for the linear inequality systems (Lotov, 1986).

Epilogue: New applications of the IDM technique on the Web

In our Epilogue we try to look forward into the future. Therefore, in contrast to the other parts of the book, where recent applications of the IDM technique are described, we discuss options that might be provided by the IDM technique sooner or later. So, this part of the book can be considered as science fiction. Nevertheless, information technology develops very fast, and so we hope that the discussed ideas will be implemented soon. In any case, we try to prove it in this part of the book.

Introduction. Involvement of non-experts in the process of environmental decision making through the Internet is the topic of our Epilogue. Usually, non-experts have minimal knowledge of environmental problems and especially of the ways to solve them. Nevertheless, non-experts want and often are involved in political actions related to such problems. It is clear that the gap between knowledge and actions of non-experts can be misused by irresponsible politicians and is dangerous. The Internet seems to be a mean that can help non-experts understand the problem and use the knowledge in their problem-related political actions in a deliberate way. We argue that Internet visualization can help to develop tools, which can be mastered by non-experts and used for obtaining knowledge of environmental strategies.

As it has already been said in this book, the screening phase is an extremely important part of the environmental decision process. At the screening phase a small number of projects are selected for subsequent discussion and final choice, while other projects are rejected forever. This is why the involvement of non-experts in the decision process is especially important at the screening phase. Here we discuss specially designed Web resources aimed at supporting of non-experts in their independent screening of possible environmental decisions. By this, such Web resources can support non-experts in their preparation for political actions related to environmental problems.

The need to support non-experts in their preparation for environmental political actions has been articulated in the paper of M. Abbott and his colleagues (Yan, Solomatine, Velickov and Abbott, 1999). This concept has the name "democratic paradigm of environmental decision making". This part of our book can be considered as an attempt to an Internet tool that can help implement the general ideas of this new paradigm.

The democratic paradigm of environmental decision making seems to be a response to complications related to the technocratic (expert-oriented) paradigm that is actually the traditional approach to environmental decision making. Roughly speaking, in the framework of the technocratic paradigm, experts develop projects, and professional decision makers approve or reject them. Mathematical models and DSSs play an important role in the technocratic paradigm; they are used by experts and sometime even by decision makers. Experts and decision makers, with the help of modelers, system analysts and computer scientists, find more or less satisfying solutions to environmental problems. In such decision processes, system analysts exploit the fact that the number of people involved in a technocratic decision process is fairly small and known in advance. Technocratic decision making, which is a norm now, will surely profit from recent development of the Internet. Professional experts and even decision makers gradually master the network tools including Web servers, distributed simulation and optimization, various forms of Internet communication (say, synchronic or asynchronic meetings), etc.

However, the technocratic approach to environmental decision making does not seem to be sufficient now. It is related to the fact that the recent situation in the field of environmental decision making differs from what it was about 50 years ago. Multiple political parties and interest groups, mass media and even particular citizens want to be involved in decision processes. The number of such new players is not known in advance, and it may happen to be very large. It is important that these players are non-experts. Usually, they have minimal knowledge of the problem and especially of ways solve it. In the framework of the technocratic paradigm, not so much can be done to involve non-experts in the decision process. Sometimes, they are informed on general features of the strategies discussed by experts and decision makers, however, they cannot influence the decision processes except by protesting against strategies selected by technocrats. Sometimes, they can stop the implementation of such a strategy.

As an illustration of a failure of the technocratic paradigm, we provide the story of a large-scale water management project based on partial diversion of the flow of Northern Russian rivers into the Volga River basin. Such a project was proposed by water managers and intensively discussed in the USSR in the 1970s–1980s. The main aim of the project was to develop the USSR grain crop production in the Volga and Don River basins. The USSR has had a permanent deficit of grain crops, which was covered by imports. Though the deficit of grain crops was related mainly to non-efficient agricultural technologies and large losses in the processes of grain transportation and grain storing, such inefficiency was considered to be inevitable (it seems that this was true for the non-market economy of the USSR). Therefore, the project, which promised to increase grain crop production, was supported by the ruling communist party. Another advantage of the project was related to the opportunity to increase the level of the Caspian Sea that was extremely low at that time (in contrast to the current situation). Additional advantages did exist, but they played a minor role in the process of discussing the project.

However, the project had many negative features. First of all, it was extremely expensive. For example, it was clear to any broad-minded person that a minor market-oriented reform of the agricultural sector of the USSR economy could improve production without such tremendous investment. Secondly, multiple negative environmental consequences of the project were articulated. In addition, unexpected environmental consequences were fairly plausible. Therefore, the project required a detailed study and long-time consideration. However, the communist party of the USSR was in a hurry to support the planned economy. In 1981 the party made a decision to approve the project and start its implementation. It resulted in mass protests of environmentalists, other researchers, writers and practically all educated people. This protest campaign lasted about five years and resulted first in suspension of the project and then in its final stopping in 1986.

This example shows how technocratic decision making processes that are not supported by non-experts may be not sufficient even in the case of a non-democratic society. However, the technocratic paradigm is still used. As the most known current example of the technocratic approach to environmental decision making, one can mention attempts to solve the problems of global climate change. Mass media inform ordinary people about threats that are likely to result from climate change. The people know that experts and politicians keep negotiating some strategies based on national restrictions on carbon dioxide emission, but nobody tries to involve non-experts in the decision process — the reasons why particular strategies were selected are not discussed with them. However, such involvement seems to be very important at least for two reasons. First, the negotiated strategies, which promise to abate the consequences of the climate change, may have negative economic consequences. So, their implementation may be related to tough political decisions, and, therefore, the involvement of ordinary people is needed. Secondly, the involvement of ordinary people can help broaden the scope of strategies under discussion — economically efficient strategies may exist that result in the same decrement of carbon dioxide emission (see Section 5 of Chapter 2).

Democratic paradigm of environmental decision making. In contrast to the technocratic paradigm, the democratic paradigm is based on the idea that "the power to make decisions must be placed as far as possible in the hands of the persons who are the most directly influenced by the decision concerned, and not in the hands of individual decision makers and their experts" (Yan, Solomatine, Velickov and Abbott, 1999). The Internet clearly must play an important role in this reallocation of power to make decisions. To be precise, the Internet must provide an environment for resources that can support activities of non-experts in decision process. For this reason, the democratic paradigm of environmental decision making needs, first of all, Internet tools that can help multiple non-experts to understand possible strategies related to environmental decision problems.

It is important to take into account the following aspect of the problem. Often it is supposed tacitly that the opportunity to get information on a conflict is sufficient to understand it and even to be involved in the related decision processes. It is clear that such opinion is not valid, especially in the case of environmental problems. A database filled with possible data on a problem cannot help a non-expert to understand how the conflict could be solved. Cunge and Erlich (Cunge and Erlich, 1999) stress the need to make a difference between *data* and *information*. As data they regard the results of various measurements, variables that describe the state of an ecosystem and its evolution, regulations and laws, etc. In short, data is anything that can be measured or collected in the field or may result from projections, extrapolations and modeling. As information they regard products elaborated from the data. Cunge and Erlich require that information must be provided under a form intelligible to the stakeholders.

Mathematical modeling provides means for transformation of raw data into information. However, it is not able to solve the problem by itself

— special tools are required that can transform results of modeling into a form intelligible to non-experts.

It is needed to stress from the very beginning that we do not hope that application of such tools will result in a coordinated decision. Normally no consensus can be expected — different groups have beliefs, values, intentions and interests that are in conflict. Therefore, Internet tools can only help to assess the problem and to facilitate preparation for subsequent negotiations and political actions (see Yan, Solomatine, Velickov and Abbott, 1999, for details).

Special tools for negotiation support have already been proposed on Internet (see, for example, Kersten and Noronha, 1999), but they are usually oriented to support of negotiations of a small number of people known in advance. Such tools try to bring the negotiators to a coordinated decision using preference-related questions. However, such structured Internet-based procedures seem to be impossible in the case of the democratic paradigm characterized by a non-fixed (and presumably large) number of people with different values, which may be not articulated at all. For this reason, we concentrate on common Internet resources that can support the development of individual judgments in the process of preparation for political actions.

A list of requirements applied to Internet resources for supporting of non-experts may be fairly large. The most important of them are related to *objectivity of tools*. According to (Cunge and Erlich, 1999), the objectivity of tools means that the tools must

1) allow "the confrontation of consequences of various potentially possible scenarios and solutions", and

2) "share information in an equitable way, i.e., so that it is identical in content and intelligible to all interested parties".

The last requirement means that information must be provided to all users under the same form. Only in this case one can hope that the transformation of data into information can be appraised by non-experts as performed in an objective way.

An important feature of objective tools is the symmetric involvement of different players in the decision preparation process. It is not a simple task to implement this requirement. Say, it is proposed fairly often that non-experts must evaluate several decision alternatives, which have already been prepared and explored by the experts. However, using of such a list of decision alternatives developed by experts results in asymmetric relations between experts and non-experts: experts can develop alternatives and non-experts cannot. This asymmetric situation is not equitable, and the objectivity principle is violated. Experts may use it to thrust their preferences on non-experts. Therefore, for the sake of objectivity it would be very important to make the situation symmetric, i.e., to help non-experts to develop the decision alternatives by themselves. This statement immediately results in the requirement to involve non-experts in the decision procedure at the screening stage.

Another group of requirements is related to the *transparency of the form*, in which information is provided. The transparent form is needed to make the information intelligible for all parties including non-experts. As it has already been discussed, visualization of information seems to be the only approach intelligible for non-experts.

Concept of Internet resources for supporting the democratic paradigm. Several concepts have already been proposed for developing of Internet resources aimed at involvement of non-experts in the environmental decision process. One of them is the concept of an Internet-based judgment engine that must help non-experts to assess environmental impact related to a number of given projects and to evaluate by this the decision making efforts (Yan, Solomatine, Velickov and Abbott, 1999). As we have already said, a given list of possible decision alternatives developed by experts in advance results in asymmetric relations between experts and non-experts. To make the situation symmetric, we propose to use the Web applications of the IDM technique that can help non-experts to select preferable decision strategies independently. Such use of the IDM technique can transform the judgment phase into decision screening and evaluation processes. By this it is made symmetric and objective.

One must remember that application of the IDM technique (as of other tools that may be used in the Web resource) requires one to collect data and develop mathematical models in advance. These problems must be solved before a tool can be used by non-experts, since we cannot assume non-experts to be able to develop models by themselves. Therefore, experts have to prepare the models and the data. This fact results in a another asymmetry of experts and non-experts. The question may arise whether non-experts would agree with such a situation. This question is closely related to the problem of transparency of the mathematical model. We do not discuss this extremely sophisticated and important problem here (see Abbott and Jonoski, 1998). However, it is clear that non-experts may be convinced of the quality of the models and data in cases when experts representing different positions in an environmental conflict agree with these models and data. In any case, objectivity of natural processes provides a basis for development of such models, however, it is a complicated problem, especially taking the conflict of interests into account.

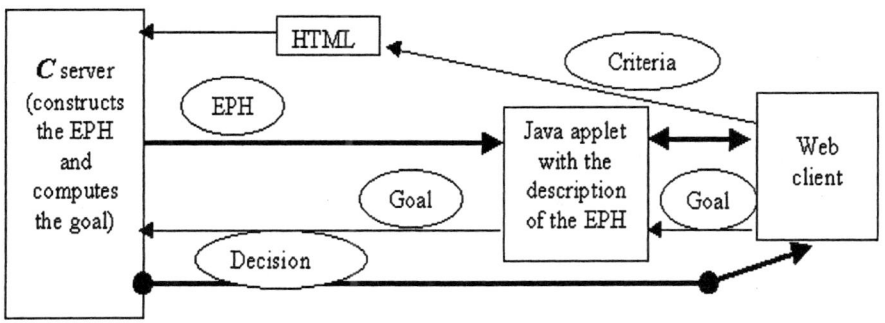

Figure 10.1. Scheme of the Web resource

This book shows that the IDM technique can be used in a broad range of multi-criteria decision problems with a large finite or infinite number of decision alternatives. In Chapter 1, Web application of the IDM technique is introduced and its demo application is described. The application proves that the IDM technique can be used on the Internet.

A possible structure of the IDM-related part of a Web resource that supports decision screening is given in (Figure 10.1). It consists of a server, which approximates the EPH for an environmental problem, of a Java applet, which provides on-line visualization of the Pareto frontier and supports goal fixation, and of middleware that helps to arrange interaction of subsystems and user dialogue with server.

First, the user has to specify the screening criteria and, perhaps, several constraints imposed on variables of the model. The criteria and constraints may be specified in large lists of possible criteria, namely performance characteristics (as it was described in Chapter 3). Then, the EPH is approximated by the server and transmitted to the user jointly with the Java applet. With the help of the applet, the user explores decision maps. By this he/she understands the opportunities that are provided by the given variety of decision alternatives. Then, the user identifies a feasible goal. The goal is transmitted back to the server where an associated decision is computed and transmitted to the user who studies the decision. Web-based multimedia techniques or GIS may be used to simplify such a study. If the user is not satisfied with the decision, he/she can specify different criteria, impose new constraints and start searching for a new decision alternative.

Due to such a Web tool, non-experts can obtain information on the whole variety of possible decisions (that is, on all potential solutions to an environmental problem) and are able to screen the variety. Note that in the framework of such a Web resource, the FGM/IDM technique can be applied for the study of linear simplified integrated models. If a nonlinear integrated model must be used for decision screening, the variety of feasible decision alternatives can be approximated by a large, but finite number of decision alternatives (many thousands or even millions of them). Sometimes, the variety of feasible decision alternatives in an environmental problem is large, but finite from the very beginning (such an example is given in Chapter 4 where a decision support system for water quality planning in a small region is considered). In this case the RGM/IDM-based Web application server for selecting preferable alternatives from large databases that uses Java-applets can be applied (see Section 5 of Chapter 4). The scheme of the application server is provided in Figure 4.18. Though the Web application server was developed to satisfy the needs of e-commerce, it can be applied in the framework of Web-based environmental decision screening, too.

Sometimes, the non-expert may want to test selected decisions using more detailed models. Simulation can help in this case, however, such detailed and complicated mathematical models must be prepared in advance. Moreover, the Web resource must contain some simple tools that can help to plan the simulation experiment. In the simplest case, standard plans prepared by experts can be used. Due to such plans, output of a decision alternative can be estimated and provided to users who have no expertise in simulation planning.

Results of the simulation experiments (as the results of decision screening) can be explored using visualization tools like GIS and other multimedia tools. Additional opportunities are related to virtual reality, which helps the user to "participate" in the life of a virtual world that would result from the simulated decision alternative (Neves and Camara, 1999). Though virtual reality requires enormous computing and information flow (and so it can not be used via the Web now), it is clear that it will be available soon. Application of these tools on the Web provides simulation results in a simple form that makes these results assessable even for non-experts. It may be important to evaluate a small number of decision alternatives, after they have been selected, simulated and explored with the help of GIS and other graphic tools. The non-expert can apply techniques for evaluation of given alternatives described, say, in (Yan, Solomatine, Velickov and Abbott, 1999). As another option, that is, a direct choice of a preferred alternative, can be applied, too.

Now we are ready to describe the structure of a possible Web resource that can help to inform non-experts in their preparation for political actions. The following subsystems can be included in the resource:

1 Subsystem for informing non-experts on the problem (it may be based on various Web multi-media tools including GIS);

2 Subsystem for specification of criteria for screening the variety of decision alternatives and, perhaps, specification of constraints imposed on other performance characteristics;

3 Subsystem for EPH or CEPH approximation;

4 Java applet that provides on-line display of decision maps and supports identification of the goal;

5 Subsystem for computing the goal-related decision alternatives;

6 Subsystem for visualization of the computed alternatives;

7 Subsystem for Web simulation of the selected alternatives;

8 Subsystem for visual exploration of the simulation output;

9 Subsystem for Web evaluation of the alternatives.

Conclusion. Concluding this section, we repeat that the IDM technique can be used on the Web for supporting environmental decision screening. Web implementation of the IDM technique can be used along with such tools as Web simulation, Web GIS and other tools for graphic exploration of particular decision alternatives. The IDM technique can support an independent search for preferable decision alternatives, helping by this to satisfy the requirements of objectivity of the Web resources. Real-life applications of the IDM technique prove that the technique is sufficiently convenient for experts. Experimental application of the existing IDM-based Web resources proves that the IDM technique can be easily implemented on the Internet. Many years of application of the software in computer educational tools for university students proves that one does not need to be expert to understand and consciously apply the IDM technique. An IDM-based environmental computer game was sporadically tested in experiments with people without university education (including teenagers). They managed to master it, too. These results make us hope that the technique can be used by any computer-literate non-expert. Due to this, development of non-expert-oriented IDM-based Web resources aimed at supporting of independent decision screening in particular environmental problems can be started immediately. The question is whether society is ready to involve non-experts in environmental decision processes.

References

Abbott, M.B., and Jonoski, A. (1998) Promoting distributed social learning and collaborative decision-making through electronic networking. In: V. Babovic and B. Larsen (eds), *Hydroinformatics'98*, Rotterdam: Balkema.

Aleksandrov, A.D. (1948) *Intrinsic geometry of convex surfaces*, Moscow: OGIZ (in Russian).

Aleksandrov, A.D. (1950) *Convex polyhedra*, Moscow: Gostekhizdat (in Russian).

Appino, P.A. (1984) A solution technique for approximating the non-inferior set of three-objective linear programs. Ph.D. dissertation, Johns Hopkins Univ., Baltimore, MD.

Ashchepkov, L.T., (1980) On constructing a maximal cube inscribed in a given domain. *Zh. Vychisl. Matem. Matem. Fiz.*, 20, 510–513 (in Russian).

Berezkin, V.E. (2002) *Analysis and implementation of the methods for Pareto frontier approximation for non-linear systems*. Moscow: Dorodnicyn Computing Centre of Russian Academy of Sciences (in Russian).

Berezkin, V., Kamenev, G., and Lotov, A. (2000) *Implementation of the feasible goals method for non-linear models in MS Excel*. Moscow: Computing Centre of Russian Academy of Sciences (in Russian).

Berezkin, V.E., Kamenev, G.K., Lotov, A.V. and Miettinen, K.M. (2003) *Application of Feasible Goals Method for Multi-Criteria Nonlinear Optimal Control Problem in Partial Derivatives*. University of Jyväskylä, Reports of the Department of Mathematical Information Systems, Series B, Scientific Computing.

Blagodatskikh, V.I., and Filippov, A.F. (1985). Differential inclusions and optimal control. In: *Topology, ordinary differential equations, dynamic systems. Proc. of Math. Inst. of Academy of Sciences of USSR*. Moscow: Nauka (in Russian).

Borcherding, K., Schmeer, S., Weber, M. (1993) Biases in multiattribute weight elicitation. In: Caverni, J.P., Bar-Hillel, M., Barron, F.N., Jungerman, H. (eds). *Contributions to Decision Research*, Amsterdam: North Holland.

Bourmistrova, L.V. (1999) *Study of an algorithm for polyhedral approximation of convex bodies*. Moscow: Computing Centre of Russian Academy of Sciences (in Russian).

Bourmistrova, L.V. (2000a) Investigation of a new method for the approximation of convex compact bodies by polyhedra. *Zh. Vychisl. Matem. i Matem. Phys.*, 40(10), 1475–1490 (in Russian, English translation in: *Comput. Maths. Math. Phys.*, 40(10), 1415–1429).

Bourmistrova, L.V. (2000b) *Analysis and application of adaptive methods for approximation of convex compact bodies by polytopes.* Ph.D. Thesis, Computing Centre of Russian Academy of Sciences (in Russian).

Bourmistrova, L.V. (2003) Experimental study of a new adaptive method for polyhedral approximation of multidimensional convex bodies. *Zh. Vychisl. Matem. i Matem. Phys.*, 43(3), 328–346 (in Russian, English translation in: *Comput. Maths. Math.Phys.*, 43(3)).

Bourmistrova, L.V., Efremov, R.V. and Lotov, A.V. (2002) Graphic method for decision support and its application in water resource management systems. *Izvestia (Proceedings) of Russian Academy of Sciences). Series "Theory and Control Systems"*, 2002, No 5 (in Russian, English translation in: *J. of Computer and Systems Sciences Int.*, 41(5), 759–769).

Bronshtein, E.M. and Ivanov, L.D. (1975) The approximation of convex sets by polyhedra. *Sibirsk. Mat. Zh.*, XVI (5), 1110–1112 (in Russian).

Burger, E. (1956) Uber homogene lineare Ungleichungs-systeme. *Z. Angew. Math. Mech.*, 36(3/4), 135–139.

Bushenkov, V.A. (1981) A numerical algorithm for construction of projections of polyhedral sets. In: *Proc. of the Conf. M.Ph.T.I. 1980. Aerophysics and Applied Mathematics,* Collected Volume, Moscow: Moscow Institute of Physics and Technology, 108–110 (in Russian).

Bushenkov, V. A, (1982) *Mathematical methods and software for analysis of linear control systems on the base of generalized reachable sets.* Ph.D. Thesis, Moscow Institute of Physics and Technology (in Russian).

Bushenkov, V.A. (1985) An iteration method of constructing orthogonal projections of covex polyhedral sets. *Zh. Vychisl. Matem. i Matem. Phys.*, 25(9), 1285–1292, (in Russian, English translation in: *U.S.S.R. Comput. Maths. Math. Phys.*, 25(5), 1–5).

Bushenkov, V.A., Ereshko, F.I., Kindler, J., Lotov A.V., and de Mare, L., (1982) Application of the GRS method to water resources problem in Southwestern Skane, Sweden. WP-82-120, Laxenburg, Austria: Int. Inst. for Applied System Analysis.

Bushenkov, V., Kaitala, V., Lotov A., and Pohjola, M., (1994) Decision and negotiation support for transboundary air pollution control between Finland, Russia, and Estonia. *Finnish Economic Papers,* 7(1), 69–80.

Bushenkov, V.A., and Lotov, A.V. (1980a) Algorithms for analysis of inequalities independence in linear systems. *Zh. Vychisl. Matem. i Matem. Phys.*, 20(3), 562–572 (in Russian, English translation in: *U.S.S.R Comput. Maths. Math. Phys.*, 20(3)).

Bushenkov, V.A., and Lotov, A.V. (1980b) Methods and algorithms for analyzing linear systems, by constructing generalized sets of attainability. *Zh. Vychisl. Matem. i Matem. Phys.*, 20(5), 1130–1141 (in Russian, English translation in: *U.S.S.R Comput. Maths. Math. Phys.*, 20(5), 38–49).

Bushenkov, V.A., and Lotov, A.V. (1982) *Methods for the construction and application of generalized reachable sets.* Moscow: Computing Center of the USSR Academy of Sciences (in Russian).

Bushenkov, V.A., and Lotov, A.V. (1983) Analysis of the potentialities of a region in multi-regional multi-industrial model of world economy. In: *Multi-regional Multi-industrial Models of World Economy,* A.Granberg and S.Menshikov (eds.), Novosibirsk: Nauka Publishing House, 202–217 (in Russian).

Bushenkov, V.A., and Lotov, A.V. (1984) POTENTIAL. applied program package. In: *Software. Applied program packages*, Moscow: Nauka, 120–132 (in Russian).

Bushenkov, V., and Smirnov, G. (1992) A new approach to the regulator design problem, *Optimization methods and software*, 1(1), 1–12.

Bushenkov, V., and Smirnov, G. (1997) *Stabilization problems with constraints: Analysis and computational aspects*. Amsterdam: Gordon and Breach Science Publishers.

Button, L., and Wilker, J.-B. (1978) Cutting exponents for polyhedral approximations to convex bodies. *Geomet. Dedic.* 7(4), 417–430.

Charnes, A., and Cooper, W.W. (1961) *Management models and industrial applications of linear programming*, vol. 1, New York: John Wiley and Sons, Inc.

Charnes, A., Cooper, W.W., and Rhodes, E. (1978) Measuring the efficiency of decision making units. *European Journal of Operational Research*, 2(6), 429–444.

Checkland, P.B. (1982) *Systems Thinking, Systems Practice*, New York: John Wiley and Sons, Inc.

Chernikov, S.N. (1965a) Convolution of finite systems of linear inequalities. *Zh. Vychisl. Matem. Matem. Fiz.*, 5(1), 1–24 (in Russian).

Chernikov, S.N. (1965b) Convolution of systems of linear inequalities, *Zh. Vychisl. Matem. Matem. Fiz.*, 5(2), 221–227 (in Russian).

Chernikov, S.N. (1968) *Linear inequalities*, Moscow: Nauka (in Russian. Translation in German: Tchernikow, S.N. *Lineare Ungleichungen*, Deutcher Verlag der Wissenschaften, Berlin, 1971).

Chernous'ko, F. L. (1988) *Estimation of the phase state of dynamical systems. The method of ellipsoids*, Moscow: Nauka, (in Russian).

Chernykh, O.L. (1986) *Algorithm for constructing the convex hull of a finite number of points that is robust in the case of rough calculations.* Moscow: Computing Centre of the USSR Academy of Sciences. Deposited at All-Union Institute for Scientific and Technical Information (VINITI) on 11.04.1986, No.2646-B (in Russian).

Chernykh, O.L. (1987) *Methods for constructing the generalized reachable sets in the case of rough computing, and their application for analysis of mathematical models of controlled systems.* Ph.D. Thesis, Computing Center of the USSR Academy of Sciences (in Russian).

Chernykh, O.L. (1988) Construction of the convex hull of a finite set of points when the computations are approximate. *Zh. Vychisl. Matem. Matem. Fiz* 28(9), 1386–1396 (in Russian; English translation in *U.S.S.R. Comput. Maths. Math. Phys.*, 28(5), 71–77).

Chernykh, O.L. (1991) Construction of the convex hull of a finite set of points on the basis of triangulation. *Zh. Vychisl. Matem. Matem. Fiz.*, 31(8), 1231–1242 (in Russian; English translation in *Comput. Maths. Math. Phys.*, 31(8), 80–86).

Chernykh, O.L. (1992) Constructing the convex hull of a point set as a system of linear inequalities, *Zh. Vychisl. Matem. Matem. Fiz.* 32(8) 1213–1228 (in Russian; English translation in *Comput. Maths. Math. Phys.*, 32(8), 1085–1096).

Chernykh, O.L. (1995) Approximation of the Pareto-hull of a convex set by polyhedral sets. *Zh. Vychisl. Matem. Matem. Fiz.* 35(8), 1285–1294 (in Russian; English translation in *Comput. Maths. Math. Phys.*, 35(8), 1033–1039).

Chernykh, O.L., and Kamenev, G.K. (1993) Linear algorithm for a series of parallel two-dimensional slices of multidimensional convex polytope. *Pattern Recognition and Image Analysis*, 3(2), 77–83.

Cohon J. (1978). *Multiobjective Programming and Planning*. Academic Press: New York.

Cunge, J.A., and Erlich, M. (1999) Hydroinformatics in 1999: what to be done? *J. of Hydroinformatics*, 1(1), 21–32.

Daniel, J.W. (1973a) On perturbations in systems of linear inequalities. *SIAM J. Numer. Analys.*, 12 , 770–772.

Daniel, J.W. (1973b) Stability of the solution of definite quadratic programs. *Math. Programming*, 5, 41–53.

Deb, K. (2001) *Multi-objective optimization using evolutionary algorithms*, Chichester, UK: Wiley.

Dorfman, R. (1965) Formal models in the design of water resource systems. *Water Resources Research*, 1(3), 329–336.

Dorfman, R. (1996) Why benefit-cost analysis is widely disregarded and what to do about it. *INTERFACES*, 26, 1–6.

Dudley, R. (1974) Metric entropy of some classes of sets with differentiable boundaries. *J. Approximation Theory*, 10, 227–236.

Dzholdybaeva, S.M., and Kamenev, G.K. (1991) *Experimental Analysis of the Approximation of Convex Bodies by Polyhedra.* Moscow: Computing Centre of the USSR Academy of Sciences (in Russian).

Dzholdybaeva, S.M., and Kamenev, G.K. (1992) Numerical study of the efficiency of an algorithm for approximating convex bodies by polyhedra. *Zh. Vychisl. Matem. Matem. Fiz.* 32(6) 857–866 (in Russian; English translation in *Comput. Maths. Math. Phys.*, 32(6), 739–746.

Dzholdybaeva, S.M., and Lotov, A.V. (1989) *Aggregated production functions of linear models.* Moscow: Computing Center of the USSR Academy of Sciences (in Russian).

Efremov, R.V. (2002) *Experimental analysis of methods for external polyhedral approximation of convex bodies.* Moscow: Dorodnicyn Computing Center of Russian Academy of Sciences (in Russian).

Efremov, R.V. (2003) An a priori best estimate of effectiveness of adaptive algorithm for external polyhedral approximation of smooth convex bodies. *Zh. Vychisl. Matem. Matem. Fiz.* 43(1), 149–160 (in Russian; English translation in *Comput. Maths. Math. Phys.*, 43(1)).

Efremov, R.V., and Kamenev, G.K. (2002) An a priori estimate for the asymptotic efficiency of a class of algorithms for the polyhedral approximation of convex bodies. *Zh. Vychisl. Matem. Matem. Fiz.* 42(1), 23–32 (in Russian; English translation in *Comput. Maths. Math. Phys.*, 42(1)).

Evtushenko, Yu.G. (1982) *Methods for solving extremal problems and their application in systems of optimization,* Moscow: Nauka (in Russian). English translation *Numerical optimization techniques*, New York: Optimization Software, Inc., 1985.

Evtushenko, Yu.G., and Grachev, N.I. (1979) A library of programs for solving optimal control problems. U.S.S.R. Comput. Maths. Math. Phys., Pergamon Press, vol.19, No.2, pp. 99–119.

Evtushenko, Yu.G., and Potapov, M.A. (1986) Methods for the numerical solution of multi-criteria problems. *Dokl. Akad. Nauk SSSR,* 291(1), 25–29 (in Russian).

Fisher, R., and Uri, W. (1983) *Getting to YES* New Jersey: Penguin Books.

Fourier, J.B. (1826) *Solution d'un question particuliere du calcul des inegalite.* Paris, p. 99. (Oeuvres II, Paris: Gauthier-Willars, 1890, 317–328).

Gass, S., and Saaty, T. (1955) The computational algorithm for the parametric objective function. *Naval Research Logistics Quarterly*, 2, p. 39.

Goicoechea, A., Hansen, D.R., and Duckstein, L. (1982) *Multiobjective Decision Analysis with Engineering and Business Applications.* New York: John Wiley and Sons.

Golany, B. (1988) An interactive MOLP procedure for the extensions of DEA to effectiveness analysis. *J. of Operational Research Society*, 39(8), 725–734.

Gordon, Y., Meyer, M., Reisner, S. (1994) Volume approximation of convex bodies by polytopes – a constructive method. *Studia Mathematica* 111 (1), 81–95.

Gruber, P.M. (1983) Approximation of convex bodies. In: *Convexity and its applications*, Basel: Birkhauser, 131–162.

Gruber, P. M. (1988) Volume approximation of convex bodies by inscribed polytopes. *Math. Ann.* 281, 2, 229–245.

Gruber, P. M. (1993a) Asymptotic estimates for best and stepwise approximation of convex bodies I. *Forum Math.* 5, 281–297.

Gruber, P. M. (1993b) Asymptotic estimates for best and stepwise approximation of convex bodies II. *Forum Math.* 5, 521–538.

Gruber, P. M. (1994) Approximation by convex polytopes. In: *POLYTOPES: Abstract, convex and computational*, 173–203. Netherlands: Kluwer Academic Publishers.

Gruber, P.M., and Kendrov, P. (1982) Approximation of convex bodies by polytopes. *Rendiconti Circolo mat. Palermo, Ser. 2*, 31(2), 195–225.

P.M. Gruber and J.M. Wills (eds.) (1993) *Handbook of convex geometry*, Amsterdam: North-Holland.

Gusev, D., and Lotov, A. (1994) Methods for decision support in finite choice problems. In: Yu. Ivanilov (ed.) *Operations Research: Models, Systems, Decisions*, Moscow: Computing Centre of RAS, 15–43 (in Russian).

Haimes, Y.Y. (1998) *Risk modeling, assessment, and management.* New York: Wiley.

Haimes, Y.V., Tarvainen, K., Shima, T., and Thadathil, J. (1990) *Hierarchical multiobjective analysis of large-scale systems.* New York: Hemisphere Publishing.

Henze, M., and Oedegaard, H. (1995) Wastewater treatment process development in Central and Eastern Europe strategies for a stepwise development involving chemical and biochemical treatment. In: *Remediation and Management of Degraded River Basins*, NATO ASI Series, Berlin: Springer-Verlag.

Hoffman, A.J., (1952) On approximate solutions of a system of linear inequalities. *J. Res. Nat. Bur. Standards*, 49, 263–265.

Horst, R., and Pardalos, P.M. (1995) *Handbook on global optimization*, Dordrecht, NL: Kluwer.

Ignizio, J.P. (1985) *Introduction to linear goal programming*, Beverly Hills: Sage Publications, Inc.

Jankowski, P., and Ewart, G. (1996) Spatial decision support system for health practitioners: selecting a location for rural health practice. *Geographical Systems,* 3, 279–299.

Jankowski, P., Lotov, A., and Gusev, D. (1999) Multiple criteria tradeoff approach to spatial decision making. In: J.-C. Thill (ed.) *Spatial Multicriteria Decision Making and Analysis: A Geographical Information Sciences Approach*, Brookfield, VT, 127–148.

Johansson, M., Tahtinen, M., and Amann, M. (1991) Optimal strategies to achieve critical loads in Finland. In: *Proceedings of the 1991 International Symposium on Energy and Environment*, Espoo, Finland.

Kaitala, V., Pohjola, M. and Tahvonen, O. (1992) Transboundary air pollution and soil acidification: A dynamic analysis of an acid rain game between Finland and the USSR. *Environmental and Resource Economics*, 2, 161–181.

Kallay, M. (1981) Convex Hull Algorithms in Higher Dimensions. Unpublished manuscript, Dept. Mathematics Univ. Oklahoma, Oklahoma: Norman.

Kamenev, G.K. (1986a) *Analysis of iterative methods of approximating convex sets by polyhedra.* Moscow: Computing Centre of the USSR Academy of Sciences (in Russian).

Kamenev, G.K. (1986b) *Methods for polyhedral approximation of convex bodies and their application for the construction and analysis of generalized reachable sets,* Ph.D. Thesis, Moscow Institute of Physics and Technology, (in Russian).

Kamenev, G.K. (1988) On one class of adaptive scheme for approximation of convex bodies by polytopes. In: *Mathematical Modeling and Discrete Optimization.* Moscow: Computing Centre of the USSR Academy of Sciences, pp. 3–9 (in Russian).

Kamenev, G.K. (1992) A class of adaptive algorithms for the approximation of convex bodies by polyhedra. *Zh. Vychisl. Matem. Matem. Fiz.* 32(1), 136–152 (in Russian; English translation in *Comput. Maths. Math. Phys.,* 32(1), 114–127).

Kamenev, G.K. (1993) Efficiency of Hausdorff algorithms for polyhedral approximation of convex bodies. *Zh. Vychisl. Matem. Matem. Fiz.* 33(5), 796–805 (in Russian; English translation in *Comput. Maths. Math. Phys.,* 33(5), 709–716).

Kamenev, G.K. (1994) Investigation of an algorithm for the approximation of convex bodies. *Zh. Vychisl. Matem. Matem. Fiz.* 34(4), 608–616 (in Russian; English translation in *Comput. Maths. Math. Phys.,* 34(4), 521–528).

Kamenev, G.K. (1996) An algorithm for approximating polyhedra *Zh. Vychisl. Matem. Matem. Fiz.* 36(4), 134–147 (in Russian; English translation in *Comput. Maths. Math. Phys.,* 36(4), 553–544).

Kamenev, G.K. (1998) A Visual Method for Parameter Estimation. *Dokl. Akad. Nauk,* 359(3), 319–322 (in Russian; translation in *Doklady Mathematics,* 57(2), 305–307).

Kamenev, G.K. (1999) Efficient algorithms for the approximation of nonsmooth convex bodies. *Zh. Vychisl. Matem. Matem. Fiz.* 39(3), 446–450 (in Russian; English translation in *Comput. Maths. Math. Phys.,* 39(3), 423–427).

Kamenev, G.K. (2000) On the approximation properties of nonsmooth convex disks. *Zh. Vychisl. Matem. Matem. Fiz.* 40(10), 1464–1474 (in Russian; English translation in *Comput. Maths. Math. Phys.,* 40(10), 1404–1414).

Kamenev, G.K. (2001) Approximation of completely bounded sets by the method of deep holes. *Zh. Vychisl. Matem. Matem. Fiz.* 41(11), 1751–1760 (in Russian; English translation in *Comput. Maths. Math. Phys.,* 41(11), 1667–1675).

Kamenev, G.K. (2002) Dual adaptive algorithms for polyhedral approximation of convex bodies. *Zh. Vychisl. Matem. Matem. Fiz.,* 42(8), 1351–1367 (in Russian; English translation in *Comput. Maths. Math. Phys.,* 42(8)).

Kamenev, G.K. (2003) A method of polyhedral approximation of convex bodies, optimal in respect to the rate of number of support and distance functions calculations. *Dokl. Akad. Nauk* (in print; in Russian).

Kamenev, G.K., and Kondrat'ev, D.L. (1992) A method for studying nonclosed nonlinear models. *Matematicheskoe Modelirovanie* (Mathematical Modeling), 4(3), 105–118 (in Russian).

Kamenev, G.K., Lotov, A.V., and van Walsum, P.E.V. (1986) *Application of the GRS method to water resources problems in the Southern Peel region of the Netherlands,* CP-86-19, Laxenburg, Austria: Int. Inst. for Applied Systems Analysis.

Karwan, M., Lotfi, V., Telgen, J., and Zionts, S. (1983) Redundancy in mathematical programming. *Lecture Notes in Economics and Mathematical Systems,* v.206, Berlin: Springer.

Kasanen, E., Wallenius, H., Wallenius, J., and Zionts, S. (2000) A study of high-managerial decision processes, with implication for MCDM research. *European Journal of Operational Research,* 120, 496–510.

Keeney, R.L., and Raiffa, H. (1976) *Decisions with multiple objectives: Preferences and value tradeoffs*, New York: Wiley.

Kersten, G.E., and Noronha, S.J. (1999) Supporting negotiations with a WWW-based system. *Decision Support Systems,* 8(3), 251–279.

Kondrat'ev, D.L., and Lotov, A.V. (1990) Exterior estimates and construction of attainable sets for nonlinear control systems. *Zh. Vychisl. Matem. Matem. Fiz.,* 30(4), 483–490 (in Russian; English translation in U.S.S.R. Comput. Maths. Math. Phys., 30(2), 93–97).

Korhonen, P., and Wallenius, J. (1988) A Pareto Race. *Naval Research Logistics Quarterly* 35, 615–623.

Korhonen, P., and Wallenius, J. (1990) A Multiple Objective Linear Programming Decision Support System. *Decision Support Systems* 6(3), 243–251.

Krasnoshchekov, P.S., Morozov, V.V., and Fedorov, V.V. (1979) Decomposition in design problems. *Proceedings (Izvestiya) of Academy of Sciences. Technical Cybernetics* No 2, 7–17 (in Russian).

Kurzhanski, A.B., and Valyi, I. (1996) *Ellipsoidal calculus for estimation and control,* Boston: Birkhauser.

Ky Fan (1956) On systems of linear inequalities. In: H.W. Kuhn and A.W. Tucker (eds.), *Linear Inequalities and Related Systems*, Princeton:Princeton Univ., 214–262.

Laitinen, E., and Neittaanmäki, P. (1988) On numerical solution of the problem connected with the control of the secondary cooling in the continuous casting process. *Control Theory and Adv. Tech.,* vol. 4, 285–305.

Larichev, O.I. (1984) Psychological Validation of Decision Methods. *J. of Applied Systems Analysis* 11, 37–46.

Larichev, O.I. (1992) Cognitive validity in design of decision aiding techniques. *J. Multi-Criteria Decision Analysis,* 1(3), 127–138.

Leichtweiss, K. (1980) *Konvex Mengen*, Berlin: VEB Deutsch.

Leschine, T., Wallenius, H., and Verdini, W. (1992) Interactive multiobjective analysis and assimilative capacity-based ocean disposal decisions. *European Journal of Operational Research*, 56, 278–289.

Levin, N., Tishler, A. and Zahavi, J., (1985) Capacity expansion of power generation system with uncertainty in the prices of primary energy resources, *Management Science*, V.31, No. 2, 175–186.

Lieberman, E. (1991) *Multi-objective programming in USSR*, New York: Academic Press.

Lotfi, V., Stewart, T.J., and Zionts, S. (1992) An aspiration level interactive model for multiple criteria decision making, *Computers and Operations Research*, 19(7), 671–681.

Lotov, A.V. (1971) Constructing of reachable sets for a linear descrete systems with constraints of bottleneck type. In: *Proc. of the Conf. M.Ph.T.I. 1970. Aerophysics and Applied Mathematics*, Collected Volume, Moscow: Moscow Institute of Physics and Technology, 113–119 (in Russian).

Lotov, A.V. (1972a) A numerical method of construction of attainability sets for a linear control system. *Zh. Vychisl. Matem. Matem. Fiz.,* 12(3), 785–788 (in Russian, English translation in: *U.S.S.R. Comput. Maths. Math. Phys.*, Pergamon Press, 1972, 12(3)).

Lotov, A.V. (1972b) Numerical method for solving the Cauchy problem for Bellmans equation in optimal–time problem for linear systems. *Zh. Vychisl. Matem. Matem.*

Fiz., 12(4), 1035–1037 (in Russian, English translation *in U.S.S.R. Comput. Maths. Math. Phys.*, 1972, 12(4)).

Lotov, A.V. (1973a) A numerical method of investigation of the continuity of the minimal time in linear systems, and a solution of the Cauchy problem for Bellman's equation. *Zh. Vychisl. Matem. Matem. Fiz.*, 13(5), 1315–1319 (in Russian, English translation in: *U.S.S.R. Comput. Maths. Math. Phys.*, Pergamon Press, 1973, 13(5)).

Lotov, A. V. (1973b) An approach to perspective planning in the case of absence of an unique objective. In: *Proc. of Conf. on Systems Approach and Perspective Planning (Moscow, 1972)*, Moscow: Computing Center of the USSR Academy of Sciences (in Russian).

Lotov, A.V. (1975a) Exploration of economic systems with the help of reachable sets. In: *Proc. Int. Conf. on the Modeling of Economic Processes (Erevan, 1974)*, Moscow: Computing Center of the USSR Academy of Sciences, 132–137 (in Russian)

Lotov, A.V. (1975b) A numerical method for the construction of attainability sets for linear controllable systems with phase constraints. *Zh. Vychisl. Matem. Matem. Fiz.*, 15(1), 67–68 (in Russian; English translation in *U.S.S.R. Comput. Maths. Math. Phys.*, 1975, 15(1)).

Lotov, A.V. (1978) On universe approximation of reachable sets for differential systems by reachable sets for their difference analogs. *Zh. Vychisl. Matem. Matem. Fiz.*, 18(1), 233–235 (in Russian; English translation in *U.S.S.R. Comput. Maths. Math. Phys.*, 1979, 18(1)).

Lotov, A.V. (1979) The convergence of methods of numerical approximation of attainability sets for linear differential systems with convex phase constraints. *Zh. Vychisl. Matem. Matem. Fiz.*, 19(1), 44–55 (in Russian; English translation in *U.S.S.R. Comput. Maths. Math. Phys.*, 1979, 19(1)).

Lotov, A.V. (1980) On the concept of the GRS and its construction for linear controlled systems. *Doklady Akademii Nauk SSSR,* 250(5) 1081–1083 (in Russian; English translation in Sov. Phys. Dokl., American Istitute of Physics, 1980, 25(2) 82–84).

Lotov, A.V. (1981a) Analysis of potentialities of economic systems. *Economika i matem. metody*, XVII, issue 2, 377–381 (in Russian).

Lotov, A.V. (1981b) Reachable sets approach to multiobjective problems and its possible applications to water resources management in the Skane Region. WP-81-145, Laxenburg, Austria: Int. Inst. for Applied System Analysis.

Lotov, A.V. (1981c) On the concept and construction of GRS for linear controlled systems described by partial differential equations. *Doklady Akademii Nauk SSSR,* 261(2), 297–300 (in Russian; English translation in *Sov. Phys. Dokl.,* American Istitute of Physics, 1981, 26(11), 1030–1031).

Lotov, A.V. (1982) Aggregation as approximation of the GRS. *Doklady Akademii Nauk SSSR,* 265(6) 1334–1337 (in Russian; English translation in *Sov. Phys. Dokl.,* American Istitute of Physics, 1982, 27(8), 593–595).

Lotov, A.V. (1983) Coordination of economic models by using feasibility sets. In: E.L. Berlyand and S.B. Barabash (eds.) *Mathematical methods for the analysis of the interaction of industrial and regional systems*, Novosibirsk: Nauka, 36–44 (in Russian).

Lotov, A.V. (1984a) *Introduction to mathematical economic modeling*, Moscow: Nauka (in Russian).

Lotov, A.V. (1984b) Estimation of stability and the conditioning number of the set of solutions of a system of linear inequalities. *Zh. Vychisl. Matem. Matem. Fiz.,*

24(12), 1763–1774 (in Russian, English translation in *U.S.S.R. Comput. Maths. Math. Phys.*, 1984, 24(6), 104–111).

Lotov, A.V. (1985a) *Methods for analysis of mathematical models of controlled systems on the basis of constructing the set of feasible values for the criteria of control quality.* Dr.Hab. Thesis, Moscow: Computing Center of the USSR Academy of Sciences (in Russian).

Lotov, A.V. (1985b) Estimation of the stability of the solutions set of systems of linear equalities and inequalities. *Zh. Vychisl. Matem. Matem. Fiz.*, 25(3) 451–455 (in Russian; English translation in *U.S.S.R. Comput. Maths. Math. Phys.*, 25 (2), 83–86).

Lotov, A.V. (1986) Estimate of the effect of round-off errors on the accuracy of elimination of variables in systems of linear inequalities. *Zh. Vychisl. Matem. Matem. Fiz.*, 26(3), 323–331 (in Russian, English translation in *U.S.S.R. Comput. Maths. Math. Phys.*, 26(2), 1–6).

Lotov, A.V. (1987) Approximation and stability of generalized attainability sets. In: V.A.Mel'nikov (ed.) *Cybernetics and computer technology*, No 3, Moscow: Nauka, 197–208 (in Russian).

Lotov, A.V. (1989) Generalized reachable sets method in multiple criteria problems. In: A. Lewandowski and I. Stanchev (eds.) *Methodology and Software for Interactive Decision Support. Lecture Notes in Economics and Mathematical Systems*, vol.337, Berlin: Springer, 65–73.

Lotov, A.V. (1994) *Integrated assessment of environmental problems*, Moscow: Computing Center of Russian Academy of Sciences (in Russian).

Lotov, A.V. (1995) An estimate of solution set perturbations for a system of linear inequalities. *Optimization Methods and Software*, 1995, 6(1), 1–24.

Lotov, A. (1998) Computer-based support for planning and negotiation on environmental rehabilitation of water resource systems. In: Loucks, D.P. (ed.) *Rehabilitation of Degraded Rivers: Challenges, Issues and Experiences*, Dordrecht: Kluwer Academic Publishers, 417–445.

Lotov, A.V. (2003) Computer visualization of production possibility set in data envelopment analysis. *Dokl. Akad. Nauk*, 388(2), 171–173 (in Russian, English translation in *Doklady Mathematics*, 67(1), **131 − −133**).

Lotov, A., Bourmistrova, L., and Bushenkov, V. (1999) Efficient strategies: an application in water quality planning. In: G. Kersten, Z. Mikolajuk, M. Rais and A. Yeh (eds.) *Decision Analysis and Support for Sustainable Development*, Dordrecht: Kluwer Academic Publishers, 145–166.

Lotov, A., Bourmistrova, L., Bushenkov, V., Efremov, R., Buber, A., Brainin, N., Maksimov, A. (1999) MIKE11 and interactive decision maps: joint application in DSS for water quality planning. In: *3d DHI Software Conference, Danish Hydraulic Institute*, Helsingor, Denmark, June 7–9, 1999.
http://www.dhi.dk/softcon/papers/027/027.htm

Lotov, A., and Bushenkov, V., (2000) Visual Market/2 for Windows, User's Guide
http://www.ccas.ru/mmes/mmeda/soft/second.htm

Lotov, A.V., Bushenkov, V.A., Chernov, A.V., Gusev, D.V., and Kamenev, G.K. (1997) Internet, GIS, and Interactive decision maps. *J. of Geographical Information and Decision Analysis*, 1(2), 119–143.
http://www.geodec.org/gida_2.htm

Lotov, A.V., Bushenkov, V.A., and Chernykh, O.L. (1992) LOTOV–LAKE. Scientific educational computer game.
http://www.ccas.ru/mmes/mmeda/soft/

Lotov, A.V., Bushenkov, V.A., and Chernykh, O.L. (1997) Multi-criteria DSS for river water quality planning. *Microcomputers in Civil Engineering,* 12(1), 57–67.

Lotov, A.V., Bushenkov, V.A., Chernykh, O.L., Wallenius, H., and Wallenius, J. (1998) Interactive decision maps, with an example illustrating ocean waste management decisions. In: Stewart, T. J., and van den Honert, R.C. (eds.) *Trends in Multicriteria Decision Making, Lecture Notes in Economics and Mathematical Systems,* v. 465, Berlin: Springer, 313–323.

Lotov, A.V., Bushenkov, V.A. and Kamenev, G.K. (1999) *Feasible Goals method. Mathematical foundations and environmental applications.* New York: Edwin Mellen Press (in Russian).

Lotov, A., Bushenkov, V. and Kamenev, G. (2001) *Feasible Goals method. Search for smart decisions.* Moscow: Computing Centre of Russian Academy of Sciences.

Lotov, A.V., Bushenkov, V.A., Kamenev, G.K., and Chernykh, O.L. (1997) *Computer search for balanced decisions,* Moscow: Nauka (in Russian).

Lotov, A., Bushenkov, V., Kistanov, A., and Chernov, A. (2000) Experimental INTERNET Resource for development of independent strategies.
http://www.ccas.ru/mmes/mmeda/resource/

Lotov, A.V., Chernykh, O.L., and Hellman, O. (1992) Multiple objective analysis of long-term development strategies for a national economy. *European Journal of Operational Research,* 56(2), 210–218.

Lotov, A.V., Kistanov, A.A., and Zaitsev, A.D. (2001) *Client support in e-commerce: Graphic search for bargains in large lists,* Arbeitsberichte N34, Siegen, Germany: Fachbereich Wirtschaftwissenschaften, Institute fuer Wirtschaftsinformatik, Universitet Siegen.

Lotov, A.V., Kamenev, G.K. and Berezkin, V.E. (2002) Approximation and visualization of Pareto frontier for nonconvex multi-objective problems. *Dokl. Akad. Nauk,* 386(6), 738–741 (in Russian, English translation in *Doklady Mathematics,* 66(2), 260–262).

Lotov, A.V., and Ognivtsev, S.B. (1980) *On preliminary resource allocation in goal-programming approach to national economy planning.* Moscow: Computing Center of the USSR Academy of Sciences (in Russian).

Lotov, A.V., and Ognivtsev, S.B. (1984) Application of methods for economic models coordination by using generalized feasible goals in goal-program approach. *Izvestia (Proceedings) of Russian Academy of Sciences). Series "Technical Cybernetics",* 1984, No 2, 77–83 (in Russian).

Lotov, A.V., and Stolyarova, H.M. (1984) Multiobjective analysis of forestry management models using the generalized reachable sets method. In: Grauer, M. and Wierzbicki, A. (eds.) *Interactive Decision Analysis. Lecture Notes in Economics and Mathematical Systems,* v.229, Springer-Verlag.

Louie, P.W.F., Yeh, W.W.-G., and Hsu, N.-S. (1984) Multiobjective water resources management planning. *J. of Water Resources Planning and Management,* 110(1), 39–56.

Maeler, K.-G. (1990) *International environmental problems,* Oxford Review of Economic Policy, 6, 80–108.

Mangasarian, O.L., (1981) A condition number for linear inequalities and linear programs. MRC Technical Summary Report No. 2185, Univ. Wisconsin, Madison, Wisconsin, March.

Mangasarian, O.L. and Shiau, T.-H., (1987) Lipschitz continuity of solutions of linear inequalities, programs, and complimentary problems. *SIAM J. Control Optim.,* 25, 583–595.

Matlin, I.S. (1978) Integrated model of national economy in the framework of the system for optimal perspective planning. *Economics and mathematical methods*, XIV, issue 6 (in Russian).

McClure, D.E. and Vitale, R.A. (1975) Polygonal approximation of plane convex bodies. *J. Math. Anal. Appl.* 51, 2, 326–358.

McMullen, P., and Shephard, G.C. (1971) *Convex polytopes and the upper bound conjecture*, Cambridge, England: Cambridge Univ.

McQuaid, M.J., Ong, T.-H., Chen, H., and Nunamaker, J.F. (1999) Multidimensional scaling for group memory visualization. *Decision Support Systems*, 27, 163–176.

Miettinen, K.M. (1999) *Nonlinear multiobjective optimization*, Boston: Kluwer Academic Publishers.

Miettinen, K., Berezkin, V., Kamenev, G.,and Lotov, A. (2000) *Graphical and interactive decision support tool for nonlinear multiobjective optimization*. University of Jyväskylä, Reports of the Department of Mathematical Information Systems, Series B. Scientific Computing, B 15.

Miettinen, K., Lotov, A.V., Kamenev, G.K., and Berezkin, V.E. (2003) Integration of Two Multiobjective Optimization Methods for Nonlinear Problems, *Optimization Methods and Software* (in print).

Miettinen, K., Mäkelä, M.M., and Männikkö, T. (1998) Optimal control of continuous casting by nondifferentiable multiobjective optimization. *Computational Optimization and Applications*, 11, 177–194.

Minkowski, H. (1903) Volumen und Oberfläche. *Math. Ann.* 57, 447–495.

Moiseev, N.N., Alexandrov, V.V., Krapivin, V.F., Lotov, A.V., Svirezhev, Ju.M., and Tarko, A.M. (1983) *Global models, the biospheric approach.* CP-83-33, Laxenburg, Austria: Int. Inst. for Applied System Analysis.

Moiseev, N.N., Alexandrov, V.V., and Tarko, A.M. (1985) *Man and the biosphere*, Moscow: Nauka.

Motzkin, T.S., Raiffa, H., Thompson, G.L., and Thrall, R.M. (1953) The double description method. In: H.W. Kuhn and A.W. Tucker (eds.), *Contributions to Theory of Games, Vol. II, Princeton: Annals of Mathematical Studies*, no. 24, 51–74.

Nefedov, V.N. (1984) *Regularization methods for multi-criteria optimization problems*, Moscow: Moscow Aviation Institute (in Russian).

Neves, J.N., and Camara, A. (1999) Virtual environments and GIS. In: M. Goodchild and D. Rhind (eds.) *Geographical Information Systems*, New York: John Wiley.

Orlovski, S.A., and van Walsum, P.E.V. (1984) *Water policies: regions with intense agriculture.* WP-84-40. Laxenburg, Austria: Int. Inst. for Applied Systems Analysis.

Pecsvaradi, T., and Narendra, K.S. (1971) Reachable sets for linear dynamic systems. *Information and Control*, 19(4).

Petrov, A.A., Pospelov I.G., and Shananin, A.A. (1999) *From GOSPLAN to non-efficient market economy.* New York: Edwin Mellen Press (in Russian).

Phelps, E. (1961) The golden rule for accumulation. *American Economic Review*, 60(4).

Pontryagin, L.S., Boltyanskii, V.G., Gamkrelidze, R.V. and Mishchenko, E.F. (1961) *The Mathematical Theory of Optimal Processes*, Moscow: Fizmatgiz (in Russian. English translation: New-York: Pergamon-Macmillan, 1964).

Pospelov, G.S., and Irikov, V.A. (1976) *Programmed-objective planning and control*, Moscow: Nauka (in Russian).

Preparata, F.P., and Shamos, M.I. (1985) *Computational geometry: An introduction*, Berlin: Springer.

Raiffa, H. (1968) *Decision analysis*. Reading, Massachusetts: Addison-Wesley.

Raiffa, H. (1982) *The Art and Science of Negotiations*. Belknap Press of Harvard University.

Robinson, S.M. (1973) Stability theory for systems of inequalities. Part I: Linear Systems. *SIAM J. Numer. Analys.*,12,754–769.

Rockafellar, R.T. (1970) *Convex analysis*, Princeton, NY: Princeton University Press.

Rogers, C.A. (1964) *Packing and covering*, Cambridge University Press.

Romero, C. (1991) *Handbook of critical issues in goal programming*, London: Pergamon Press.

Rosenhead, J., and Mingers, J. (eds.) (2001) *Rational analysis for a Problematic World Revisited*, Chichister: Wiley.

Roy, B. (1972) Decisions avec criteres multiples. Problems et methodes. *Metra International* 11(1), 121–151.

Saaty, T. (1996) *The analytic network process*, Pittsburg: RWS Publications.

Samsonov, S.P. (1983) Reconstruction of a convex set from its supporting function with prescribed accuracy. *Vestnik MGU, Ser. 15, Vychisl. Matem. i Kiber.* 1, 68–71.

Sawaragi, Y., Nakayama, H., and Tanino, T. (1985) *Theory of multiobjective optimization*, Orlando: Academic Press.

Schneider, R. (1983) Zur optimalen Approximation konvexer Hyperflachen durch Polyeder. *Math. Ann.* 256, 3, 289–301.

Schneider, R. (1987) Polyhedral approximation of smooth convex bodies. *J. Math. Anal. Appl.* 128, 2, 470–474.

Schneider, R., and Wieacker, J.A. (1981) Approximation of convex bodies by polytopes. *Bull. London Math. Soc.*, 13, pt. 2, no. 41, 149–156.

Seidel, R. (1981) *A Convex hull algorithm optimal for points in even dimensions,* M.S. Thesis, Tech. Rep. 81–14, Vancouver, Canada: Dept. Comput. Sci., Univ. British Columbia.

Simon, H. (1960) *The New Science of Management Decision*, New York: Harper and Row.

Shiryaev, A.N. (1989) *Probability*, Moscow: Nauka (in Russian).

Solanki, R.S., Appino, P.A., and Cohon, J.L. (1993) Approximating the noninferior set in multiobjective linear programming problems. *European Journal of Operational Research*, 68(3), 356–373.

Soloveitchik, D., Ben-Aderet, N., Grinman, M., Lotov, A. (2002) Multiobjective optimization and marginal pollution abatement cost in the electricity sector – An Israeli case study. *European Journal of Operational Research*, 140(3), 571–583.

Solso, R.L. (1988) *Cognitive psychology*, Boston: Allyn and Bycon, Inc.

Sonnevend, G. (1977) On the optimization of algorithms for function minimization. *Zh. vychisl. Matem. mat. Fiz.* 17, 591–609.

Sonnevend, G. (1980) Asymptotically optimal, sequential methods for the approximation of convex, compact sets in $\mathbb{R}^n$ in the Hausdorff metrics. *Colloq. Math. Soc. Janos Bolyai* 35(2), 1075–1089.

Sonnevend, G. (1983) An optimal sequential algorithm for uniform approximation of convex functions on $[0,I]^2$. *Appl. Math. Optimiz.*, 1983, no. 10, 127–142.

Stadler, W. (1986) Initiators of multicriteria optimization. In: J. Jahn and W. Krabs (eds.) *Recent Advances and Historical Development of Vector Optimization*, Berlin: Springer-Verlag, 3–47.

Statnikov, R.B., and Matusov, J. (1995) *Multicriteria optimization and engineering*, Chapman and Hall.

Steuer, R. (1986) *Multiple criteria optimization*, New York: John Wiley.

Stewart, T.J. (1996) Relationships between DEA and MCDM. *J. of Operational Research Society*, 47(5), 654–665.

Stoleru, L. (1967) *L'equilibre et la croissance economique*, Paris: Dunod.

Thanassoulis, E., and Dyson, R.G. (1992) Estimating preferred target input-output levels using DEA. *European J. of Operational Research*, 56, 80–97.

Thanassoulis, E. (2000) DEA and its use in the regulation of water companies. *European J. of Operational Research*, 127(1), 1–13.

Tuovinen, J.-P., Kangas, L., and Nordlund, G. (1990) Model calculations of sulphur and nitrogen depositions in Finland. In: P. Kauppi, P. Anttila and K. Kenttamies (eds.) *Acidification in Finland*, Berlin: Springer-Verlag, 167–197.

United Nations (1988) *Assessment of Multiple Objective Water Resources Projects*, New York: United Nations.

Vasil'yev, N.S. (1983) On finding of global minimum of qusi-concave function. *Zh. Vychisl. Matem. Matem. Fiz.* 23(2), 307–313 (in Russian).

Vasil'yev, N.S. (1988) On nonimprovable bounds of approximation of strongly convex bodies. *Vopr. Kiber.* 136, 49–56 (in Russian).

Voevodin V.V. (1977) *Numerical foundations of linear algebra*, Moscow: Nauka (in Russian).

Wallenius, H., Leschine, T.M., and Verdini, W. (1987) *Multiple criteria decision methods in formulating marine pollution policy: A comparative investigation*. Research Paper No.126, Vaasa, Finland: Proceedings of the University of Vaasa.

Wierzbicki, A. (1981) A mathematical basis for satisficing decision making. In: J. Morse (ed.) *Organizations: Multiple Agents with Multiple Criteria*, Berlin: Springer, 465–485.

Yan, H., Solomatine, D.P., Velickov, S., and Abbott, M.B. (1999) Distributed environmental impact assessment using Internet. *J. of Hydroinformatics*, 1(1), 59–70.

Zeleny, M. (1974) *Linear multiobjective programming*, Berlin: Springer-Verlag.

Zhigljavsky, A.A. (1991) *Theory of global random search*, Dordrecht, NL: Kluwer.

Author Index

Index